GW00708383

The SLL Code for Lighting

The Society of
Light and Lighting

222 Balham High Road, London SW12 9BS
+44 (0)20 8675 5211
www.cibse.org

This document is based on the best knowledge available at the time of publication. However, no responsibility of any kind for any injury, death, loss, damage or delay however caused resulting from the use of these recommendations can be accepted by the Chartered Institution of Building Services Engineers, The Society of Light and Lighting, the authors or others involved in its publication. In adopting these recommendations for use each adopter by doing so agrees to accept full responsibility for any personal injury, death, loss, damage or delay arising out of or in connection with their use by or on behalf of such adopter irrespective of the cause or reason therefore and agrees to defend, indemnify and hold harmless the Chartered Institution of Building Services Engineers, The Society of Light and Lighting, the authors and others involved in their publication from any and all liability arising out of or in connection with such use as aforesaid and irrespective of any negligence on the part of those indemnified.

The rights of publication or translation are reserved.

No part of this publication may be reproduced, stored in a retrieval system or transmitted in any form or by any means without the prior permission of the publisher.

Note from the publisher

This publication is primarily intended to give guidance. It is not intended to be exhaustive or definitive, and it will be necessary for users of the guidance given to exercise their own professional judgement when deciding whether to abide by or depart from it.

© March 2012 The Society of Light and Lighting

The Society of
Light and Lighting

The Society is part of CIBSE, which is a registered charity, number 278104.

ISBN 978-1-906846-21-3

Print management and typesetting by The Charlesworth Group

Printed in Great Britain on FSC certified mix source paper by Page Bros (Norwich) Ltd., Norwich, Norfolk, NR6 6SA

Main cover image: The Royal Pavilion, Brighton; photograph by Liz Peck
(LPA Photography)

Foreword

The Society of Light and Lighting (SLL) and its predecessors, the Lighting Division of the Chartered Institution of Building Services Engineers (CIBSE) and the Illuminating Engineering Society (IES), have published recommendations on lighting practice since 1936. From the beginning, these recommendations, called Codes for Lighting, have all contained details of the illuminances required for use in different applications together with qualitative guidance on how to implement these recommendations. Over the years, as the understanding of how lighting conditions can affect visual performance and cause visual discomfort has increased, additional quantitative lighting criteria have been added, notably those concerned with the level of colour rendering required and with limiting discomfort glare. For many years, the *IES Code for Lighting* was the *de facto* standard for lighting provision in the United Kingdom. However, in 2002, the Committee for European Standardisation (CEN) took on the task of providing lighting recommendations, and, since then the British Standards Institution has adopted the CEN recommendations for use in the United Kingdom. As a result, there are now a range of British Standards that specify the quantitative lighting requirements for a wide range of applications. Consequently, the role of the *SLL Code for Lighting* has shifted from being the only source for quantitative lighting recommendations to being a guide on how to interpret the British Standard recommendations and how to implement them in practice.

This edition of the *SLL Code for Lighting* takes the changes in lighting guidance a step further by a process of separation and concentration. The separation involves moving the details of vision, lighting technology and lighting applications into another publication, the *SLL Lighting Handbook*. The concentration occurs because this *SLL Code for Lighting* provides information on three fundamental matters of relevance to lighting practice. These matters are:

● A summary of what is known about the effects of lighting on task performance, behaviour, safety, perception and health as well as its financial and environmental costs.

● A compendium of all the lighting recommendations relevant to the United Kingdom with suggestions as to how these should be interpreted. This compendium covers recommendations for both interior and exterior lighting in normal conditions.

● A detailed description of all the calculations required for quantitative lighting design. While it is a fact that, today, most lighting calculations are done using software that simply implements the fundamental calculations described here, without knowledge of these calculations, it is difficult to assess the meaning and merit of the results produced by software.

Principal author
Peter Raynham BSc Msc CEng FSLL MCIBSE MILP

Contributors
Peter Boyce PhD FSLL FIESNA
John Fitzpatrick BSc PhD

Editor
The Charlesworth Group

SLL Secretary
Liz Peck FSLL

CIBSE Editorial Manager
Ken Butcher

CIBSE Head of Knowledge
Nicolas Peake

Acknowledgements
The Society of Light and Lighting gratefully acknowledges the following for reviewing the draft prior to publication: Robert Bean, Lou Bedocs, Brian Glynn, Ruth Kelly, Paul Littlefair, David Loe, Michael Pointer, Paul Ruffles, Anthony Slater, Peter Thorns, Alan Tulla and Bob Venning.

The Society also wishes to acknowledge the following for permission to reproduce illustrations: Illuminating Engineering Society of North America (Figure 1.4), Naomi Miller Lighting Design LLC (Figure 1.8), Commission Internationale de l'Eclairage (Figure 1.11), Electronic Healing (Figure 1.12), NCS Colour AB (Figures 16.14, 16.15 and 16.16), British Standards Institution (Figure 16.19). Other illustrations courtesy of the author and contributors.

NCS-Natural Colour System®© is the property of and used on licence from NCS Colour AB Stockholm 2012. References to NCS®© in this publication are used with permission from NCS Colour AB.

Permission to reproduce extracts from BS 667, BS 5489, BS 5252, BS 8206, BS EN 12464, BS EN 12665, BS EN 13201, BS EN 13032, BS EN 15193 and BS ISO 23539 is granted by BSI. British Standards can be obtained in PDF or hard copy formats from the BSI online shop: www.bsigroup.com/Shop or by contacting BSI Customer Services for hard copies only: Tel: +44 (0)20 8996 9001, Email: cservices@bsigroup.com.

Contents

Chapter 6: Energy

Chapter 7: Construction (Design and Management) Regulations

Chapter 8: Basic energy and light

Chapter 9: Luminous flux, intensity, illuminance, luminance and their interrelationships

Chapter 10: Direct lighting

Chapter 11: Indirect lighting

Chapter 12: Photometric datasheets

Chapter 13: Indoor lighting calculations

Chapter: 14 Outdoor lighting calculations

Chapter 15: Measurement of lighting installations and interpreting the results

Chapter 16: Colour

Chapter 1: The balance of lighting

1.1 Lighting quality

The objective of anyone concerned with providing lighting should be to produce good quality lighting, but what constitutes a good quality lighting installation? The answer is one that meets the objectives and constraints set by the client and the designer. Depending on the context, the objectives can include facilitating desirable outcomes, such as enhancing the performance of relevant tasks, creating specific impressions and generating a desired pattern of behaviour, as well as ensuring visual comfort and safety. The constraints are usually the maximum allowed financial and power budgets, a maximum time for completion of the work and, sometimes, restrictions on the design approach to be used.

Such a definition of good quality lighting has its limitations. It is not expressed in terms of photometric measures, but rather in terms of the impact lighting has on more distant outcomes. There are three arguments in favour of such an outcome-based definition of lighting quality rather than one based directly on photometric measures. The first is that lighting is usually designed and installed as a means to an end, not as an end in itself, so the extent to which the end is achieved becomes the measure of success. The second is that what is desirable lighting depends very much on the context. Almost all of the aspects of lighting that are considered undesirable in one context are attractive in another. The third is that there are many physical and psychological processes that can influence the perception of lighting quality (Veitch, 2001a,b). It is this inherent variability that makes a single, universally applicable recipe for good quality lighting based on photometric quantities an unreal expectation.

So what role do lighting recommendations have to play in ensuring good quality lighting? A simple concept that offers a place for lighting recommendations is that lighting installations can be divided into three classes of quality: the good, the bad and the indifferent:

- Bad quality lighting is lighting that does not allow you to see what you need to see, quickly and easily and/or causes visual discomfort.

- Indifferent quality lighting is lighting that does allow you to see what you need to see quickly and easily and does not cause visual discomfort but does nothing to lift the spirit.

- Good quality lighting is lighting that allows you to see what you need to see quickly and easily and does not cause visual discomfort but does raise the human spirit.

On this scale, lighting recommendations are useful for eliminating bad lighting. Following lighting recommendations is usually enough to ensure that indifferent quality lighting is achieved. This is no mean achievement. Indeed, it may be the best that can be expected from the use of guidelines and quantitative lighting criteria. It may be that once bad lighting is avoided, the difference between indifferent lighting and good lighting is a matter of fashion and opportunity. Fashion is important because we often crave the new to provide interest and variety in our lives. There is no reason to suppose that lighting should be any different in this respect than most other aspects of life. As for opportunity, that is partly a matter of technology and partly a matter of being in the right place at the right time. And what is the right place? An eminent lighting designer, J.M. Waldram, once said "If there is nothing worth looking at, there is nothing worth lighting" so the right place is presumably, a place which contains something worth looking at. Also, given that to be really good, the lighting has to be matched in some way to the particular environment, each lighting solution would be specific and not generally applicable. This combination of fashion and specificity suggests that the conditions necessary for good lighting quality are liable to change over time and space and hence will not be achievable through the use of lighting recommendations alone. At the moment, good quality lighting most frequently occurs at the conjunction of a talented architect and a creative lighting designer, neither of whom is given to slavishly following lighting recommendations.

By now it should be apparent that the writers of lighting recommendations do not have an easy task. They have to strike the right balance between a number of conflicting aims. They have to make recommendations that are precise and preferably quantitative but not so precise that they lose credibility. Equally, they have to avoid making recommendations that are so vague as to be meaningless. The recommendations have to be technically and economically feasible and simple enough to be implemented, although they should also reflect the complexity of the subject. Yet lighting recommendations are needed. Advice is needed on appropriate lighting by people who buy lighting installations and by some who design them. To such people it does not matter that they represent a balance between conflicting aims. What does matter is that the recommendations made produce reasonable results in practice. If they do, then the recommendations will be accepted and the judgements of the people who write the recommendations vindicated. If they do not, then no amount of contrary evidence will convince anyone that the recommendations are correct. It is this test of practice that is the ultimate justification for many lighting recommendations (Jay, 1973). The lighting recommendations given in Chapters 2 to 4 of this *SLL Code for Lighting* are known to produce reasonable results in practice.

1.2   The place of lighting in the modern world

Lighting is vital to the modern world – it enables a 24-hour society to exist. When first introduced, electric lighting was expensive and available to few. Today, it is ubiquitous and cheap. Lighting is used for many different purposes – to ensure visual work can be done accurately, quickly, safely and in comfort, to make places attractive and interesting, to generate business activity, to enhance security and to promote human health. Together these functions make a real contribution to the quality of life of millions.

But lighting comes at a cost, both financial and environmental. The financial cost involves first costs, the cost of the electricity consumed and disposal costs. The environmental cost takes three forms: the consequences of generating the electricity required to power lighting, the chemical pollution upon disposal and the presence of light pollution at night.

This means that lighting recommendations are a balance between the benefits and costs. Lighting recommendations reflect this balance and are inevitably a consensus view of what is reasonable for the conditions prevailing when they are written (Boyce, 1996). That consensus will be different in different countries and different at different times in the same country, depending on the state of knowledge about lighting, the technical and economic situation, and the interests of the people contributing to the consensus. This is evident in the history of illuminance recommendations which show considerable variations between and within countries over time (Mills and Borg, 1999). The following sections discuss what we know about both the benefits and costs of lighting.

1.3   An overview of the effects of light on human performance

Light can affect what people can do and what they choose to do via three different routes; through the visual system, through non-visual effects on human physiology and through perception. The outcome in any particular case is human performance in its widest sense. Figure 1.1 shows a conceptual framework for considering the factors that influence progress down each route and the interactions between them.

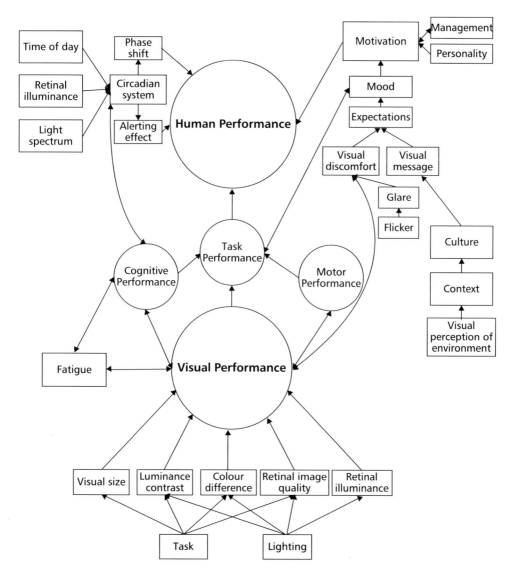

Figure 1.1 A conceptual framework illustrating the routes via which lighting can affect human performance (from Boyce, 2003)

The effect of lighting on vision is the most obvious impact of light on humans. With light we can see, without light we cannot. The visual system is an image processing system. The optics of the eye form an image of the outside world on the retina of the eye. At the retina, some image processing occurs. Different aspects of the retinal image are processed through two different channels up to the visual cortex of the brain. The magnocellular channel processes information rapidly but with little detail or colour information, while the parvocellular channel provides detail of brightness, colour and texture but at a slower rate. In addition, the visual system is organised spatially into two parts, the fovea of the retina, where fine detail is available, and the periphery, which is basically a detection system indicating where in the visual field the fovea should be directed. When there is a lot of light available, e.g. in daytime, the whole of the retina is active. When there is very little light, e.g. outside on a moonless night, the fovea is blind and only the peripheral retina operates. A more detailed discussion of the visual system and the parameters considered below can be found in the *SLL Lighting Handbook* (The Society of Light and Lighting, 2009).

Any stimulus to the visual system can be described by five parameters, its visual size, luminance contrast, colour difference, retinal image quality and retinal illumination. These parameters are important in determining the extent to which the visual system can detect and identify the stimulus, i.e. the extent to which the stimulus is visible.

The visual size of a stimulus for detection is usually given by the solid angle the stimulus subtends at the eye. The solid angle is given by the quotient of the areal extent of the object and the square of the distance from which it is viewed. The larger the solid angle, the easier the stimulus is to detect. The visual size for resolution is usually given as the angle the critical dimension of the stimulus subtends at the eye. The larger the visual size of detail in a stimulus, the easier it is to resolve that detail. Lighting can do little to change the visual size of two-dimensional objects but shadows can be used to enhance the effective visual size of some three-dimensional objects.

The luminance contrast of a stimulus expresses its luminance relative to its immediate background. The higher the luminance contrast, the easier it is to detect the stimulus. Lighting can change the luminance contrast of a stimulus by producing disability glare in the eye or veiling reflections from the stimulus.

Luminance quantifies the amount of light emitted from a stimulus and ignores the combination of wavelengths making up that light. It is the wavelengths emitted from the stimulus that influence its colour. It is possible to have a stimulus with zero luminance contrast that can still be detected because it differs from its background in colour (Eklund, 1999; O'Donell et al., 2011). Lighting can alter the colour difference between the object and its background when light sources with different spectral power distributions are used.

As with all image processing systems, the visual system works best when it is presented with a sharp image. The sharpness of the retinal image is determined by the stimulus itself, the extent to which the medium through which light from the stimulus is transmitted scatters light, and the ability of the visual system to focus the image on the retina. Lighting can do little to alter any of these factors, although it has been shown that light sources that are rich in the short wavelengths produce smaller pupil sizes for the same luminance than light sources that are deficient in the short wavelengths (Berman, 1992). A smaller pupil size produces a better quality retinal image because it implies a greater depth of field and less spherical and chromatic aberrations.

The illuminance on the retina determines the state of adaptation of the visual system and therefore alters the capabilities of the visual system. At higher states of visual adaptation, visual acuity and contrast sensitivity are enhanced and colour discrimination is finer (see the *SLL Lighting Handbook*). The amount of light entering the eye is mainly determined by the luminances in the field of view. For interiors, these luminances are determined by the reflectances of the surfaces in the field of view and the illuminances on them. For exteriors, the relevant luminances are those of reflecting surfaces, such as the ground, and of self-luminous sources, such as the sky.

What these five parameters imply is that it is the interaction between the object to be seen, the background against which it is seen and the lighting of both object and background that determine the stimulus the object presents to the visual system and the operating state of the visual system. It is the stimulus and the operating state of the visual system that largely determine the level of visual performance possible. The other factors that can influence visual performance are concerned with the presentation conditions, specifically, movement and presentation time. When the object to be seen is in motion, particularly when the movement is not predictable so the object cannot be fixated or when the movement is very fast so the presentation time is short, visual performance will be worse than what is possible for a static object presenting the same stimuli.

Another route whereby lighting conditions can affect human performance is through the non-visual effects of light entering the eye. That there are such effects is made most evident by the role of a regular alternating pattern to light and darkness in entraining the human circadian system (Dijk et al., 1995). The most obvious evidence for the existence of a circadian system in humans is the occurrence of the sleep/wake cycle although there are many other variations in hormonal and behavioural rhythms over a 24-hour period. The organ that controls these cycles in humans is the suprachiasmatic nuclei (SCN) set deep in the brain. The SCN is linked directly to the retina, receiving signals from a recently discovered photoreceptor, the intrinsically photosensitive retinal ganglion cell (ipRGC) (Berson et al., 2002). When signals are transmitted from the retina to the SCN, no attempt is made to preserve their original location. Rather, the ipRGCs supplying the SCN with a signal act like a simple photocell discriminating between light and dark. Signals from the SCN are, in turn, transmitted to many parts of the brain, many of which have not yet been investigated (CIE, 2004a). This means that the aspects of lighting that influence the state of the SCN are the spectrum and amount of radiation reaching the retina, which in turn depend on the spectrum of the light source used, the distribution of the resulting radiation, the spectral reflectances of the surfaces in the space, the spectral transmittance of the optic media and where the observer is looking.

Lighting conditions can affect human performance through the circadian system in two different ways; a shifting effect in which the phase of the circadian rhythm can be advanced or delayed by exposure to bright light at specific times (Dijk et al., 1995); and an acute effect related to the suppression of the hormone melatonin at night (Campbell et al., 1995). Both of these effects can be expected to enhance human performance in the right circumstances. Attempts have been made to use the phase shift to more quickly adapt people to nightshift work but for that to work requires control over light exposure over the whole 24 hours (Eastman et al., 1994). As for the acute effect, there is clear evidence that exposure to bright light increases alertness at night (Badia et al., 1991) and that this can enhance the performance of complex cognitive tasks (Boyce et al., 1997). But the circadian system is only the most well explored of the non-visual effects of light on human physiology and hence on human performance. There are known to be other effects of light exposure, such as increased vitality during the day (Partonen and Lönnqvist, 2000), but the mechanisms through which these effects occur are unknown.

The third route whereby lighting conditions can affect human performance is through the perceptual system. The perceptual system takes over once the retinal image has been processed by the visual system. The simplest output of the perceptual system is a sense of visual discomfort, which may change the observer's mood and motivation, particularly if the work is prolonged.

Lighting conditions in which achieving a high level of visual performance is difficult will be considered uncomfortable as will conditions in which the lighting leads to distraction from the task, as can occur when glare and flicker are present. But perception is much more sophisticated than just producing a feeling of visual discomfort. In a sense, every lighting installation sends a 'message' about the people who designed it, who bought it, who work under it, who maintain it and about the place where it is located. Observers interpret the 'message' according to the context in which it occurs and their own culture and expectations. According to what the 'message' is, the observer's mood and motivation can be changed. Every lighting designer appreciates the importance of 'message' but it is only in the context of retailing and entertainment that the 'message' a lighting installation sends is given the importance its potential to influence behaviour deserves.

The effect of lighting on mood and motivation has not been the subject of extensive study but what has been done has shown that lighting can be used to draw attention to objects (LaGiusa and Perney, 1974), to modify an observer's mood (Baron et al., 1992; McCloughan et al., 1999) and to move people in a desired direction. There is also some evidence that lighting can be used to generate desirable behaviour. For example, one study has shown a correlation between the presence of skylights and the value of sales in a supermarket – the presence of skylights leads to higher sales (Heschong Mahone Group, 1999). There is still much to learn about using the 'message' lighting sends to good effect.

While the visual, non-visual and perceptual routes have been discussed separately, it is important to appreciate that they can interact and extend the range of effects of lighting to all tasks, even those that do not require vision. For example, working at night when your body is telling you to go to sleep will affect the performance of both cognitive and visual tasks. Another example would be a situation where the lighting provides poor task visibility, so that visual performance is poor. If the worker is aware of the poor level of performance and it fails to meet his or her expectations, then the worker's mood and motivation may be altered. To further complicate the picture, it is necessary to appreciate that while visual performance for a given task is determined by lighting conditions alone, a worker's motivation can be influenced by many physical and social factors, lighting conditions being just one of them (CIBSE, 1999).

What this overview demonstrates is that lighting conditions can influence our lives in many different ways, sometimes being the primary factor and at other times being only one factor amongst many. Lighting recommendations for different applications are produced with this diversity in mind, some applications giving priority to the ability to see detail and others focusing on the 'message' delivered through the perception of the space and the people in it. Only rarely have the non-visual effects been considered but that may change in the future as more knowledge in this area is developed.

1.4   Lighting and visual task performance

One of the main benefits of lighting is its ability to enhance the performance of visual tasks by increasing the visibility of critical details. A visual task is one that requires the use of vision. Virtually all visual tasks actually have three components: a visual component, a cognitive component and a motor component (see Figure 1.1). Consider driving as an example. The visual component is seeing the road ahead and what is on it. The cognitive component is understanding what the scene ahead implies for the control of the vehicle. The motor component is the movement of the vehicle's controls. Task performance is the performance of the whole task. Visual performance is the performance of the visual component.

Lighting only directly affects the visual component by making the task details more or less visible. How important lighting is to the performance of a specific task depends on the place of the visual component in the task structure. Different tasks have different structures. Assembling small electronic components has a large visual component; mixing concrete does not. It is the diversity of task structure that makes it impossible to generalise about the impact of lighting on task performance for all tasks from measurements on one. What is possible is to measure the impact of lighting on the performance of tasks in which the visual component has been maximised and the other components minimised. This is the approach used in the study of visual performance. The effect of lighting conditions on visual performance is one factor that contributes to lighting recommendations.

1.4.1   Visual performance

Visual performance can be considered at two levels, threshold and suprathreshold. Threshold visual performance is the performance of a visual task close to the limits of what is possible. Suprathreshold visual performance is the performance of tasks that are easily visible because all of the details required to perform the task are well above threshold. This raises the question as to why lighting makes a difference to visual performance once what has to be seen is clearly visible. The answer is that even when all the necessary details are clearly visible, lighting influences the speed with which the visual information can be processed.

Both threshold and suprathreshold visual performance measures are useful, but in different ways. Threshold measures are useful for determining whether or not a specific lighting condition will be seen, e.g. will a given fluctuation in light output be seen as flicker? However, suprathreshold visual performance is usually more relevant to lighting practice because lighting is usually designed to make sure what needs to be seen can be easily seen and this means that the relevant aspects of the task are well above threshold.

The most widely used form of lighting recommendation is the illuminance on the task. One of the first people to systematically investigate the effect of illuminance on visual task performance was H.C. Weston (1935, 1945). He used a matrix of Landolt Cs (Figure 1.2) as a standard task. The advantage of this approach is that the difficulty of the task can be varied by changing the size and luminance contrast of the Landolt Cs. In this task, what the observer has to do is to identify all the Cs with a gap in a specified direction. The time taken to examine the matrix and the number of errors made are combined to provide a measure of visual performance. Figure 1.3 shows the results obtained from Landolt C matrices of different sizes and luminance contrasts. From such data, it is possible to identify four qualitative features of the effect of illuminance on visual performance.

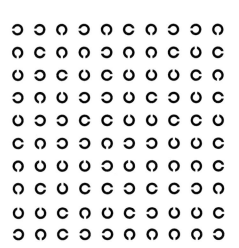

Figure 1.2 A matrix of Landolt Cs as used in the measurement of visual performance

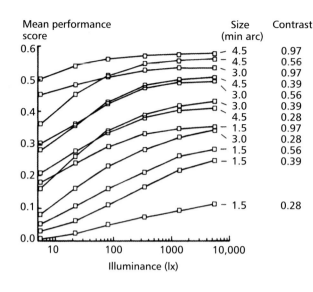

Size (min arc)	Contrast
4.5	0.97
4.5	0.56
3.0	0.97
4.5	0.39
3.0	0.56
3.0	0.39
4.5	0.28
1.5	0.97
3.0	0.28
1.5	0.56
1.5	0.39
1.5	0.28

Figure 1.3 Mean performance scores for Landolt ring charts of different critical size and contrast, plotted against illuminance (after Weston, 1945)

These are:

● Increasing illuminance follows a law of diminishing returns, i.e. equal increments in illuminance lead to smaller and smaller changes in visual performance until saturation occurs.

● The point where saturation occurs is different for different sizes and luminance contrasts of critical detail.

● Larger improvements in visual performance can be achieved by changing the task than by increasing the illuminance, at least over any illuminance range of practical interest.

● It is not possible to make a visually difficult task reach the same level of performance as a visually easy task simply by increasing the illuminance over any reasonable range.

While such understanding is useful, it is not enough to make quantitative predictions of the effect of illuminance on visual performance. This can be done using the Relative Visual Performance (RVP) model derived from measurements of reaction time to the onset of a target and the time taken to compare lists of numbers (Rea, 1986; Rea and Ouellette, 1991). Figure 1.4 shows the shape of a RVP surface for a target of a fixed size over a range of luminance contrasts and adaptation luminances. Similar surfaces have been found for more realistic visual tasks (Eklund et al., 2001). The overall shape of the relative visual performance surface has been described as a plateau and an escarpment (Boyce and Rea, 1987). In essence, what it shows is that the visual system is capable of a high level of visual performance over a wide range of sizes, luminance contrasts and adaptation luminances (the plateau) but at some point, either size or luminance contrast or adaptation luminance will become insufficient and visual performance will rapidly collapse (the escarpment). The existence of a plateau of visual performance implies that for a wide range of visual conditions, visual performance changes very little with changes in the lighting conditions. This is why a high level of precision is not necessary in the provision of illuminance in most lighting installations. This lack of precision is evident in two aspects of lighting recommendations. The first is the fact that the illuminances recommended are arranged along a scale with significant gaps between adjacent steps, e.g. the illuminances recommended for offices are either 300 or 500 lx. The second is that associated with each recommended maintained illuminance is an illuminance uniformity criterion. This criterion allows for considerable variation in illuminance across the relevant surface.

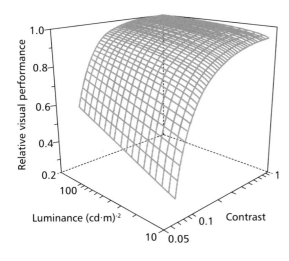

Figure 1.4 A Relative Visual Performance (RVP) surface (after Rea, 1986)

The RVP model has been developed using achromatic targets, i.e. targets without colour. When the target or the background or both are coloured, then the spectrum of the light illuminating the task is important. This importance also takes two forms. The first is simply when the colour has a meaning, as with fruit or vegetables where the colour can indicate the degree of ripeness, or where there is a desired appearance, as in clothes retailing. For these applications, the light source has to be chosen to give the objects the required colour appearance. The ability of a light source to render colours accurately is expressed through the CIE General Colour Rendering Index (CRI), although this has limitations when dealing with white light constructed from narrow band light sources such as LEDs. In general, the higher the CRI of a light source, the more accurately colours are rendered. The second is where the colour difference between the task and its immediate background contributes to the visibility of the task. This is particularly important where luminance contrast is low because then, a clear colour difference can help maintain visual performance (O'Donell et al., 2011).

It might be thought that when the task is achromatic, i.e. involving only black, white and grey, the light spectrum would not have an effect on visual performance. However, when visual performance is on the escarpment of visual performance, it does. In this situation, a light source with a high proportion of power in the short wavelength part of the visible spectrum will enhance visual acuity (Berman et al., 2006) and that can be beneficial where performance is limited by the size of detail that needs to be seen (Liebel et al., 2010) but when the plateau is reached the effect of light source spectrum on the performance of achromatic tasks disappears (Boyce et al., 2003).

1.4.2 Visual search

So far, all the knowledge presented has been based on tasks in a known position so the task can be viewed directly using the fovea. But not all tasks are like this. In some tasks, the object to be detected is in an unknown location. Such tasks involve visual search. Visual search is typically undertaken through a series of eye fixations, the fixation pattern being guided either by expectations about where the object to be seen is most likely to appear or by what part of the visual scene is most important (Figure 1.5).

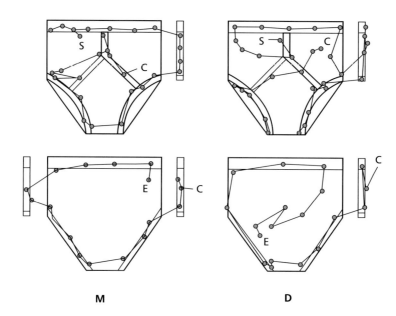

M   D

Figure 1.5 The pattern of fixations made by two inspectors examining men's briefs held on a frame. S = start of scan path, C = end of scan of front and one side, rotation of frame and continuation of scan across back and sides, E = end of scan. Inspector M examines only the seams while Inspector D examines the fabric as well (after Megaw and Richardson, 1979)

Typically, the object to be detected is first seen away from the visual axis and then confirmed or resolved by an on-axis fixation. The speed with which a visual search task is completed depends on the visibility of the object to be found, the presence of other objects in the search area, and the extent to which the object to be found is different from the other objects. The simplest visual search task is one in which the object to be found appears somewhere in an otherwise empty field, e.g. an air bubble in a pane of glass. The most difficult visual search task is one where the object to be found is situated in a cluttered field, and the clutter is very similar to the object to be found, e.g. searching for a face in a crowd. The lighting required for fast visual search has to be matched to the physical characteristics of the object to be found so as to maximise the luminance contrast or colour difference between the target, its immediate background and any clutter present.

1.4.3   Mesopic conditions

The human visual system has three operating states; photopic, mesopic and scotopic. In the photopic state, the cone photoreceptors of the retina are dominant and the rod photoreceptors are saturated. In the scotopic state, only the rod photoreceptors are active. In the mesopic state, both rod and cone photoreceptors contribute. By convention, photopic vision occurs when the adaptation luminance is above 3 cd/m². Scotopic vision occurs when the adaptation luminance is below 0.001 cd/m². Mesopic vision occurs when the adaptation luminance is between 0.001 and 3 cd/m². These boundaries are approximate because the sensitivity of individuals varies. What these values imply is that virtually all interior lighting ensures that the visual system is in the photopic state but some exterior lighting, such as road lighting, is only sufficient to ensure operation in the mesopic state.

Mesopic vision is different from photopic vision in that different parts of the retina have different spectral sensitivities. The fovea maintains photopic spectral sensitivity through the mesopic region until vision fails because it contains only cone photoreceptors. However, the rest of the retina has a different spectral sensitivity, one in which the peak sensitivity steadily shifts towards the short wavelength end of the visible spectrum as the adaptation luminance decreases. This occurs because the rest of the retina contains both rod and cone photoreceptors and the rods become increasingly dominant as the adaptation luminance decreases. This divergence in spectral sensitivity matters because all the photometric quantities used in lighting recommendations assume a photopic spectral sensitivity. As a result, two light sources that both meet the same outdoor photopic lighting recommendations may allow different levels of off-axis performance when used outdoors at night. To be sure of a good level of off-axis performance when following the exterior lighting recommendations, it is necessary to select a light source with a high scotopic/photopic (S/P) ratio, i.e. a light source that provides a high level of stimulation to the rod photoreceptors as well as the cone photoreceptors. Most light sources have been designed to maximise their photopic light output/unit power but some also provide good levels of scotopic light output. Table 1.1 shows the S/P ratios for some common light sources. Light sources with high S/P ratios have been shown to improve the detection of off-axis targets in mesopic conditions (Akashi et al., 2007; Fotios and Cheal, 2009).

Table 1.1 Scotopic/photopic ratios for some common light sources

Light source	Scotopic/photopic ratio
High pressure sodium	0.6
Fluorescent – 3000 K	1.3
Tungsten halogen	1.4
Quartz metal halide	1.5
Ceramic metal halide	1.7
White LED – 6000 K	2.0
Fluorescent – 6500 K	2.2

These values are indicative only. Exact values should be obtained from the manufacturer of the light source.

1.4.4   A discrepancy

While the study of how lighting conditions affect visual performance has made significant progress over the years, one of the outcomes of this study has been rather disconcerting. This is that the illuminances required to reach the plateau of visual performance are, for most tasks done in interiors, much less than the illuminances currently recommended. This discrepancy can be justified in three ways. First, the ability to perform a task is not the only factor to be taken into account when deciding on the appropriate illuminance. The illuminance has to be sufficient to ensure that people are comfortable and can perform the task with ease. Judgments of the illuminance required to perform a task are consistently higher than that identified as necessary by models of visual performance (Newsham and Veitch, 2001; Boyce et al., 2006b). In a sense, this is a difference between what people want and what they need. Second, it is rarely known exactly which tasks will occur in a given location. Third, different people have different visual capabilities. The unknown nature of the range of tasks that can occur and the differences among people imply that a safety factor needs to be applied to the illuminance determined from models of visual performance to ensure that all people can easily do all the tasks likely to occur in a given situation.

1.4.5   Improving visual performance

The illuminance recommendations for applications where task performance is of primary importance represent a consensus as to the amount of light required to be sure of a high level of visual performance by people of working age for all the tasks likely to be found in the specific application. However, it is always possible for new and more visually difficult tasks to occur or for people with limited visual capabilities to be employed. If this should happen and people are experiencing difficulty doing a specific visual task, there are a number of steps that can be taken to improve visual performance. Not all of these steps will be possible in every situation and not all are appropriate for every task.

A task can be made visually easier by:

- increasing the size of the detail in the task

- increasing the luminance contrast of the detail in the task

- ensuring that the target can be looked at directly without visual search

- ensuring that the worker can focus the object, using corrective lenses if necessary

- if luminance contrast is low and cannot be increased, changing the colour of the task to increase the colour difference against the immediate background

- if the target is moving, reducing its speed so as to make it easier to track

- if the target is presented for a limited time, increasing the presentation time.

The lighting can be improved by:

- increasing the illuminance so that the adaptation luminance is increased

- selecting a light source with more appropriate colour properties

- ensuring that the lighting is free from disability glare and veiling reflections as these both reduce luminance contrast.

1.5   Lighting and behaviour

Lighting can certainly be used to enhance the visual performance of tasks requiring the resolution of detail but there are also other activities that benefit when lighting is used to modify behaviour. Lighting can be used to attract attention, to direct movement and to facilitate communication.

1.5.1 Attracting attention

Flashing lights are widely used to attract attention to signs and vehicles but they can also cause discomfort. Theatre lighting uses spotlights to create small areas of high luminance to direct audience attention to important characters. This is the foundation of display lighting where the aim is to attract attention without causing discomfort. Practice in display lighting is to provide a luminance ratio between where the attention is to be directed and the rest of the space of at least 5:1 for a definite effect and more than 30:1 for a very dramatic effect (Figure 1.6). Accent lighting in a shop can increase the time that consumers spend at the display (Summers and Hebert, 2001) and lighting design that focuses the light on the merchandise can improve customer and staff perception of the store (Cuttle and Brandston, 1995). The benefits of accent lighting extend to educational and office settings. Studies in classrooms have found that pupils tend to pay more attention to instructional materials and perform better on tests about the material when visual aids are spotlighted (LaGiusa and Perney, 1973, 1974).

Figure 1.6 Accent lighting used to attract attention to merchandise

1.5.2 Directing movement

Designers of places where large numbers of people move about, such as museums and shopping centres, use lighting to direct traffic. It is possible to influence the direction of movement by increasing the luminance of the desired direction of movement by a factor of 10–16 over the less-desired alternative (Kang, 2004; Taylor and Sucov, 1974).

1.5.3 Communication

Work frequently requires verbal communication, especially in classrooms, conference rooms and at service counters. Understanding speech is primarily an auditory task but intelligibility is improved when one can see the face of the speaker. This is particularly true when there is interference from background noise or when the speaker has an unfamiliar accent. Further, many people with subtle hearing impairments depend on lip reading to supplement their understanding of speech. The rise of videoconferencing has also increased the importance of understanding how to use lighting to facilitate speech intelligibility. Lighting solutions to facilitate communication must combine lighting the speaker's mouth with providing acceptable facial modelling for a pleasant appearance (Zhou and Boyce, 2001).

1.6 Lighting and safety

Lighting can be used to enhance safety, both indoors and outdoors. The recommendations for workplace lighting made in Chapters 2 and 3 of this *SLL Code for Lighting* take the hazards posed by specific working situations into account. However, one situation in which safety may be compromised is when there is a power failure, particularly if the power failure is associated with fire and smoke. In this situation, emergency lighting comes into play. Outdoors, one of the objectives of road lighting is to enhance road safety, while street lighting has a role to play in crime prevention.

1.6.1 Emergency escape lighting

Emergency escape lighting requires a system that will provide information about where to leave the building in case of an emergency (exit signs), information about how to get to the exit, and sufficient light to enable people to move along the path to the exit. As would be expected from knowledge of visual performance, the speed with which people can move along the path varies as a function of the illuminance on the floor and the person's age as well as the complexity of the path and the individual's knowledge of the route (Figure 1.7).

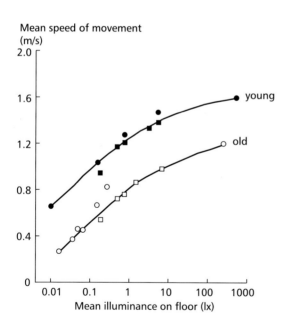

Figure 1.7 Mean speed of movement in cluttered and furnished spaces in a clear atmosphere by young and old people, plotted against mean illuminance on the floor (after Ouellette and Rea, 1989)

Further advice on emergency lighting can be found in the *SLL Lighting Handbook* and in the *SLL Lighting Guide 12*. Well designed emergency lighting will encourage people to leave a building promptly when the alarm sounds.

1.6.2 Road lighting

Road lighting is designed to enhance road safety. How successful road lighting is in enhancing road safety can be judged by what happens in its absence. A meta-analysis of multiple studies of the effect of road lighting on accidents has led to the conclusion that introducing road lighting to previously unlit roads should lead to a 65% reduction in night-time fatal accidents, a 30% reduction in night-time injury accidents and a 15% reduction in night-time property damage accidents (Elvik, 1995). These are overall figures and offer little guidance as to where introducing road lighting might be most effectively employed. An alternative approach based on the sudden change in light level at the same clock time that occurs at the daylight savings time change has been used to examine the consequences of reduced visibility (Sullivan and Flannagan, 2007). The results indicate that some types of accident are more sensitive to the reduction in visibility that follows the end of the day than others. For example, adult pedestrians are almost seven times more likely to be killed after dark than during daytime, but fatalities associated with overturning the vehicle are less likely after dark. Further, the pattern of sensitivity to reduced visibility conforms to common sense. The accident types with the highest sensitivity to reduced visibility are those involving unlighted objects, such as pedestrians and animals, or where objects appear unexpectedly in the road, or where the road suddenly changes direction. Unlighted objects will have a low visibility after dark compared to lighted objects. Unexpected objects and unexpected road configurations require a response within a limited time. Improving visibility through better road lighting allows more time to make a response. There can be little doubt that road lighting has a role to play in improving road safety through greater visibility.

1.6.3 Lighting and crime

The probability of a crime being committed is influenced by many factors, the possibility of being seen being just one of them. Nonetheless, a series of studies of increasing sophistication leave little doubt that lighting has a role to play in crime prevention (Painter and Farrington, 1999, 2001). Improving street lighting can lead to a decrease in crime, but it may not. This is because lighting, *per se*, does not have a direct effect on the level of crime. Rather, lighting can affect crime by two indirect mechanisms. The first is the obvious one of facilitating surveillance by people on the street after dark, by the community in general and by the authorities. If such increased surveillance is perceived by criminals as increasing the effort and risk and decreasing the reward for a criminal activity, then the incidence of crime is likely to be reduced. Where increased surveillance is perceived by the criminally inclined not to matter, then better lighting will not be effective. The second indirect mechanism by which an investment in better lighting might affect the level of crime is by enhancing community confidence and hence increasing the degree of informal social control. This mechanism can be effective both day and night.

What constitutes better lighting for crime prevention is unclear. Studies which demonstrate that better lighting reduces crime usually involve the use of more light sources with higher light output and better colour rendering, more closely spaced. From such information and basic knowledge of how to make it easier to see details at night, it can be concluded that the important factors are the illuminance provided, the illuminance uniformity, the control of glare and the light spectrum. Such lighting should allow anyone on the street to detect and recognise a threatening situation while there is still time to do something about it, and any witnesses to provide accurate information about the perpetrators.

1.7 Lighting and perception

Lighting affects the perception of spaces and objects. In many applications, it is the perception of the people and the space around them that are matters of primary concern. Examples are the lighting of homes, hotels, shops, restaurants, parks and plazas. The perceptions influenced by lighting can be divided into simple and complex, although both are influenced by the amount, colour and distribution of light. Simple perceptions, such as brightness and form, tend to be governed by the performance of the visual system and hence are somewhat consistent across cultures. Complex perceptions, such as attractiveness, interest and safety are also related to previous experience and culture. A more extensive discussion of lighting and perception can be found in the *SLL Lighting Handbook* (SLL, 2009).

1.7.1 Brightness

Strictly, the simple perception of brightness only occurs for a self-luminous source, such as a computer screen, and is linked to the luminance of that source by a power law with an exponent of 0.33. However, brightness is also commonly used to describe the perception of spaces, both indoors and outdoors, although then it is the luminance of the surfaces in the space that influence the perception. In such situations, the brightness of the space is influenced by the amount and distribution of light, the reflectances of the surfaces, the luminance of the luminaire and the spectrum of the light. This means that for a given set of surface reflectances, increasing the illuminance on those surfaces will increase the perception of brightness. But there are ways to increase brightness other than simply increasing illuminance. For example, for the same surface luminance, choosing a light source with a spectrum containing a higher proportion of short wavelength light or one which makes colours appear more saturated will increase the brightness. As for luminaire luminance, depending on the luminance and area of any bright patches on the luminaire, the brightness of a room can be enhanced or diminished. Balancing the luminance and area so that the bright patch of the luminaire is perceived as sparkling will also enhance the brightness of a room (Akashi et al., 2006). Increasing the luminance further so that the luminaire becomes glaring will diminish the brightness of the room. The simple perception of colour appearance is linked to the spectrum of the light source and the luminance. How strong an effect the choice of light source has depends on whether the space is essentially achromatic or one containing many coloured surfaces. The effect of light source will be much greater for the latter conditions than the former because chromatic adaptation can offset some of the difference due to different light spectra in an achromatic room but cannot offset the effect of the light spectrum on the saturation of colours in the room. Light sources with a higher CIE General Colour Rendering Index increase the saturation of surface colours which is why they are recommended for places where the appearance of the space and the objects in it is of primary importance.

1.7.2 Form

Form is primarily influenced by light distribution. Different light distributions create different patterns of highlight and shadow, patterns that can be used to reveal or mask features of the object. Highlights are important in revealing the specular nature of materials such as silver and glass. One of the most important objects to be lit in many locations is the human face. Figure 1.8 shows how the appearance of a face can be changed by altering the light distribution.

Figure 1.8 The modelling of a face by different light distributions. From left to right, the lighting is completely diffuse, strong down-lighting and a combination of diffuse and side accent lighting (courtesy Naomi Miller)

The importance attached to avoiding the extremes of light distribution, i.e. totally diffuse and totally directional, where the appearance of faces matters, is evident in the recommendations about the range of preferred surface reflectances and the minimum illuminances that should fall on those surfaces. By using high reflectances for walls and ceilings (high reflectance floors are not a realistic proposition) and ensuring that at least a minimum illuminance falls on them, the contribution of diffuse inter-reflected light to the lighting of the space is increased, particularly for small rooms. Increasing the amount of inter-reflected light serves to diminish veiling reflections, soften shadows and reduce glare.

One metric that can be used to measure the light distribution in a space is the cylindrical illuminance. Cylindrical illuminance is the average illuminance falling on the surface of a small cylinder. When the cylinder is aligned vertically and is positioned at an appropriate height, cylindrical illuminance is related to the appearance of people in the space, much more so than the conventional illuminance on a horizontal plane. The appearance of people matters because social interaction is governed in part by our judgments of the attractiveness of the people with whom we interact. Attractive people are judged to be more intelligent and pleasant and are more likely to be helped by others (Langlois et al., 2000), and enjoy better job-related outcomes such as likelihood of being employed or promoted (Hosoda et al., 2003).

Of course, when it comes to the appearance of people, the spectrum of the light source also matters. People are very sensitive to skin tone as it is used as an indicator of health. Similar considerations apply to any other object where the colour has meaning. Again, this is the reason why it is recommended to use a light source with a high CIE General Colour Rendering Index in places where the colour appearance of people and objects in the space is important.

1.7.3 Higher order perceptions

The effect of the lighting on higher order perceptions such as attractiveness, interest and safety is much less certain than for simple perceptions. This is because higher order perceptions are influenced by the whole of the environment, not just the lighting, as well as the context of the space and the culture of the observer. What we do know is that functional spaces, such as offices, are evaluated on the dimensions of brightness and visual interest, the former being related to the amount of light in the space; the latter being enhanced by a non-uniform distribution of light away from the work area. This should not be taken to mean that lighting has only a limited effect on the perception of spaces and objects. There is clear evidence that by changing the lighting, the perception of objects can be changed from drab and boring to eye-catching and dramatic (Mangum, 1998) a finding that is important for many places including shops, hotels and museums. The problem with functional spaces such as offices is that the possible lighting effects and the materials they have to work with are often limited by the need to provide good visibility for work over a large portion of the space. In such spaces, lighting is most likely to have an effect on higher order perceptions when the architect has generated an attractive space and the lighting designer has produced lighting that provides sufficient brightness in the task area and enhances the architecture to provide some visual interest elsewhere. Unfortunately, this is not something that can be done through lighting recommendations. Where recommendations are given, they should be treated as a baseline from which to elaborate rather than the beginning and end of design.

By manipulating light to change perception, lighting can influence feelings. Changes in feelings can influence the performance of all types of tasks, not just visual, by altering mood and motivation. In this way, lighting can influence many outcomes that are important to individuals, organisations and society. One conceptual model of how people evaluate a space (Kaplan and Kaplan, 1982) suggests that they seek an illuminance that is sufficiently high to make it easy to see what needs to be seen and a pattern of light that reflects the hierarchy of objects in the space. Lighting can vary in many ways: amount, distribution, colour appearance, colour rendering, light source type, luminaire appearance, controllability, temporal variation, etc. Many of these dimensions interact with room surface characteristics, with the culture and expectations of the people and with the architecture and purpose of the space. This makes the study of preference for lighting very complex. Consequently, the only application that has been extensively studied is office lighting. For offices, we know that:

- People prefer to have a window nearby, both for daylight as a light source and for a view out (Farley and Veitch, 2001).

- People want to be protected from direct sun as a source of glare and heat gain. If available, they will use window blinds to block direct sun (Galasiu and Veitch, 2006).

- People prefer a combination of direct and indirect lighting (Boyce et al., 2006a; Houser et al., 2002; Veitch and Newsham, 2000).

- There are large differences between individuals in light level preferences. The central tendency for offices with computer work hovers in the range of 300–500 lx on the desk (Newsham and Veitch, 2001; Boyce et al., 2006b).

- Desktop uniformity (minimum/maximum illuminance) should be within the range 0.5–0.7 (Slater and Boyce, 1990).

- Spaces with an average vertical luminance in the field of view of at least 30 cd/m^2 are judged to be more attractive and comfortable (Loe et al., 1994; Newsham et al., 2005).

- People prefer a lit environment with a moderate degree of variability (Loe et al., 1994), although there is some evidence that too much variability is undesirable (Newsham et al., 2005).

To what extent these preferences apply to other applications such as classrooms, shops and hospitals remains to be determined.

When one experiences one's preferred environmental conditions, a pleasant mood called 'positive affect' results (Baron, 1990; Baron and Thomley, 1994). People experiencing positive affect are more likely to respond cooperatively than competitively to interpersonal conflict, are more creative and perform better on intellectual tasks (Isen and Baron, 1991). Interestingly, a link has been found between preferred lighting conditions and positive feelings of health and well-being (Veitch et al., 2008). Such findings mean it is important for the lighting designer to match the lighting to the occupants' preferences. Following the lighting recommendations in Chapter 2 will ensure that dramatic departures from the conditions that people prefer are avoided but will not guarantee that the most preferred lighting is achieved.

A higher order perception of concern to people outdoors after dark is safety. Figure 1.9 shows the mean rating of how safe it was perceived to be to walk alone through car parks by day and night in two urban and suburban areas (Boyce et al., 2000). It is evident from Figure 1.9 that the perceived safety for walking alone in the car parks during the day is higher in the suburban area than in the urban area. As for perceived safety when walking alone at night, Figure 1.9 shows that, for both urban and suburban car parks, lighting can bring that perception close to what it is during the day but cannot exceed it. Figure 1.10 shows the difference in ratings of safety when walking alone by day and night plotted against the median illuminance in the car park at night, for the urban and suburban car parks. These results indicate that at a sufficiently high illuminance, the difference in ratings of safety for day and night approach zero. However, the approach to zero difference is asymptotic. For illuminances in the range 0 to 10 lx, small increases in illuminance produce a large increase in perceived safety. Illuminances in the range 10 and 50 lx show a law of diminishing returns. The illuminances recommended in Chapter 4 for lighting residential streets cover a range of 2 to 15 lx. Of course, there is much more to successful lighting for a perception of safety than illuminance. Attention has also to be given to the uniformity of illuminance, glare control and light source colour rendering. Nonetheless, it is interesting that there is some agreement between a measure of the perception of safety at night and the main recommendation for lighting residential streets.

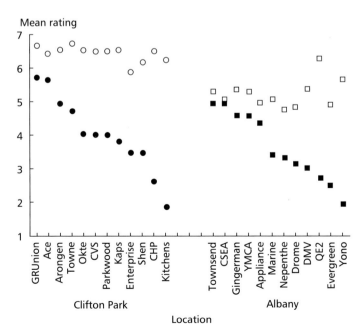

Figure 1.9 Mean ratings of perceived safety for walking alone in a car park, by day and night, for the parking lots in Albany, NY (urban) and Clifton Park, NY (suburban). The car parks are presented in order of decreasing perceived safety at night (1 = very dangerous; 7 = very safe; filled symbols = night, open symbols = day) (after Boyce et al., 2000)

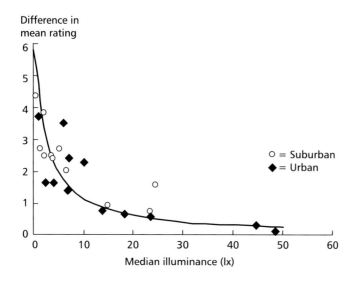

Figure 1.10 Difference in mean ratings of perceived safety for walking alone in a car park, by day and night (day–night) plotted against median pavement illuminance for the car parks in Albany, NY (urban) and Clifton Park, NY (suburban) (after Boyce et al., 2000)

1.8   Lighting and health

Although illuminance on the working plane is the most widely used lighting recommendation, simply providing that illuminance is not enough to ensure good quality lighting. Depending on how the illuminance is delivered, the result can either be comfortable or uncomfortable. Visual discomfort occurs when the lighting makes it difficult to see what needs to be seen, causes distraction or pushes one to the limits of the visual system, all of which are likely to affect visual performance negatively. Aspects of lighting that can commonly cause visual discomfort are insufficient light, excessive light, shadows, veiling reflections, glare and flicker. Following the illuminance recommendations given in Chapter 2 will almost always ensure that the extremes of insufficient light and too much light are avoided. More details of the situations in which shadows, veiling reflections, various forms of glare and flicker occur can be found in the *SLL Lighting Handbook* (The Society of Light and Lighting, 2009).

1.8.1   Eyestrain

The consequence of prolonged exposure to uncomfortable lighting conditions is likely to be eyestrain. The symptoms of eyestrain vary in form and magnitude from one individual to another but headache, blurred vision, dry or watery eyes, tense muscles and burning or itching eyes are common. Sleep or resting the eyes for a time usually alleviates the condition. While eyestrain can be annoying, it rarely indicates a serious eye problem. Nonetheless, anyone who frequently experiences eyestrain cannot be said to be enjoying the best of health.

There are a number of lighting recommendations designed to minimise the occurrence of eyestrain. The illuminance on the working plane and the illuminance uniformity across that plane are intended to ensure that all the tasks likely to occur in the space under consideration will be highly visible. The luminance ratios recommended between the task area, the immediate surround and the background seek to maximise the visual acuity and contrast sensitivity of the visual system and minimise the likelihood of discomfort glare occurring. The ranges of room reflectances have a role to play in increasing the amount of inter-reflected light in a room and hence reducing the magnitude of any veiling reflections. The Unified Glare Rating recommendations have been specifically developed to set a maximum allowed level of discomfort glare for many applications. There are no recommendations for minimising flicker, this being seen as a property of the light source. This does not mean that flicker does not matter. Rather it means that light source manufacturers are aware of the problems caused by the perception of flicker and seek to minimise it in their products so that flicker rarely occurs in practice.

To summarise, eyestrain can be avoided by following these recommendations:

- Provide the recommended illuminance on the task.

- Ensure the distribution of illuminance on the task is uniform.

- Keep the illuminance on the area surrounding the task similar to that on the task.

- Avoid confusing shadows in the detail of the task.

- Ensure there are no veiling reflections on the task. For work on a computer, this can be achieved by using a high luminance background for the display and a diffusely reflecting screen.

- Eliminate disability glare from luminaires or windows.

- Reduce discomfort glare from luminaires or windows.

- Choose light sources that are free of flicker.

1.8.2 Non-visual effects

Although being able to see is the most obvious and immediate consequence of light entering the eye, there are other non-visual consequences. This is because, in addition to the rod and cone photoreceptors that serve vision, there is another type of photoreceptor in the retina, the intrinsically photosensitive retinal ganglion cell that signals the presence of light or dark to parts of the brain concerned with many basic physiological functions. The one that has been most extensively studied is the circadian system.

The circadian system produces circadian rhythms. Circadian rhythms are a basic part of life and can be found in virtually all plants and animals, including humans. The most obvious of these rhythms is the sleep/wake cycle but there are many others some relevant to task performance such as alertness, mood, memory recall and cognitive throughput. The human circadian system involves three basic components:

- an internal oscillator, the suprachiasmatic nuclei, located in the brain

- a number of external oscillators that can entrain the internal oscillator

- a messenger hormone, melatonin, which carries the internal 'time' information to all parts of the body through the bloodstream.

In the absence of light, and other cues, the internal oscillator continues to operate but with a period longer than 24 h. External stimuli are necessary to entrain the internal oscillator to a 24-hour period and to adjust for the seasons. The light–dark cycle is one of the most potent of these external stimuli.

Given that the circadian system is fundamental for a lot of human physiology, it should not be a surprise that when it fails or is disrupted for a long time, there are negative implications for human health. A failing circadian system is a common feature of old age, a failing that is associated with sleep problems. A disrupted circadian system is a common feature of rapidly rotating night shift work. The resulting sleep deprivation is, in turn associated with an increased number of accidents and reduced productivity (Lockley et al., 2007; Rosekind et al., 2010). Shift work over many years is linked to an enhanced risk of major health hazards such as heart disease, cancer and diabetes (Rosa and Colligan, 1997; Jasser et al., 2006). Further, there is growing support for the hypothesis that repeated exposure to sufficient light to suppress melatonin from its normal concentration has some role to play in the incidence and development of breast cancer but there may be other necessary conditions yet to be established (Figueiro et al., 2006).

Another group of people with sleep problems are those suffering from Alzheimer's disease. This is a degenerative disease of the brain and is the most common cause of dementia. People with Alzheimer's disease and other forms of dementia often demonstrate fragmented sleep/wake patterns throughout the day and night (van Someren et al., 1996).

Light exposure patterns have a positive role to play in the alleviation or prevention of health problems. Exposure to bright light immediately after awakening is effective for treating delayed sleep phase disorder, as is exposure in the evening for advanced sleep phase disorder and sleep maintenance insomnia (Czeisler et al., 1988; Lack and Schumacher, 1993; Campbell et al., 1993). Exposing Alzheimer's patients to bright light during the day and little light at night, thereby increasing the signal strength for entrainment, has been shown to help to make their rest/activity patterns more stable, bright light being an illuminance of about 1200 lx (van Someren et al., 1997). As for circadian disruption, it is possible to identify some actions that can be taken to minimise it. They are:

- be exposed to high illuminances during the day and low illuminances at night

- do not do shift work

- if shift work has to be done, avoid rapidly-rotating shifts; physiologically it is preferable to stay on one schedule as long as possible. Rosa and Colligan (1997) give advice on how to assess any proposed shift system

- when working at night use a low illuminance provided by a light source that has a spectrum dominated by the long wavelength end of the visible spectrum

- avoid frequent jet travel across multiple time zones. When travelling, expect to need about 1 day to adjust per 2–3 h of time shift.

All the above has been concerned with the circadian system but as is apparent in Figure 1.11, there are many links between the retina of the eye and various parts of the brain, many of these still waiting to be explored.

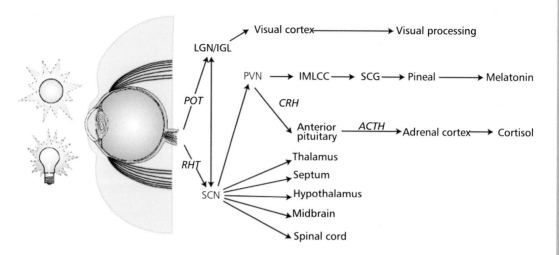

Figure 1.11 A simplified schematic diagram of two eye-brain pathways. Light received by the eye is converted to neural signals that pass via the optic nerve to these visual and non-visual pathways. POT = Primary optic tract; RHT = retino-hypothalamic tract; LGN/IGL = lateral geniculate nucleus/intergeniculate leaflet; SCN = suprachiasmatic nucleus of the hypothalamus; PVN = paraventricular nucleus of the hypothalamus; IMLCC = intermediolateral cell column of the spinal cord; SCG = superior cervical ganglion; CRH = corticotropic releasing hormone; ACTH = adrenocorticotropic hormone (after CIE, 2004a)

This means there is still much to learn about the non-visual effects of light exposure (Boyce, 2006). It would be unwise to attempt to manipulate the circadian system with light too much or too often until all the possible consequences have been explored and understood which is why the effects of light exposure on human health are discussed in BS EN 12464-1 but no explicit recommendations are given. However, it is worth pointing out that both the visual system and the circadian system have evolved under a natural regime of daylight days and dark nights. The alternative electric light sources have only been available for use by day and night for about a hundred years, a very short time in evolutionary terms. It may be that the main impact of a greater understanding of the role of light exposure on human health will be to return attention to the better daylighting of buildings.

A proven benefit for light exposure is in the treatment of seasonal mood disorders, such as depression and bipolar disorder, although the exact mechanism is unknown. The most common form is winter depression, better known as Seasonal Affective Disorder (SAD). Seasonal depression symptoms include increased appetite, carbohydrate craving and unacceptable weight gain as well as increased sleep (Lam and Levitt, 1998). Estimates vary as to the prevalence of seasonal mood disorders and there are several theories as to their cause, but there is little controversy concerning the value of light therapy as an effective treatment (Ravindran et al., 2009). Standard light therapy involves the delivery of up to ~10 000 lx, measured at the eye, for 30 min daily, usually in the early morning delivered either by daylight or by a light box (Figure 1.12).

Figure 1.12 A light box being used to deliver a high light level to the eye (courtesy Electronic Healing)

1.8.3   Tissue damage by optical radiation

Light can have an effect on human health simply as optical radiation incident on the eye and skin, regardless of any signals produced by the retina that enter the brain. When considering light as optical radiation, the definition of light is usually extended to include ultra-violet (UV) and infrared (IR) radiation as many light sources produce all three types of radiation. In sufficient doses, light can cause tissue damage to both the eye and skin, quickly and over many years. In the short term, UV radiation can cause photokeratitis of the eye and erythema of the skin. Prolonged exposure to UV radiation can lead to cataract in the lens of the eye as well as skin aging and skin cancer. In sufficient quantities, visible radiation can produce photoretinitis of the retina. Visible and short wavelength IR radiation can cause thermal damage to the retina and burns to the skin. Prolonged exposure to IR radiation can lead to cataract and burns. Guidance setting out threshold limit values that should be observed to avoid these detrimental effects on health, and a lamp classification system based on these threshold limit values are available (IESNA, 2005, 2007, 2009; ACGIH, 2010; CIE 2006) as is advice on methods of measurement of exposure (BSI, 2005a). When evaluated according to these threshold limit values, most light sources, when used and viewed in a conventional manner, pose no hazard to health. Nonetheless, cases of tissue damage can still occur in special circumstances where either exposure is excessive (O'Hagan et al., 2011) or when individuals with extreme sensitivities are present so care should be taken to consider each case on its merits.

1.9   Lighting costs

Lighting has both financial and environmental costs. Both need to be considered if the lighting design is to be of good quality. There are no lighting recommendations in Chapters 2, 3 or 4 for the financial cost of a lighting installation, although the lighting recommendations that are made there certainly affect the cost of a lighting installation. There are also lighting recommendations relevant to the environmental costs.

1.9.1 Financial costs

The financial costs of lighting are easy to quantify. The costs of buying the required equipment; lamps, luminaires, control systems, and the costs of installing the equipment are readily obtained. This total constitutes the first cost of the lighting installation. Having installed the lighting, there are then the operating and maintenance costs to be considered. Operating costs will depend on the number of hours of use and the price of electrical energy; the latter may be fixed or might vary with time and market conditions depending on the form of the contract with the supplier. As for maintenance costs, these will depend on the maintenance schedule adopted as part of the lighting design as well as the costs of replacement equipment and the costs of gaining access to and the cleaning of the installation. Finally, there are disposal costs to be considered. It is no longer possible to simply dispose of electrical equipment by throwing it into a landfill. Under the Waste Electrical and Electronic Equipment (WEEE) Regulations, all lighting equipment, other than filament lamps and that in domestic premises, is considered to be hazardous waste and has to be disposed of appropriately.

What combinations of these costs are considered will depend on who is paying. For an owner-occupier, it is likely that all will be combined to give an estimate of life cycle costs. This measure involves an adjustment for the fact that the operating and maintenance costs are spread out over several years while the disposal costs occur many years in the future. For the developer who plans to lease the property, it is likely that only the first cost and possibly the disposal costs will be considered. The operating and maintenance costs will usually be paid by the tenant although if the tenant is wise, he will estimate the operating cost of the lighting system before signing the lease. The accuracy with which operating costs can be estimated will vary depending on the amount of daylight available, the sophistication of the lighting control system and the pattern of use of the space. The most accurate estimates will be possible where there is little daylight, the control system is a simple on–off switch and the space is continuously occupied throughout the working day. The least accurate estimates will occur when there are large amounts of daylight available, the electric lighting is designed to be switched or dimmed depending on the amount of daylight present and when the occupancy of the space is intermittent and occupancy sensors are employed. It will also be difficult to estimate operating costs accurately when there is an element of individual control of the illuminances provided, such as when a task/ambient lighting system is in place.

1.9.2 Electricity consumption

The most obvious environmental consequence of the provision of lighting is the pollution resulting from the consumption of electricity. Depending on the fuel mix used, the generation of electricity can involve the burning of fossil fuels such as coal, oil and gas which will lead to carbon emissions. In an age of global warming, the reduction of carbon emissions is a policy objective of many governments. To give an idea of the magnitude of the problem, it has been estimated that in the State of California, the fifth biggest economy in the world and for many countries a model for the future, lighting uses 23% of the electricity generated; 14% in commercial buildings, 8% in homes and 1% for road lighting. Over the last quarter century, governments have made attempts to reduce or at least slow the growth in electricity demand with little success. Improving the efficiency of lighting installations is an attractive means to achieve this end, because more energy efficient lighting technology is available and the life of lighting installations is a matter of years rather than decades. Consequently, a number of governments have introduced laws either specifying a maximum lighting power density to be used in buildings or outlawing the use of specific energy-inefficient lighting equipment. The result has been a steady reduction in the lighting power densities in commercial buildings but little change for homes. But this is now changing. The most common light source used in homes is the incandescent lamp, a light source with the lowest luminous efficacy of those in common use.

Many countries, including both the United States and the countries of the European Union, are in the process of legally removing the incandescent lamp from the market. The expectation is that householder will adopt more energy efficient means to light their homes, such as the compact fluorescent lamp and the light emitting diode.

1.9.3 Chemical pollution

While the consumption of electricity is the most obvious environmental impact of lighting, there are also two forms of pollution; chemical pollution and light pollution. Lighting can generate chemical pollution either directly or indirectly. Direct chemical pollution comes about when lamps and control gear are scrapped. Some old control gear that may still be found in use contains toxic materials such as polychlorinated biphenyls (PCBs) and/or di(2-ethylhexyl)phthalate (DEHP). Modern control gear, both electromagnetic and electronic, does not contain such toxic materials. As for lamps, a toxic material commonly found in lamps is mercury (Begley and Linderson, 1991; Clear and Berman, 1994). Mercury is used in many discharge lamps. In fluorescent and metal halide lamps, some mercury is essential because the mechanism for generating light is the creation of an electric discharge through a mercury atmosphere. The lighting industry has responded to pressure from governments and environmentalists by dramatically reducing the amount of mercury used in fluorescent lamps.

Indirect chemical pollution occurs in the generation of the electricity consumed by the operation of the lamp. Given the concern with mercury, it might be thought that eliminating the incandescent lamp from the market and replacing it with the compact fluorescent lamp would lead to an increase in mercury deposits as the compact fluorescents are scrapped. However, the amount of mercury released by lamp disposal is very small compared to the amounts of mercury released into the atmosphere naturally, through volcanoes, and, more controllably, by the burning of fossil fuels to generate electricity (EPA, 1997). Thus, as long as the wider use of compact fluorescents leads to lower electricity consumption by lighting, the amount of mercury in the environment will be reduced as will the amounts of sulphur dioxide, nitrogen oxides and carbon dioxide produced by electricity generation. These chemicals are considered key indicators of pollution, sulphur dioxide, nitrogen oxides and carbon dioxide being associated with air quality, acid rain and global climate change, respectively. There can be little doubt that maximising the efficiency of lighting installations and thereby reducing the electricity consumed to provide the desired lighting conditions, would have a beneficial effect on many forms of chemical pollution.

1.9.4 Light pollution

Light, itself, can be considered a form of pollution. This is apparent by the identification of light as a potential statutory nuisance in the Clean Neighbourhoods and Environment Act 2005. Complaints about light pollution at night, or obtrusive light as it is also known, can be divided into two categories: light trespass and sky glow. Light trespass is local in that it is associated with complaints from individuals in a specific location, typically about light from a road lighting luminaire shining into a bedroom. Light trespass can be avoided by the careful selection, positioning and aiming of luminaires with appropriate luminous intensity distributions. If that fails then some form of shielding can usually be devised. Recommendations for limiting light trespass in the form of maximum illuminances on windows are given in Chapter 3 of this *Code*.

Sky glow is more diffuse than light trespass in that it can affect people over great distances and is more difficult to deal with. Complaints about sky glow originate from many people, ranging from those who have a professional interest in a dark sky, i.e. optical astronomers (McNally, 1994), to those who simply like to be able to see the stars at night. Light pollution is caused by the multiple scattering of light in the atmosphere, resulting in a diffuse distribution of luminance called sky glow. The problem this sky glow causes is twofold. The more obvious is that it reduces the luminance contrast of all the features of the night sky. A reduction in luminance contrast means that features that are naturally close to the visual threshold will be taken below threshold by the addition of the sky glow. As a result, as sky glow increases, the number of stars and other astronomical phenomena that can be seen is much reduced. This means that in most cities, it is difficult to see anything at night other than the moon and a few stars. The second is that sky glow can adversely affect the lives of flora and fauna (Rich and Longcore, 2006). For example, mammals and reptiles can be attracted or repulsed by light at night, birds can become disorientated and plants can be induced to flower at inappropriate times.

Sky glow has two components, one natural and one unnatural. Natural sky glow is light from the sun, moon, planets and stars that is scattered by interplanetary dust, and by air molecules, dust particles, water vapour and aerosols in the Earth's atmosphere, and light produced by a chemical reaction of the upper atmosphere with ultra-violet radiation from the sun. The luminance of the natural sky glow at zenith is of the order of 0.0002 cd/m². The unnatural form of light pollution is produced by light generated on Earth traversing the atmosphere and being scattered by air molecules and by water vapour, dust and aerosols in the atmosphere.

The problem in dealing with sky glow is not in measuring or predicting its effects on the visibility of the stars, but rather in agreeing what to do about it. The problem is that what constitutes the astronomer's pollution is often the business owner's commercial necessity and sometimes the citizen's preference. Residents of cities like their streets to be lit at night for the feeling of safety the lighting provides. Similarly, many roads are lit at night to enhance the safety of travel. Businesses use light to identify themselves at night and to attract customers. Further, the floodlighting of buildings and the lighting of landscapes are methods used to create an attractive environment at night. The problem of sky glow is how to strike the right balance between these conflicting desires.

One solution to this problem is to classify different areas into different environmental zones. CIE (1997) have identified four major zones ranging from areas of intrinsically dark landscapes to city centres (Table 1.2). Then a maximum upward light output ratio for any luminaires used in each zone is given, ranging from zero in areas of intrinsically dark landscapes to 25% in city centres. This approach offers some flexibility in that it recognises it is ridiculous to give the same priority to sky glow limitation in a city centre as it is in an area of intrinsically dark landscapes.

Table 1.2 The environmental zoning system of the CIE (after CIE, 1997)

Zone	Zone description
E1	Areas with intrinsically dark landscapes: National Parks, areas of outstanding natural beauty (where roads are usually unlit)
E2	Areas of 'low district brightness': Outer urban and rural residential areas (where roads are lit to residential road standard)
E3	Areas of 'middle district brightness': Generally urban residential areas (where roads are lit to traffic route standard)
E4	Areas of 'high district brightness': Generally, urban areas having mixed recreational and commercial land use with high night-time activity

Of course, light output ratio is a metric of relative light distribution and does nothing to control the total amount of light emitted. To effectively limit the contribution of human activity to sky glow, there are two complementary options. The first option is to limit the amount of light used at night. The second option is to pay careful attention to the timing of the use of light. Light pollution is unlike chemical pollution in that when the light source is extinguished, the pollution goes away very rapidly. This suggests that a curfew defining the times when lighting can and cannot be used could have a dramatic effect on the prevalence of sky glow.

1.10 The future

Our understanding of the effects of light on people rests on lighting research which is, currently, very active. This is for three reasons. The first is the rapid introduction of solid state light sources to the lighting market. The properties of this technology have raised a number of questions about colour metrics, glare perception and flicker that require answering. The second is the exploration of the non-visual effects of light. The possibility that exposure to light can have significant effects on human health and functioning beyond the obvious visual effect implies another basis for making lighting recommendations. However, more knowledge is required before non-visual effects can be applied with confidence, particularly about the effects on healthy people exposed to light by day. The third is the renewed emphasis placed on energy consumption by the threat of global warming. Lighting is a major user of electrical energy, a use that needs to be reduced. Together, these influences are driving lighting research forward. The findings from such research are likely to influence lighting recommendations in the future.

Chapter 2: Indoor workplaces

This chapter of the *Code* is based on BS EN 12464-1: 2011: *Light and lighting – Lighting of work places – Part 1: Indoor work places* (BSI, 2011a). All of the text in this chapter that is in italics and the tables of lighting requirements have been taken directly from the standard. Whilst the standard is an important document that sets out what it considers to be a sensible code of practice, it does not have any direct statutory weight, and thus it should be treated as advice rather than an absolute set of requirements. There are some occasions when standards and codes may be written into contracts and in those situations, it is sensible for any deviations from its requirements to be agreed by the designer and the client.

The other material in this chapter is advice on how best to apply the standard whilst at the same time following what the Society of Light and Lighting regards to be good lighting practice. It should be noted that this chapter does not include all of BS EN 12464-1 (BSI, 2011a); the scope, normative references, terms and definitions together with annexes A and B have been omitted as they are either not appropriate to this document or they are covered elsewhere in this *Code*.

2.1 Lighting design criteria

This section covers the material in section 4 of BS EN 12464-1 (BSI, 2011a).

2.1.1 Luminous environment
For good lighting practice it is essential that as well as the required illuminances, additional qualitative and quantitative needs are satisfied.

Lighting requirements are determined by the satisfaction of three basic human needs:

- *visual comfort, where the workers have a feeling of well-being; in an indirect way this also contributes to a higher productivity level and a higher quality of work;*

- *visual performance, where the workers are able to perform their visual tasks, even under difficult circumstances and during longer periods;*

- *safety.*

Main parameters determining the luminous environment with respect to artificial light and daylight are:

- *luminance distribution;*

- *illuminance;*

- *directionality of light, lighting in the interior space;*

- *variability of light (levels and colour of light);*

- *colour rendering and colour appearance of the light;*

- *glare;*

- *flicker.*

Values for illuminance and its uniformity, discomfort glare and colour rendering index are given in section 2.2; other parameters are described in the rest of section 2.1.

Note: In addition to the lighting there are other visual ergonomic parameters which influence visual performance, such as:

- *the intrinsic task properties (size, shape, position, colour and reflectance properties of detail and background),*

- *ophthalmic capacity of the person (visual acuity, depth perception, colour perception),*

- *intentionally improved and designed luminous environment, glare-free illumination, good colour rendering, high contrast markings and optical and tactile guiding systems can improve visibility and sense of direction and locality. See CIE Guidelines for Accessibility: Visibility and Lighting Guidelines for Older Persons and Persons with Disabilities.*

Attention to these factors can enhance visual performance without the need for higher illuminance.

2.1.2   Luminance distribution
2.1.2.1 General

The luminance distribution in the visual field controls the adaptation level of the eyes which affects task visibility.

A well balanced adaptation luminance is needed to increase:

- *visual acuity (sharpness of vision);*

- *contrast sensitivity (discrimination of small relative luminance differences);*

- *efficiency of the ocular functions (such as accommodation, convergence, pupillary contraction, eye movements, etc).*

The luminance distribution in the visual field also affects visual comfort. The following should be avoided for the reasons given:

- *too high luminances which can give rise to glare;*

- *too high luminance contrasts which will cause fatigue because of constant re-adaptation of the eyes;*

- *too low luminances and too low luminance contrasts which result in a dull and non-stimulating working environment.*

To create a well balanced luminance distribution the luminances of all surfaces shall be taken into consideration and will be determined by the reflectance and the illuminance on the surfaces. To avoid gloom and to raise adaptation levels and comfort of people in buildings, it is highly desirable to have bright interior surfaces particularly the walls and ceiling.

The lighting designer shall consider and select the appropriate reflectance and illuminance values for the interior surfaces based on the guidance below.

2.1.2.2 Reflectance of surfaces
Recommended reflectances for the major interior diffuse surfaces are:

- *ceiling: 0.7 to 0.9;*

- *walls: 0.5 to 0.8;*

- *floor: 0.2 to 0.4.*

Note: The reflectance of major objects (like furniture, machinery, etc) should be in the range of 0.2 to 0.7.

The reflectance of surfaces is often a parameter that is outside the control of the lighting designer; however, where possible, the designer should try to persuade those responsible to aim for reflectances in the above range. The values given in this edition of the *Code* are slightly tighter than the 2009 edition which gave the following values:

- Ceiling: 0.6–0.9

- Walls: 0.3–0.8

- Working planes: 0.2–0.6

- Floor: 0.1–0.5

The current recommendations for reflectances have increased the lower limits for walls and ceilings; this will help to create lighter spaces and will reduce energy consumption as the amount of inter-reflected light will increase. Note that the concept of working planes is not used in this edition of the *Code*.

2.1.2.3 Illuminance on surfaces
In all enclosed places the maintained illuminances on the major surfaces shall have the following values:

- $\bar{E}_m > 50\ lx$ *with* $U_o \geq 0.10$ *on the walls and*

- $\bar{E}_m > 30\ lx$ *with* $U_o \geq 0.10$ *on the ceiling.*

Note 1:   It is recognised that, in some places such as racked storage places, steelworks, railway terminals, etc, due to the size, complexity and operational constraints, the desired light levels on these surfaces will not be practical to achieve. In these places reduced levels of the recommended values are accepted. Whilst this note acknowledges it may be difficult to achieve in some complex areas it is vital that measures are taken to ensure that the correct illuminance is achieved in all areas where there is a visual task. Example of such areas include the vertical face of goods stored in a racking system and control valves and gauges located on complex plant structures.

Note 2:   In some enclosed places such as offices, education, health care and general areas of entrance, corridors, stairs, etc, the walls and ceiling need to be brighter. In these places it is recommended that the maintained illuminances on the major surfaces should have the following values: $\bar{E}_m > 75\ lx$ with $U_o \geq 0.10$ on the walls and $\bar{E}_m > 50\ lx$ with $U_o \geq 0.01$ on the ceiling.

The previous edition of the *Code* provided requirements for the illuminance of the ceiling and walls as a fraction of the working plane illuminance. The requirements were:

- Ceiling: 0.3–0.9

- Walls: 0.5–0.6

Now that the concept of a working plane has been removed, these fractions relate to the average floor illuminance. In general, the use of these old ratios results in higher wall and ceiling illuminance values and hence a better visual environment; whilst these higher values are not required by BS EN 12464-1 (BSI, 2011a), the higher values do in general provide a brighter interior and their application should be looked on as best practice.

2.1.3 Illuminance

2.1.3.1 General

The illuminance and its distribution on the task area and on the surrounding area have a great impact on how quickly, safely and comfortably a person perceives and carries out the visual task. All values of illuminances specified in this European Standard are maintained illuminances and fulfil visual comfort and performance needs.

All maintained illuminance and uniformity values are dependent upon the grid definition (see sections 2.1.4 and 15.2.2).

Maintained illuminance is the value of illuminance achieved just before maintenance is carried out. There are a number of factors that cause the illuminance delivered by a lighting system to fall with time. Chapter 18 explains the various factors and gives tables to predict their magnitude.

2.1.3.2 Scale of illuminance

To give a perceptual difference the recommended steps of illuminance (in lx) are according to EN 12665:

$$20–30–50–75–100–150–200–300–500–750–1000–1500–2000–3000–5000$$

The terms and definitions from BS EN 12665 (BSI, 2011b) are given in the glossary of this *Code*.

2.1.3.3 Illuminances on the task area

The values given in section 2.2 are maintained illuminances over the task area on the reference surface which can be horizontal, vertical or inclined. The average illuminance for each task shall not fall below the value given in section 2.2, regardless of the age and condition of the installation. The values are valid for normal visual conditions and take into account the following factors:

- *psycho-physiological aspects such as visual comfort and well-being;*

- *requirements for visual tasks;*

- *visual ergonomics;*

- *practical experience;*

- *contribution to functional safety;*

- *economy.*

The value of illuminance may be adjusted by at least one step in the scale of illuminances (see 2.1.3.2), if the visual conditions differ from the normal assumptions.

The required maintained illuminance should be increased when:

- *visual work is critical;*

- *errors are costly to rectify;*

- *accuracy, higher productivity or increased concentration is of great importance;*

- *task details are of unusually small size or low contrast;*

- *the task is undertaken for an unusually long time;*

- *the visual capacity of the worker is below normal.*

The required maintained illuminance may be decreased when:

- *task details are of an unusually large size or high contrast;*

- *the task is undertaken for an unusually short time.*

Note: For visually impaired people special requirements can be necessary with regard to illuminances and contrasts.

The size and position of the task area should be stated and documented.

For work stations where the size and/or location of the task area(s) is/are unknown, either:

- *the whole area is treated as the task area or*

- *the whole area is uniformly ($U_o \geq 0.40$) lit to an illuminance level specified by the designer; if the task area becomes known, the lighting scheme shall be re-designed to provide the required illuminances.*

If the type of the task is not known the designer has to make assumptions about the likely tasks and state task requirements.

Where the location of a task area within a space is unknown it is very wasteful of energy to light the whole space just for one particular task carried out over a relatively small area. Possible solutions include the use of individually dimmable luminaires to give flexibility to the lighting system, or the provision of task lighting where needed.

2.1.3.4 Illuminance of the immediate surrounding area

Large spatial variations in illuminances around the task area can lead to visual stress and discomfort.

The illuminance of the immediate surrounding area shall be related to the illuminance of the task area and should provide a well-balanced luminance distribution in the visual field. The immediate surrounding area should be a band with a width of at least 0.5 m around the task area within the visual field.

The illuminance of the immediate surrounding area may be lower than the illuminance on the task area but shall be not less than the values given in Table 2.1.

In addition to the illuminance on the task area the lighting shall provide adequate adaptation luminance in accordance with section 2.1.2.

The size and position of the immediate surrounding area should be stated and documented.

Table 2.1 Relationship of illuminances on immediate surroundings to the illuminance on the task area

Illuminance on the task area E_{task} / lx	Illuminance on immediate surrounding areas / lx
≥ 750	500
500	300
300	200
200	150
150	E_{task}
100	E_{task}
< 50	E_{task}

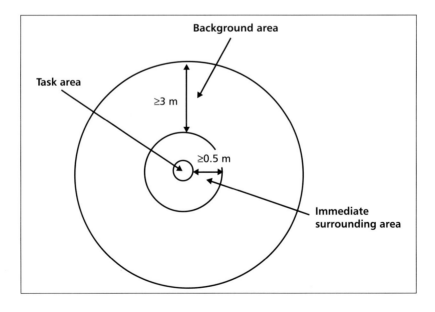

Figure 2.1 Task, immediate surrounding and background areas

In most spaces, there are various visual tasks with differing degrees of difficulty. Although it is possible to use general lighting systems to give flexibility of task location, the average illuminance is determined by the needs of the most exacting task. It is almost always very wasteful to illuminate all areas to the same level and non-uniform lighting should be provided by local or localised lighting systems. If control systems are used, individuals may be able to adjust their levels of supplementary task lighting, furthermore; absence detection may also switch off luminaires in unoccupied areas. Whatever lighting system is used, excessive variations of horizontal illuminance across an interior should be avoided and thus it is important to ensure that around each task area, there is a surrounding area when the illuminance is only slightly less.

2.1.3.5 Illuminance on the background area

In indoor work places, particularly those devoid of daylight, a large part of the area surrounding an active and occupied task area needs to be illuminated. This area known as the 'background area' should be a border at least 3 m wide adjacent to the immediate surrounding area within the limits of the space and shall be illuminated with a maintained illuminance of ⅓ of the value of the immediate surrounding area.

The size and position of the background area should be stated and documented.

2.1.3.6 Illuminance uniformity

In the task area, the illuminance uniformity (U_o) shall be not less than the minimum uniformity values given in the tables of section 2.2.

For lighting from artificial lighting or roof lights the illuminance uniformity:

- *in the immediate surrounding area shall be $U_o \geq 0.40$;*

- *on the background area shall be $U_o \geq 0.10$.*

For lighting from windows:

- *in larger areas, activity areas and background areas the available daylight decreases rapidly with the distance from the window; the additional benefits of daylight (see 2.1.12) can compensate for the lack of uniformity.*

2.1.4   Illuminance grid

Grid systems shall be created to indicate the points at which the illuminance values are calculated and verified for the task area(s), immediate surrounding area(s) and background area(s).

Grid cells approximating to a square are preferred; the ratio of length to width of a grid cell shall be kept between 0.5 and 2. (The same method of grid specification is used in other parts of this *Code* and there is a worked example of how to apply the formulae given in section 15.2.2.) *The maximum grid size shall be:*

$$p = 0.2 \times 5^{\log d} \tag{2.1}$$

where $p \leq 10$ m

d is the longer dimension of the calculation area (m), however if the ratio of the longer to the shorter side is 2 or more then d becomes the shorter dimension of the area, and

p is the maximum grid cell size (m).

The number of points in the relevant dimension is given by the nearest whole number that is equal to or greater than d/p.

The resulting spacing between the grid points is used to calculate the nearest whole number of grid points in the other dimension. This will give a ratio of length to width of a grid cell close to 1.

When the area of a grid is a room or part of a room, then a band of 0.5 m from the walls is excluded from the calculation area except when a task area is in or extends into this border area.

An appropriate grid size shall be applied to walls and ceiling and a border of 0.5 m may be applied also.

Note: The grid point spacing should not coincide with the luminaire spacing.

2.1.5   Glare
2.1.5.1 General
Glare is the sensation produced by bright areas within the visual field, such as lit surfaces, parts of the luminaires, windows and/or roof lights. Glare shall be limited to avoid errors, fatigue and accidents. Glare can be experienced either as discomfort glare or as disability glare. In interior work places disability glare is not usually a major problem if discomfort glare limits are met.

Glare caused by reflections in specular surfaces is usually known as veiling reflections or reflected glare.

Note: Special care is needed to avoid glare when the direction of view is above horizontal.

2.1.5.2 Discomfort glare
For the rating of discomfort glare from windows there is currently no standardized method.

The rating of discomfort glare caused directly from the luminaires of an indoor lighting installation shall be determined using the CIE Unified Glare Rating (UGR) tabular method, based on the formula 2.2:

$$UGR = 8\log_{10}\left[\sum \frac{0.25}{L_b} \frac{L^2\omega}{p^2}\right] \tag{2.2}$$

where

L_b is the background luminance in $cd \cdot m^{-2}$, calculated as $E_{ind} \cdot \pi^{-1}$, in which E_{ind} is the vertical indirect illuminance at the observer's eye in $cd \cdot m^{-2}$,

L is the luminance of the luminous parts of each luminaire in the direction of the observer's eye in $cd \cdot m^{-2}$,

ω is the solid angle in steradian of the luminous parts of each luminaire at the observer's eye,

p is the Guth position index for each individual luminaire which relates to its displacement from the line of sight.

All assumptions made in the determination of UGR shall be stated in the scheme documentation. The UGR value of the lighting installation shall not exceed the value given in section 2.2.

The recommended limiting values of the UGR form a series whose steps indicate noticeable changes in glare.

The series of UGR is: 10, 13, 16, 19, 22, 25, 28.

Note 1:   The variations of UGR within the room can be determined using the comprehensive tables for different observer positions, as detailed in CIE 117-1995 (CIE, 1995a).

Note 2:   If the maximum UGR value in the room is higher than the UGR limit given in section 2.2, information on appropriate positions for work stations within the room should be given.

Note 3:   If the tabular method is not applicable and the observer position and the viewing directions are known the UGR value can be determined by using the formula. However limited research has been done, to determine the applicability of existing limiting values. Limits for this condition are under consideration.

The discomfort experienced when some elements of an interior have a much higher luminance than others can be immediate but sometimes may only become evident after prolonged exposure. The degree of discomfort experienced will depend on the luminance and size of the glare source, the luminance of the background against which it is seen and the position of the glare source relative to the line of sight. A high source luminance, large source area, low background luminance and a position close to the line of sight all increase discomfort glare. Unfortunately, most of the variables available to the designer alter more than one factor. For example, changing the luminaire to reduce the source luminance may also reduce the background luminance. These factors could counteract each other, resulting in no reduction in discomfort glare. However, as a general rule, discomfort glare can be avoided by the choice of luminaire layout and orientation, and the use of high reflectance surfaces for the ceiling and upper walls. Thus if a proposed lighting scheme has a glare rating in excess of the limiting value then the three options open to the designer are to change either the luminaire type, the luminaire orientation or increase the reflectance of the walls and ceiling. In general, glare in a room is worse where the field of view contains a lot of luminaires so if the field of view of people performing a glare-sensitive task is such that they can only see a few luminaires then a scheme that nominally breaks the limiting value of glare rating may be acceptable, however, there is no standardised way to predict this effect.

As discomfort glare is worse in a room where the walls and ceiling have low reflectance, if the reflectances of a space are below the values used by the lighting designer there may well be problems with discomfort glare.

The use and production of glare tables are discussed in Chapter 12 *Photometric Datasheets*.

2.1.5.3 Shielding against glare
Bright sources of light can cause glare and can impair the vision of objects. It shall be avoided for example by suitable shielding of lamps and roof lights, or suitable shading from bright daylight through windows.

For luminaires, the minimum shielding angles (see Figure 2.2) in the visual field given in Table 2.2 shall be applied for the specified lamp luminances.

Note: The values given in Table 2.2 do not apply to up-lighters or to luminaires with a downward component only mounted below normal eye level.

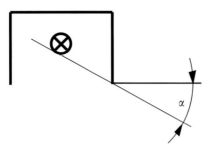

Figure 2.2 Shielding angle α

Table 2.2 Minimum shielding angles at specified lamp luminances

Lamp luminance / kcd·m⁻²	Minimum shielding angle α
20 to < 50	15°
50 to < 500	20°
≥500	30°

Whilst it is important to ensure that the lamp is shielded from direct view, it is also important to ensure that any high luminance areas within the luminaire are equally well shielded from view.

2.1.5.4 Veiling reflections and reflected glare

High brightness reflections in the visual task can alter task visibility, usually detrimentally. Veiling reflections and reflected glare can be prevented or minimised by the following measures:

- *arrangement of work stations with respect to luminaires, windows and roof lights;*

- *surface finish (matt surfaces);*

- *luminance restriction of luminaires, windows and roof lights;*

- *bright ceiling and bright walls.*

Veiling reflections are high luminance reflections which overlay the detail of the task. Such reflections may be sharp-edged or vague in outline, but regardless of form they may affect task performance and cause discomfort. Task performance will be affected because veiling reflections usually reduce the contrast of a task, making task details difficult to see and may give rise to discomfort.

Two conditions have to be met before veiling reflections occur:

- part of the task, task detail or background, or both, has to be glossy to some degree

- part of the interior, called the 'offending zone', which specularly reflects towards the observer has to have a high luminance.

The most common sources of veiling reflections are windows and luminaires. Generally applicable methods of avoiding veiling reflections are to use matt materials in task areas, to arrange the geometry of the viewing situation so that the luminance of the offending zone is low or reduce the luminance by, for example, using curtains or blinds on windows. Figure 2.3 illustrates the visual problems caused by veiling reflections.

Figure 2.3 Veiling reflections

2.1.6 Lighting in the interior space
2.1.6.1 General
In addition to task lighting the volume of space occupied by people should be lit. This light is required to highlight objects, reveal texture and improve the appearance of people within the space. The terms 'mean cylindrical illuminance', 'modelling' and 'directional lighting' describe the lighting conditions.

The use of mean cylindrical illuminance in a lighting code is new, and it represents a big step forward in recognising the importance of the visibility of objects, particularly people's faces, within a space. The calculation of cylindrical illuminance from a point source is discussed in section 10.1.2 and the derivation of cylindrical illuminance from the illuminance on six faces of a cube at the same point is given in section 13.4.

2.1.6.2 Mean cylindrical illuminance requirement in the activity space
Good visual communication and recognition of objects within a space require that the volume of space in which people move or work shall be illuminated. This is satisfied by providing adequate mean cylindrical illuminance, $\bar{E}_z$, in the space.

The maintained mean cylindrical illuminance (average vertical plane illuminance) in the activity and interior areas shall be not less than 50 lx with $U_o \geq 0.10$, on a horizontal plane at a specified height, for example 1.2 m for sitting people and 1.6 m for standing people above the floor.

Note: In areas, where good visual communication is important, especially in offices, meeting and teaching areas, $\bar{E}_z$ should be not less than 150 lx with $U_o \geq 0.10$.

2.1.6.3 Modelling

The general appearance of an interior is enhanced when its structural features, the people and objects within it are lit so that form and texture are revealed clearly and pleasingly.

The lighting should not be too directional or it will produce harsh shadows, neither should it be too diffuse or the modelling effect will be lost entirely, resulting in a very dull luminous environment. Multiple shadows caused by directional lighting from more than one position should be avoided as this can result in a confused visual effect.

Modelling describes the balance between diffuse and directed light and should be considered.

Note 1:   The ratio of cylindrical to horizontal illuminance at a point is an indicator of modelling. The grid points for cylindrical and horizontal illuminances should coincide.

Note 2:   For uniform arrangement of luminaires or roof lights a value between 0.30 and 0.60 is an indicator of good modelling.

Note 3:   Daylight is distributed predominantly horizontally from windows. The additional benefits of daylight (see 2.1.12) can compensate for its effect on modelling values, and modelling values from daylight can be extended from the range indicated.

The values indicated in *Note 2* are purely intended for spaces where people's faces are being lit and other values may be needed elsewhere; for example, in applications such as retail and display lighting, it is often desirable to go outside the range of modelling suggested in *Note 2* above when putting particular emphasis on to a given object.

2.1.6.4 Directional lighting of visual tasks

Lighting from a specific direction can reveal details within a visual task, increasing their visibility and making the task easier to perform. Unintended veiling reflections and reflected glare should be avoided, see 2.1.5.4.

Harsh shadows that interfere with the visual task should be avoided. But some shadows help to increase the visibility of the task.

2.1.7   Colour aspects

2.1.7.1 General

The colour qualities of a near-white lamp or transmitted daylight are characterised by two attributes:

● *the colour appearance of the light;*

● *its colour rendering capabilities, which affect the colour appearance of objects and persons.*

These two attributes shall be considered separately.

2.1.7.2 Colour appearance

The colour appearance of a lamp refers to the apparent colour (chromaticity) of the light emitted. It is quantified by its correlated colour temperature (T_{CP}).

Colour appearance of daylight varies throughout the day.

Colour appearance of artificial light can also be described as in Table 2.3.

Note: Colour appearance does not uniquely specify the colour appearance of a light source. It is possible for two sources with the same colour temperature to have different appearances, one looking slightly purple and the other looking greenish. See section 16.2 for details on how to characterise the colour properties of a light source.

Table 2.3 Lamp colour appearance groups

Colour appearance	Correlated colour temperature T_{CP}
Warm	Below 3300 K
Intermediate	3300 to 5300 K
Cool	Above 5300 K

The choice of colour appearance is a matter of psychology, aesthetics and what is considered to be natural. The choice will depend on illuminance level, colours of the room and furniture, surrounding climate and the application. In warm climates generally a cooler light colour appearance is preferred, whereas in cold climates a warmer light colour appearance is preferred.

In section 2.2, for specific applications a restricted band of suitable colour temperatures is given. These are applicable for daylighting as well as artificial lighting.

2.1.7.3 Colour rendering
For visual performance and the feeling of comfort and well being colours in the environment, of objects and of human skin, shall be rendered naturally, correctly and in a way that makes people look attractive and healthy.

To provide an objective indication of the colour rendering properties of a light source the general colour rendering index Ra is used. The maximum value of Ra is 100.

The minimum values of colour rendering index for distinct types of interiors (areas), tasks or activities are given in the tables in section 2.2.

Safety colours according to BS ISO 3864-1 (ISO, 2009) shall always be recognisable as such.

Note 1: Colour rendering of light from a light source may be reduced by optics, glazing and coloured surfaces.

Note 2: For accurate rendition of colours of objects and human skin the appropriate individual special colour rendering index (R_i) should be considered.

Full details of the basis of colour rendering are given in section 16.2.4. The *Ra* index used above is based on eight colour samples and is often referred to as Ra_8. Where colour quality is particularly important, for example in art galleries, the Ra_{14} index is sometimes used. There are limitations on how effective a single number can be at describing the colour rendering qualities of a light source, as a source may be very good at rendering most colours but be very poor with one particular colour. Thus for particular applications where the colour of certain objects is critical it is recommended that a visual appraisal of the object under the light source is carried out.

2.1.8   Flicker and stroboscopic effects

Flicker causes distraction and can give rise to physiological effects such as headaches.

Stroboscopic effects can lead to dangerous situations by changing the perceived motion of rotating or reciprocating machinery.

Lighting systems should be designed to avoid flicker and stroboscopic effects.

2.1.9   Lighting of work stations with Display Screen Equipment (DSE)
2.1.9.1 General

The lighting for the DSE work stations shall be appropriate for all tasks performed at the work station, e.g. reading from the screen, reading printed text, writing on paper, keyboard work.

For these areas the lighting criteria and system shall be chosen in accordance with type of area, task or activity from the schedule in section 2.2.

Reflections in the DSE and, in some circumstances, reflections from the keyboard can cause disability and discomfort glare. It is therefore necessary to select, locate and arrange the luminaires to avoid high brightness reflections.

The designer shall determine the offending mounting zone and shall choose equipment and plan mounting positions which will cause no disturbing reflections.

In general, it is a much better strategy to arrange the luminaires around a work station in such a way that it is unlikely that they cause reflections in a screen rather than relying on the luminance limits set out in section 2.1.9.2. This is because by restricting the light at higher angles, there is less light able to reach the walls of a room, and the space may start to appear dark. Also, with restricted light at high angles, it may be difficult to achieve necessary mean cylindrical illuminance for the space (see section 2.1.6.2).

Figure 2.4 shows a method of determining which luminaires may cause problems with reflections in a screen.

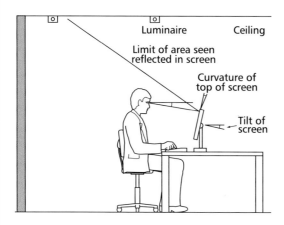

Figure 2.4 Typical geometry for eye, screen and luminaires

2.1.9.2 Luminaire luminance limits with downward flux

Light can lower the contrast of the presentation on a DSE by:

- *veiling reflection caused by the illuminance on the display's surface and*

- *luminances from luminaires and bright surfaces reflecting in the display.*

BS EN ISO 9241-307 (ISO, 2008a) gives requirements for the visual qualities of displays concerning unwanted reflections.

This subclause describes luminance limits for luminaires which can be reflected in DSE for normal viewing directions.

Table 2.4 gives the limits of the average luminaire luminance at elevation angles of 65° and above from the downward vertical, radially around the luminaires, for work stations where display screens which are vertical or inclined up to 15° tilt angle are used.

Note: Section 12.2.7 gives information on normalised luminance tables for luminaires that may be used to check if a given luminaire complies with the requirements in Table 2.4.

Table 2.4 Average luminance limits of luminaires, which can be reflected in flat screens

Screen high state luminance	High luminance screen $L > 200$ cd·m^{-2}	Medium luminance screen $L \leq 200$ cd·m^{-2}
Case A *(positive polarity and normal requirements concerning colour and details of the shown information, as used in office, education, etc)*	*≤ 3000 cd·m^{-2}*	*≤ 1500 cd·m^{-2}*
Case B *(negative polarity and/or higher requirements concerning colour and details of the shown information, as used for CAD colour inspection, etc)*	*≤ 1500 cd·m^{-2}*	*≤ 1000 cd·m^{-2}*

Note: *Screen high state luminance (see BS EN ISO 9241-302 (ISO, 2008b)) describes the maximum luminance of the white part of the screen and this value is available from the manufacturer of the screen.*

If a high luminance screen is intended to be operated at luminances below 200 cd·m^{-2} the conditions specified for a medium luminance screen shall be considered.

Some tasks, activities or display screen technologies, particularly high gloss screens, require different lighting treatment (e.g. lower luminance limits, special shading, individual dimming, etc).

In areas of industrial activities and crafts screens are sometimes protected by additional front glasses. The unwanted reflections on these protection glasses have to be reduced by suitable methods (such as antireflection treatment, tilting of the protection glass or by shutters).

2.1.10   Maintenance factor

The lighting scheme should be designed with an overall maintenance factor (MF) calculated for the selected lighting equipment, environment and specified maintenance schedule.

The recommended illuminance for each task is given as maintained illuminance. The maintenance factor depends on the maintenance characteristics of the lamp and control gear, the luminaire, the environment and the maintenance programme.

The lighting scheme should be designed with the overall MF for the selected lamp(s), luminaire(s), surface reflectances, environment and specified maintenance schedule.

For daylight calculations, reduction of glazing transmittance due to dirt deposition should be taken into account.

The designer shall:

● *state the MF and list all assumptions made in the derivation of the value,*

● *specify lighting equipment suitable for the application environment and prepare a comprehensive maintenance schedule to include frequency of lamp replacement, luminaire, room and glazing cleaning intervals and cleaning method.*

The MF has a great impact on energy efficiency. The assumptions made in the derivation of the MF shall be optimized in a way that leads to a high value. Guidance on the derivation of MF for artificial lighting systems can be found in CIE 97-2005 (CIE, 2005).

Chapter 18 of this *Code* gives a comprehensive method for the evaluation of maintenance factor; it is based on CIE 97: 2005 and other similar documents.

In selecting lighting equipment and a maintenance programme, the lighting designer should seek to keep the overall maintenance as high as possible without imposing too onerous a maintenance schedule on the building owner.

2.1.11   Energy efficiency requirements

Lighting should be designed to meet the lighting requirements of a particular task or space in an energy efficient manner. It is important not to compromise the visual aspects of a lighting installation simply to reduce energy consumption. Light levels as set in this European Standard are minimum average illuminance values and need to be maintained.

Energy savings can be made by harvesting daylight, responding to occupancy patterns, improving maintenance characteristics of the installation, and making full use of controls.

The amount of daylight varies throughout the day depending on climate conditions. In addition, in interiors with side windows the available daylight decreases rapidly with the distance from the window. Supplementary lighting may be needed to ensure the required illuminance levels at the work station are achieved and to balance the luminance distribution within the room. Automatic or manual switching and/or dimming can be used to ensure appropriate integration between artificial lighting and daylight.

A procedure for the estimation of the energy requirements of a lighting installation is given in BS EN 15193 (BSI, 2007a). It gives a methodology for the calculation of a lighting energy numeric indicator (LENI), representing the energy performance of lighting of buildings. This indicator may be used for single rooms on a comparative basis only, as the benchmark values given in the BS EN 15193 are drawn up for a complete building.

Providing energy efficient lighting is of utmost importance and using a metric like LENI, which expresses the energy used by lighting in terms of kilowatt hours per year, is the best way to characterise the energy performance of a lighting system. However, there are several other metrics in use and some of them are used in documents such as the Building Regulations, thus they must also be used. See Chapter 6 for more information.

2.1.12 Additional benefits of daylight

Daylight can supply all or part of the lighting for visual tasks, and therefore offers potential energy savings. Additionally, it varies in level, direction and spectral composition with time and provides variable modelling and luminance patterns, which is perceived as being beneficial for people in indoor working environments. Windows are strongly favoured in work places for the daylight they deliver, and for the visual contact they provide with the outside environment. However, it is also important to ensure windows do not cause visual or thermal discomfort, or a loss of privacy.

There is more information on the use of daylight in Chapter 5.

2.1.13 Variability of light

Light is important to people's health and wellbeing. Light affects the mood, emotion and mental alertness of people. It can also support and adjust the circadian rhythms and influence people's physiological and psychological state. Up to date research indicates that these phenomena, in addition to the lighting design criteria defined in BS EN 12464-1, can be provided by the so-called 'non-image forming' illuminances and colour appearance of light. Varying lighting conditions in time by higher illuminance, luminance distribution and wider range of colour temperature than specified in this European Standard with daylight and/or dedicated artificial lighting solutions can stimulate people and enhance their wellbeing. The recommended bands of variation are under consideration.

Whilst there may be beneficial effects due to changing the colour and level of light during the course of a day, research in this area is very limited and the existence of these effects has yet to be unequivocally demonstrated. Whilst it is clear that high levels of light (usually from daylight) have an important role in the process of circadian entrainment, the biological effects of lower levels of light normally associated with electric lighting are less clear.

2.2 Schedule of lighting requirements

2.2.1 Composition of the tables

Column 1 *lists the reference number for each interior area, task or activity.*

Column 2 *lists those areas, tasks or activities for which specific requirements are given. If the particular interior area, task area or activity area is not listed, the values given for a similar, comparable situation should be adopted.*

Column 3 *gives the maintained illuminance* $\bar{E}_{m}$ *on the reference surface (see 2.1.3) for the interior (area),* task or activity given in Column 2.

Note 1:   The maintained illuminance in some circumstances may need to be increased (see 2.1.3.3).

Note 2:   Lighting control can be required to achieve adequate flexibility for the variety of tasks performed.

Column 4 *gives the maximum UGR limits (Unified Glare Rating limit, UGR$_L$) that are applicable to the situation listed in Column 2.*

Column 5 *gives the minimum illuminance uniformity U$_o$ on the reference surface for the maintained illuminance given in Column 3.*

Column 6 *gives the minimum colour rendering indices (Ra) (see 2.1.7.3) for the situation listed in Column 2.*

Column 7 *gives specific requirements for the situations listed in Column 2.*

The majority of the specific requirements in Column 7 are taken from BS EN 12464-1 (BSI, 2011a), however, there are a number of additional recommendations from the SLL also given in this column. The requirements of BS EN 12464-1 are given italics whilst the SLL recommendations are not.

2.2.2   Schedule of interior areas, tasks and activities

Table 2.5 – *Traffic zones inside buildings*

Table 2.6 – *General areas inside buildings – Rest, sanitation and first aid rooms*

Table 2.7 – *General areas inside buildings – Control rooms*

Table 2.8 – *General areas inside buildings – Store rooms, cold stores*

Table 2.9 – *General areas inside buildings – Storage rack areas*

Table 2.10 – *Industrial activities and crafts – Agriculture*

Table 2.11 – *Industrial activities and crafts – Bakeries*

Table 2.12 – *Industrial activities and crafts – Cement, cement goods, concrete, bricks*

Table 2.13 – *Industrial activities and crafts – Ceramics, tiles, glass, glassware*

Table 2.14 – *Industrial activities and crafts – Chemical, plastics and rubber industry*

Table 2.15 – *Industrial activities and crafts – Electrical and electronic industry*

Table 2.16 – *Industrial activities and crafts – Food stuffs and luxury food industry*

Table 2.17 – *Industrial activities and crafts – Foundries and metal casting*

Table 2.18 – *Industrial activities and crafts – Hairdressers*

Table 2.19 – *Industrial activities and crafts – Jewellery manufacturing*

Table 2.5 Traffic zones inside buildings

Ref No.	Type of area, task or activity	$\bar{E}_m$ / lx	UGR_L	U_o	R_a	Specific requirements
2.5.1	Circulation areas and corridors	100	28	0.40	40	• Illuminance at floor level • Ra and UGR similar to adjacent areas • 150 lx if there are vehicles on the route • The lighting of exits and entrances shall provide a transition zone to avoid sudden changes in illuminance between inside and outside by day or night • Care should be taken to avoid glare to drivers and pedestrians
2.5.2	Stairs, escalators, travelators	100	25	0.40	40	• Requires enhanced contrast on the steps • For escalators below step lighting may be effective in providing contrast between the steps and risers • Increased illuminance may be necessary at the entrances and exits of escalators and travelators
2.5.3	Elevators, lifts	100	25	0.40	40	Light level in front of the lift should be at least $\bar{E}_m = 200$ lx
2.5.4	Loading ramps/bays	150	25	0.40	40	Avoid glare to drivers of vehicles approaching the loading bay. Light and mark clearly the edge of the loading bay

Table 2.6 General areas inside buildings – Rest, sanitation and first aid rooms

Ref No.	Type of area, task or activity	$\bar{E}_m$ / lx	UGR_L	U_o	R_a	Specific requirements
2.6.1	Canteens, pantries	200	22	0.40	80	• The lighting should aim to provide a relaxed but interesting atmosphere • In food storage area luminaires should be capable of being washed or hosed down in safety
2.6.2	Rest rooms	100	22	0.40	80	Lighting should be different in style from the work areas
2.6.3	Rooms for physical exercise	300	22	0.40	80	
2.6.4	Cloakrooms, washrooms, bath-rooms, toilets	200	25	0.40	80	• In each individual toilet if these are fully enclosed • In bathrooms luminaires must be suitable for damp and humid situations
2.6.5	Sick bay	500	19	0.60	80	
2.6.6	Rooms for medical attention	500	16	0.60	90	$4000\ K \leq T_{CP} \leq 5000\ K$

Table 2.7 General areas inside buildings – Control rooms

Ref No.	Type of area, task or activity	$\bar{E}_m$ /lx	UGR_L	U_o	R_a	Specific requirements
2.7.1	*Plant rooms, switch gear rooms*	*200*	*25*	*0.40*	*60*	• Localised lighting of control display and control desks may be appropriate • Care should be taken to avoid shadows and veiling reflections on the instruments and VDT screens
2.7.2	*Telex, post room, switchboard*	*500*	*19*	*0.60*	*80*	In switchboard areas, avoid veiling reflections from controls. Too high an illuminance may reduce the visibility of signal lights. Supplementary local lighting may be desirable where directories are used

Table 2.8 General areas inside buildings – Store rooms, cold stores

Ref No.	Type of area, task or activity	$\bar{E}_m$ /lx	UGR_L	U_o	R_a	Specific requirements
2.8.1	*Store and stockrooms*	*100*	*25*	*0.40*	*60*	• *200 lx if continuously occupied* • If small items that are visually difficult to identify are stored then 300 lx or supplementary local lighting may be needed
2.8.2	*Dispatch packing handling areas*	*300*	*25*	*0.60*	*60*	

Table 2.9 General areas inside buildings – Storage rack areas

Ref No.	Type of area, task or activity	$\bar{E}_m$ /lx	UGR_L	U_o	R_a	Specific requirements
2.9.1	*Gangways: unmanned*	20	–	0.40	40	• *Illuminance at floor level* • Supplementary lighting may be required for maintenance
2.9.2	*Gangways: manned*	150	22	0.40	60	*Illuminance at floor level*
2.9.3	*Control stations*	150	22	0.60	80	Avoid glare to operator, local lighting should be considered
2.9.4	*Storage rack face*	200	–	0.40	60	*Vertical illuminance, portable lighting may be used*

Table 2.10 Industrial activities and crafts – Agriculture

Ref No.	Type of area, task or activity	$\bar{E}_m$ /lx	UGR_L	U_o	R_a	Specific requirements
2.10.1	*Loading and operating of goods, handling equipment and machinery*	200	25	0.40	80	
2.10.2	*Buildings for livestock*	50	–	0.40	40	
2.10.3	*Sick animal pens; calving stalls*	200	25	0.60	80	A lower illuminance is acceptable in the absence of the stockman
2.10.4	*Feed preparation; dairy; utensil washing*	200	25	0.60	60	Luminaires suitable for being hosed down may be required in some areas

See Table 3.9 in section 3.2.3 for information on the requirements for outdoor agriculture.

Table 2.11 Industrial activities and crafts – Bakeries

Ref No.	Type of area, task or activity	$\bar{E}_m$ / lx	UGR_L	U_o	R_a	Specific requirements
2.11.1	Preparation and baking	300	22	0.60	80	
2.11.2	Finishing, glazing, decorating	500	22	0.70	80	

Table 2.12 Industrial activities and crafts – Cement, cement goods, concrete, bricks

Ref No.	Type of area, task or activity	$\bar{E}_m$ / lx	UGR_L	U_o	R_a	Specific requirements
2.12.1	Drying	50	28	0.40	20	Safety colours shall be recognisable
2.12.2	Preparation of materials; work on kilns and mixers	200	28	0.40	40	
2.12.3	General machine work	300	25	0.60	80	
2.12.4	Rough forms	300	25	0.60	80	

Table 2.13 Industrial activities and crafts – Ceramics, tiles, glass, glassware

Ref No.	Type of area, task or activity	$\bar{E}_{m}$ / lx	UGR_L	U_o	R_a	Specific requirements
2.13.1	Drying	50	28	0.40	20	Safety colours shall be recognisable
2.13.2	Preparation, general machine work	300	25	0.60	80	
2.13.3	Enamelling, rolling, pressing, shaping simple parts, glazing, glass blowing	300	25	0.60	80	
2.13.4	Grinding, engraving, glass polishing, shaping precision parts, manufacture of glass instruments	750	19	0.70	80	
2.13.5	Grinding of optical glass, crystal, hand grinding and engraving	750	16	0.70	80	
2.13.6	Precision work e.g. decorative grinding, hand painting	1000	16	0.70	90	$4000\ K \le T_{CP} \le 6500\ K$
2.13.7	Manufacture of synthetic precious stones	1500	16	0.70	90	$4000\ K \le T_{CP} \le 6500\ K$

Table 2.14 Industrial activities and crafts – Chemical, plastics and rubber industry

Ref No.	Type of area, task or activity	$\bar{E}_m$ /lx	UGR_L	U_o	R_a	Specific requirements
2.14.1	Remote-operated processing installations	50	–	0.40	20	• Safety colours shall be recognisable • Supplementary local lighting may be needed for maintenance work
2.14.2	Processing installations with limited manual intervention	150	28	0.40	40	
2.14.3	Constantly manned work stations in processing installations	300	25	0.60	80	
2.14.4	Precision measuring rooms, laboratories	500	19	0.60	80	
2.14.5	Pharmaceutical production	500	22	0.60	80	
2.14.6	Tyre production	500	22	0.60	80	
2.14.7	Colour inspection	1000	16	0.70	90	$4000\ K \leq T_{CP} \leq 6500\ K$
2.14.8	Cutting, finishing, inspection	750	19	0.70	80	

Table 2.15 Industrial activities and crafts – Electrical and electronic industry

Ref No.	Type of area, task or activity	$\bar{E}_m$ / lx	UGR_L	U_o	R_a	Specific requirements
2.15.1	Cable and wire manufacture	300	25	0.60	80	With large machines, some obstruction is likely, portable or local lighting may be needed
2.15.2	Winding: • large coils • medium–sized coils • small coils	300 500 700	25 22 19	0.60 0.60 0.70	80 80 80	With large machines, some obstruction is likely, portable or local lighting may be needed
2.15.3	Coil impregnating	300	25	0.6	80	With large machines, some obstruction is likely, portable or local lighting may be needed
2.15.4	Galvanising	300	25	0.6	80	With large machines, some obstruction is likely, portable or local lighting may be needed
2.15.5	Assembly work: • rough, e.g. large transformers • medium, e.g. switchboards • fine, e.g. telephones, radios, IT equipment (computers) • precision, e.g. measuring equipment, printed circuit boards	300 500 750 1000	25 22 19 16	0.60 0.60 0.70 0.70	80 80 80 80	With large machines, some obstruction is likely, portable or local lighting may be needed
2.15.6	Electronic workshops, testing, adjusting	1500	16	0.70	80	Local lighting may be appropriate

Table 2.16 Industrial activities and crafts – Food stuffs and luxury food industry

Ref No.	Type of area, task or activity	$\bar{E}_{m}$ / lx	UGR_L	U_o	R_a	Specific requirements
2.16.1	Work stations and zones in: • breweries, malting floor, • for washing, barrel filling, cleaning, sieving, peeling, • cooking in preserve and chocolate factories, • work stations and zones in sugar factories, • for drying and fermenting raw tobacco, fermentation cellar	200	25	0.40	80	
2.16.2	Sorting and washing of products, milling, mixing, packing	300	25	0.60	80	
2.16.3	Work stations and critical zones in slaughter houses, butchers, dairies, mills, on filtering floor in sugar refineries	500	25	0.60	80	• Damp conditions may be present and hosing down may be part of the cleaning process • Areas containing a dust explosion hazard may be present, appropriate luminaires should be chosen
2.16.4	Cutting and sorting of fruit and vegetables	300	25	0.60	80	
2.16.5	Manufacture of delicatessen foods, kitchen work, manufacture of cigars and cigarettes	500	22	0.60	80	
2.16.6	Inspection of glasses and bottles, product control, trimming, sorting, decoration	500	22	0.60	80	
2.16.7	Laboratories	500	19	0.60	80	
2.16.8	Colour inspection	1000	16	0.70	90	$4000\ K \leq T_{CP} \leq 6500\ K$

General note: Luminaires should be constructed so that no part of the luminaire can fall into the foodstuffs, even when the luminaire is opened for lamp changing. The luminaires should be capable of being washed or hosed down in safety. Lamps suitable for operation at low temperatures will be necessary for some food storage areas. Lamps and luminaires suitable for hot and humid conditions may be required for some other areas.

Table 2.17 Industrial activities and crafts – Foundries and metal casting

Ref No.	Type of area, task or activity	$\bar{E}_m$ /lx	UGR_L	U_o	R_a	Specific requirements
2.17.1	Man-sized underfloor tunnels, cellars, etc	50	–	0.40	20	Safety colours shall be recognisable
2.17.2	Platforms	100	25	0.40	40	
2.17.3	Sand preparation	200	25	0.40	80	If blast cleaning is used, luminaires should be positioned away from the work area. Where metal castings are cleaned by abrasive wheels or bands then the dust produced may present an explosion hazard; luminaires should be chosen appropriately
2.17.4	Dressing room	200	25	0.40	80	
2.17.5	Work stations at cupola and mixer	200	25	0.40	80	
2.17.6	Casting bay	200	25	0.40	80	
2.17.7	Shake out areas	200	25	0.40	80	If blast cleaning is used, luminaires should be positioned away from the work area. Where metal castings are cleaned by abrasive wheels or bands then the dust produced may present an explosion hazard; luminaires should be chosen appropriately
2.17.8	Machine moulding	200	25	0.40	80	If blast cleaning is used, luminaires should be positioned away from the work area. Where metal castings are cleaned by abrasive wheels or bands then the dust produced may present an explosion hazard; luminaires should be chosen appropriately
2.17.9	Hand and core moulding	300	25	0.60	80	Light distribution needs to be diffused and flexible to ensure good lighting of deep moulds
2.17.10	Die casting	300	25	0.60	80	
2.17.11	Model building	500	22	0.60	80	Light distribution needs to be diffused and flexible to ensure good lighting of deep moulds

Table 2.18 Industrial activities and crafts – Hairdressers

Ref No.	Type of area, task or activity	$\bar{E}_m$ /lx	UGR_L	U_o	R_a	Specific requirements
2.18.1	Hairdressing	500	19	0.60	90	

Table 2.19 Industrial activities and crafts – Jewellery manufacturing

Ref No.	Type of area, task or activity	$\bar{E}_m$ /lx	UGR_L	U_o	R_a	Specific requirements
2.19.1	Working with precious stones	1500	16	0.70	90	$4000\ K \leq T_{CP} \leq 6500\ K$
2.19.2	Manufacture of jewellery	1000	16	0.70	90	
2.19.3	Watch making (manual)	1500	16	0.70	80	
2.19.4	Watch making (automatic)	500	19	0.60	80	

Table 2.20 Industrial activities and crafts – Laundries and dry cleaning

Ref No.	Type of area, task or activity	$\bar{E}_m$ /lx	UGR_L	U_o	R_a	Specific requirements
2.20.1	Goods in, marking and sorting	300	25	0.60	80	
2.20.2	Washing and dry cleaning	300	25	0.60	80	
2.20.3	Ironing, pressing	300	25	0.60	80	
2.20.4	Inspection and repairs	750	19	0.70	80	

General note: Luminaires may be subject to a warm humid atmosphere.

Table 2.21 Industrial activities and crafts – Leather and leather goods

Ref No.	Type of area, task or activity	$\bar{E}_\mathrm{m}$ / lx	UGR_L	U_o	R_a	Specific requirements
2.21.1	*Work on vats, barrels, pits*	200	25	0.40	40	
2.21.2	*Fleshing, skiving, rubbing, tumbling of skins*	300	25	0.40	80	
2.21.3	*Saddlery work, shoe manufacture: stitching, sewing, polishing, shaping, cutting, punching*	500	22	0.60	80	
2.21.4	*Sorting*	500	22	0.60	90	$4000\ K \leq T_\mathrm{CP} \leq 6500\ K$
2.21.5	*Leather dyeing (machine)*	500	22	0.60	80	
2.21.6	*Quality control*	1000	19	0.70	80	
2.21.7	*Colour inspection*	1000	16	0.70	90	$4000\ K \leq T_\mathrm{CP} \leq 6500\ K$
2.21.8	*Shoe making*	500	22	0.60	80	
2.21.9	*Glove making*	500	22	0.60	80	

Table 2.22 Industrial activities and crafts – Metal working and processing

Ref No.	Type of area, task or activity	$\bar{E}_m$ / lx	UGR_L	U_o	R_a	Specific requirements
2.22.1	Open die forging	200	25	0.60	80	
2.22.2	Drop forging	300	25	0.60	80	
2.22.3	Welding	300	25	0.60	80	Care is necessary to prevent exposure of eyes and skin to radiation. Welding screens will be used so considerable obstruction is likely. Portable lighting may be useful
2.22.4	Rough and average machining: tolerances ≥ 0.1 mm	300	22	0.60	80	
2.22.5	Precision machining; grinding: tolerances < 0.1 mm	500	19	0.70	80	
2.22.6	Scribing; inspection	750	19	0.70	80	Care should be taken to avoid multiple shadows
2.22.7	Wire and pipe drawing shops; cold forming	300	25	0.60	80	
2.22.8	Plate machining: thickness ≥ 5 mm	200	25	0.60	80	Some obstruction is likely. Care should be taken to minimise stroboscopic effects on rotating machinery
2.22.9	Sheet metalwork: thickness < 5 mm	300	22	0.60	80	Some obstruction is likely. Care should be taken to minimise stroboscopic effects on rotating machinery
2.22.10	Tool making; cutting equipment manufacture	750	19	0.70	80	
2.22.11	Assembly: • rough • medium • fine • precision	200 300 500 750	25 25 22 19	0.60 0.60 0.60 0.70	80 80 80 80	Some obstruction is likely. Care should be taken to minimise stroboscopic effects on rotating machinery
2.22.12	Galvanising	300	25	0.60	80	
2.22.13	Surface preparation and painting	750	25	0.70	80	
2.22.14	Tool, template and jig making, precision mechanics, micro-mechanics	1000	19	0.70	80	

Table 2.23 Industrial activities and crafts – Paper and paper goods

Ref No.	Type of area, task or activity	$\bar{E}_m$ / lx	UGR_L	U_o	R_a	Specific requirements
2.23.1	Edge runners, pulp mills	200	25	0.40	80	
2.23.2	Paper manufacture and processing, paper and corrugating machines, cardboard manufacture	300	25	0.60	80	
2.23.3	Standard bookbinding work, e.g. folding, sorting, gluing, cutting, embossing, sewing	500	22	0.60	80	

Table 2.24 Industrial activities and crafts – Power stations

Ref No.	Type of area, task or activity	$\bar{E}_m$ / lx	UGR_L	U_o	R_a	Specific requirements
2.24.1	Fuel supply plant	50	–	0.40	20	Safety colours shall be recognisable
2.24.2	Boiler house	100	28	0.40	40	
2.24.3	Machine halls	200	25	0.40	80	Additional local lighting of instruments and inspection points may be required
2.24.4	Side rooms, e.g. pump rooms, condenser rooms, etc; switchboards (inside buildings)	200	25	0.40	60	In areas such as ash handling plants, settling pits and battery rooms there may corrosive and hazardous atmospheres
2.24.5	Control rooms	500	16	0.70	80	1. Control panels are often vertical 2. Dimming may be required 3. DSE-work, see 2.1.9

See Table 3.15 in section 3.2.3 for information on the requirements for outdoor power, gas and heat plants

Table 2.25 Industrial activities and crafts – Printers

Ref No.	Type of area, task or activity	$\bar{E}_m$ /lx	UGR_L	U_o	R_a	Specific requirements
2.25.1	Cutting, gilding, embossing, block engraving, work on stones and platens, printing machines, matrix making	500	19	0.60	80	
2.25.2	Paper sorting and hand printing	500	19	0.60	80	
2.25.3	Type setting, retouching, lithography	1000	19	0.70	80	Local lighting may be appropriate
2.25.4	Colour inspection in multicoloured printing	1500	16	0.70	90	$5000\ K \leq T_{CP} \leq 6500\ K$
2.25.5	Steel and copper engraving	2000	16	0.70	80	For directionality; see 2.1.6.4

Table 2.26 Industrial activities and crafts – Rolling mills, iron and steel works

Ref No.	Type of area, task or activity	$\bar{E}_m$ /lx	UGR_L	U_o	R_a	Specific requirements
2.26.1	*Production plants without manual operation*	50	–	0.40	20	• *Safety colours shall be recognisable* • Supplementary lighting may be required for maintenance work
2.26.2	*Production plants with occasional manual operation*	150	28	0.40	40	Supplementary lighting may be required for maintenance work
2.26.3	Production plants with continuous manual operation	200	25	0.60	80	
2.26.4	Slab store	50	–	0.40	20	*Safety colours shall be recognisable*
2.26.5	Furnaces	200	25	0.40	20	*Safety colours shall be recognisable*
2.26.6	Mill train; coiler; shear line	300	25	0.60	40	
2.26.7	Control platforms; control panels	300	22	0.60	80	
2.26.8	Test, measurement and inspection	500	22	0.60	80	
2.26.9	*Underfloor man-sized tunnels; belt sections, cellars, etc*	50	–	0.40	20	• *Safety colours shall be recognisable* • Supplementary lighting may be required for maintenance work

Table 2.27 Industrial activities and crafts – Textile manufacture and processing

Ref No.	Type of area, task or activity	$\bar{E}_m$ /lx	UGR_L	U_o	R_a	Specific requirements
2.27.1	Work stations and zones in baths, bale opening	200	25	0.60	60	
2.27.2	Carding, washing, ironing, devilling machine work, drawing, combing, sizing, card cutting, pre-spinning, jute and hemp spinning	300	22	0.60	80	
2.27.3	Spinning, plying, reeling, winding	500	22	0.60	80	Prevent stroboscopic effects
2.27.4	Warping, weaving, braiding, knitting	500	22	0.60	80	Prevent stroboscopic effects
2.27.5	Sewing, fine knitting, taking up stitches	750	22	0.70	80	
2.27.6	Manual design, drawing patterns	750	22	0.70	90	
2.27.7	Finishing, dyeing	500	22	0.60	80	
2.27.8	Drying room	100	28	0.40	60	
2.27.9	Automatic fabric printing	500	25	0.60	80	
2.27.10	Burling, picking, trimming	1000	19	0.70	80	
2.27.11	Colour inspection, fabric control	1000	16	0.70	90	$4000\ K \leq T_{CP} \leq 6500\ K$
2.27.12	Invisible mending	1500	19	0.70	90	$4000\ K \leq T_{CP} \leq 6500\ K$
2.27.13	Hat manufacturing	500	22	0.60	80	

Table 2.28 Industrial activities and crafts – Vehicle construction and repair

Ref No.	Type of area, task or activity	$\bar{E}_m$ / lx	UGR_L	U_o	R_a	Specific requirements
2.28.1	Body work and assembly	500	22	0.60	80	
2.28.2	Painting, spraying chamber, polishing chamber	750	22	0.70	80	
2.28.3	Painting: touch-up, inspection	1000	19	0.70	90	$4000\ K \le T_{CP} \le 6500\ K$
2.28.4	Upholstery manufacture (manned)	1000	19	0.70	80	
2.28.5	Final inspection	1000	19	0.70	80	
2.28.6	General vehicle services, repair and testing	300	22	0.60	80	Consider local lighting

Table 2.29 Industrial activities and crafts – Wood working and processing

Ref No.	Type of area, task or activity	$\bar{E}_m$ / lx	UGR_L	U_o	R_a	Specific requirements
2.29.1	Automatic processing e.g. drying, plywood manufacturing	50	28	0.40	40	Dust from sanding and similar operations may represent an explosion hazard; luminaires should be chosen appropriately
2.29.2	Steam pits	150	28	0.40	40	
2.29.3	Saw frame	300	25	0.60	60	Prevent stroboscopic effects
2.29.4	Work at joiner's bench, gluing, assembly	300	25	0.60	80	
2.29.5	Polishing, painting, fancy joinery	750	22	0.70	80	Dust from sanding and similar operations may represent an explosion hazard; luminaires should be chosen appropriately
2.29.6	Work on wood working machines, e.g. turning, fluting, dressing, rebating, grooving, cutting, sawing, sinking	500	19	0.60	80	• Prevent stroboscopic effects • Dust from sanding and similar operations may represent an explosion hazard; luminaires should be chosen appropriately
2.29.7	Selection of veneer woods	750	22	0.70	90	4000 K ≤ T_{CP} ≤ 6500 K
2.29.8	Marquetry, inlay work	750	22	0.70	90	4000 K ≤ T_{CP} ≤ 6500 K
2.29.9	Quality control, inspection	1000	19	0.70	90	4000 K ≤ T_{CP} ≤ 6500 K

See Table 3.17 in section 3.2.3 for information on the requirements for outdoor wood working and saw mills.

Table 2.30 Offices

Ref No.	Type of area, task or activity	$\bar{E}_{m}$ /lx	UGR_{L}	U_{o}	R_{a}	Specific requirements
2.30.1	Filing, copying, etc	300	19	0.40	80	
2.30.2	Writing, typing, reading, data processing	500	19	0.60	80	DSE work, see 2.1.9
2.30.3	Technical drawing	750	16	0.70	80	
2.30.4	CAD work stations	500	19	0.60	80	DSE work, see 2.1.9
2.30.5	Conference and meeting rooms	500	19	0.60	80	Lighting should be controllable
2.30.6	Reception desk	300	22	0.60	80	
2.30.7	Archives	200	25	0.40	80	For filing, the vertical surfaces are especially important

Table 2.31 Retail premises

Ref No.	Type of area, task or activity	$\bar{E}_{m}$ /lx	UGR_{L}	U_{o}	R_{a}	Specific requirements
2.31.1	Sales area	300	22	0.40	80	
2.31.2	Till area	500	19	0.60	80	
2.31.3	Wrapper table	500	19	0.60	80	

Table 2.32 Places of public assembly – General areas

Ref No.	Type of area, task or activity	$\bar{E}_m$ /lx	UGR_L	U_o	R_a	Specific requirements
2.32.1	Entrance halls	100	22	0.40	80	UGR only if applicable
2.32.2	Cloakrooms	200	25	0.40	80	
2.32.3	Lounges	200	22	0.40	80	
2.32.4	Ticket offices	300	22	0.60	80	

Table 2.33 Places of public assembly – Restaurants and hotels

Ref No.	Type of area, task or activity	$\bar{E}_m$ /lx	UGR_L	U_o	R_a	Specific requirements
2.33.1	Reception/cashier desk, porters desk	300	22	0.60	80	Localised lighting may be appropriate
2.33.2	Kitchen	500	22	0.60	80	There should be a transition zone between kitchen and restaurant
2.33.3	Restaurant, dining room, function room	–	–	–	80	The lighting should be designed to create the appropriate atmosphere
2.33.4	Self-service restaurant	200	22	0.40	80	
2.33.5	Buffet	300	22	0.60	80	
2.33.6	Conference rooms	500	19	0.60	80	Lighting should be controllable
2.33.7	Corridors	100	25	0.40	80	During night-time lower levels are acceptable

Table 2.34 Places of public assembly – Theatres, concert halls, cinemas, places for entertainment

Ref No.	Type of area, task or activity	$\bar{E}_m$ /lx	UGR_L	U_o	R_a	Specific requirements
2.34.1	Practice rooms	300	22	0.60	80	
2.34.2	Dressing rooms	300	22	0.60	90	Lighting at mirrors for make-up shall be 'glare-free'. Disability glare should be avoided at mirrors for make-up
2.34.3	Seating areas – maintenance, cleaning	200	22	0.50	80	Illuminance at floor level
2.34.4	Stage area – rigging	300	25	0.40	80	Illuminance at floor level
2.34.5	Projection rooms	150	22	0.60	40	Lighting should be provided on the working side of the projector. The lighting should not detract from the view into the auditorium. Dimming facilities may be desirable

Table 2.35 Places of public assembly – Trade fairs, exhibition halls

Ref No.	Type of area, task or activity	$\bar{E}_m$ /lx	UGR_L	U_o	R_a	Specific requirements
2.35.1	General lighting	300	22	0.40	80	

Table 2.36 Places of public assembly – Museums

Ref No.	Type of area, task or activity	$\bar{E}_\mathrm{m}$ / lx	UGR_L	U_o	R_a	Specific requirements
2.36.1	*Exhibits, insensitive to light*					*Lighting is determined by the display requirements*
2.36.2	*Exhibits, sensitive to light*					1. *Lighting is determined by the display requirements* 2. *Protection against damaging radiation is paramount*

Table 2.37 Places of public assembly – Libraries

Ref No.	Type of area, task or activity	$\bar{E}_\mathrm{m}$ / lx	UGR_L	U_o	R_a	Specific requirements
2.37.1	*Bookshelves*	200	19	0.40	80	The illuminance should be provided on the vertical face at the bottom of the bookshelf
2.37.2	*Reading area*	500	19	0.60	80	• Local or localised lighting may be appropriate • DSE work, see 2.1.9
2.37.3	*Counters*	500	19	0.60	80	• Local or localised lighting may be appropriate • DSE work, see 2.1.9

Table 2.38 Places of public assembly – Public car parks (indoor)

Ref No.	Type of area, task or activity	$\bar{E}_m$ /lx	UGR_L	U_o	R_a	Specific requirements
2.38.1	In/out ramps (during the day)	300	25	0.40	40	1. Illuminances at floor level 2. Safety colours shall be recognisable
2.38.2	In/out ramps (at night)	75	25	0.40	40	1. Illuminances at floor level 2. Safety colours shall be recognisable
2.38.3	Traffic lanes	75	25	0.40	40	1. Illuminances at floor level 2. Safety colours shall be recognisable
2.38.4	Parking areas	75	–	0.40	40	1. Illuminances at floor level 2. Safety colours shall be recognisable 3. A high vertical illuminance increases recognition of people's faces and therefore the feeling of safety
2.38.5	Ticket office	300	19	0.60	80	1. Reflections in the windows shall be avoided 2. Glare from outside shall be prevented

See Table 3.13 in section 3.2.3 for information on the requirements for outdoor car parks.

Table 2.39 Educational premises – Nursery school, play school

Ref No.	Type of area, task or activity	$\bar{E}_{m}$ /lx	UGR_{L}	U_{o}	R_{a}	Specific requirements
2.39.1	Play room	300	22	0.40	80	*High luminances should be avoided in viewing directions from below by use of diffuse covers*
2.39.2	Nursery	300	22	0.40	80	*High luminances should be avoided in viewing directions from below by use of diffuse covers*
2.39.3	Handicraft room	300	19	0.60	80	

Table 2.40 Educational premises – Educational buildings

Ref No.	Type of area, task or activity	$\bar{E}_{m}$ /lx	UGR_{L}	U_{o}	R_{a}	Specific requirements
2.40.1	*Classrooms, tutorial rooms*	300	19	0.60	80	*Lighting should be controllable*
2.40.2	*Classroom for evening classes and adults education*	500	19	0.60	80	*Lighting should be controllable*
2.40.3	*Auditorium, lecture halls*	500	19	0.60	80	*Lighting should be controllable to accommodate various A/V needs*
2.40.4	*Black, green and white boards*	500	19	0.70	80	*Specular reflections shall be prevented. Presenter/teacher shall be illuminated with suitable vertical illumination*
2.40.5	*Demonstration table*	500	19	0.70	80	*In lecture halls 750 lx*
2.40.6	*Art rooms*	500	19	0.60	80	

Table 2.40 Continued

Ref No.	Type of area, task or activity	$\bar{E}_m$ /lx	UGR_L	U_o	R_a	Specific requirements
2.40.7	Art rooms in art schools	750	19	0.70	90	$5000\ K \le T_{CP} \le 6500\ K$
2.40.8	Technical drawing rooms	750	16	0.70	80	
2.40.9	Practical rooms and laboratories	500	19	0.60	80	
2.40.10	Handicraft rooms	500	19	0.60	80	
2.40.11	Teaching workshop	500	19	0.60	80	
2.40.12	Music practice rooms	300	19	0.60	80	
2.40.13	Computer practice rooms (menu driven)	300	19	0.60	80	DSE work, see 2.1.9
2.40.14	Language laboratory	300	19	0.60	80	
2.40.15	Preparation rooms and workshops	500	22	0.60	80	
2.40.16	Entrance halls	200	22	0.40	80	
2.40.17	Circulation areas, corridors	100	25	0.40	80	
2.40.18	Stairs	150	25	0.40	80	
2.40.19	Student common rooms and assembly halls	200	22	0.40	80	

Table 2.40 Continued

Ref No.	Type of area, task or activity	$\bar{E}_m$ /lx	UGR_L	U_o	R_a	Specific requirements
2.40.20	Teachers rooms	300	19	0.60	80	
2.40.21	Library: bookshelves	200	19	0.60	80	
2.40.22	Library: reading areas	500	19	0.60	80	
2.40.23	Stock rooms for teaching materials	100	25	0.40	80	
2.40.24	Sports halls, gymnasiums, swimming pools	300	22	0.60	80	See BS EN 12193 (BSI, 2007b) or SLL Lighting Guide 4: Sports (SLL, 2006) for training conditions
2.40.25	School canteens	200	22	0.40	80	
2.40.26	Kitchen	500	22	0.60	80	

For more information see SLL Lighting Guide 5: Lighting for education (SLL, 2011).

Table 2.41 Health care premises – Rooms for general use

Ref No.	Type of area, task or activity	$\bar{E}_\mathrm{m}$ / lx	UGR_L	U_o	R_a	Specific requirements
						Too high luminances in the patients' visual field shall be prevented
2.41.1	Waiting rooms	200		0.40	80	
2.41.2	Corridors: during the day	100	22	0.40	80	Illuminance at floor level
2.41.3	Corridors: cleaning	100	22	0.40	80	Illuminance at floor level
2.41.4	Corridors: during the night	50	22	0.40	80	Illuminance at floor level
2.41.5	Corridors with multi-purpose use	200	22	0.60	80	Illuminance at task/activity level
2.41.6	Day rooms	200	22	0.60	80	
2.41.7	Elevators, lifts for persons and visitors	100	22	0.60	80	Illuminance at floor level
2.41.8	Service lifts	200	22	0.60	80	Illuminance at floor level

For more information see SLL Lighting Guide 2: *Hospitals and health care buildings* (SLL, 2008).

Table 2.42 Health care premises – Staff rooms

Ref No.	Type of area, task or activity	$\bar{E}_m$ /lx	UGR_L	U_o	R_a	Specific requirements
2.42.1	Staff office	500	19	0.60	80	
2.42.2	Staff rooms	300	19	0.60	80	

For more information see SLL Lighting Guide 2: *Hospitals and health care buildings* (SLL, 2008).

Table 2.43 Health care premises – Wards, maternity wards

Ref No.	Type of area, task or activity	$\bar{E}_m$ /lx	UGR_L	U_o	R_a	Specific requirements
						Too high luminances in the patients' visual field shall be prevented
2.43.1	General lighting	100	19	0.40	80	Illuminance at floor level
2.43.2	Reading lighting	300	19	0.70	80	
2.43.3	Simple examinations	300	19	0.60	80	
2.43.4	Examination and treatment	1000	19	0.70	90	Examination luminaire may be required
2.43.5	Night lighting, observation lighting	5	–	–	80	
2.43.6	Bathrooms and toilets for patients	200	22	0.40	80	

For more information see SLL Lighting Guide 2: *Hospitals and health care buildings* (SLL, 2008).

Table 2.44 Health care premises – Examination rooms (general)

Ref No.	Type of area, task or activity	$\bar{E}_m$ /lx	UGR_L	U_o	R_a	Specific requirements
2.44.1	General lighting	500	19	0.60	90	$4000\ K \leq T_{CP} \leq 5000\ K$
2.44.2	Examination and treatment	1000	19	0.70	90	Examination luminaire may be required

For more information see SLL Lighting Guide 2: *Hospitals and health care buildings* (SLL, 2008).

Table 2.45 Health care premises – Eye examination rooms

Ref No.	Type of area, task or activity	$\bar{E}_m$ /lx	UGR_L	U_o	R_a	Specific requirements
2.45.1	General lighting	500	19	0.60	90	$4000\ K \leq T_{CP} \leq 5000\ K$
2.45.2	Examination of the outer eye	1000	–	–	90	Examination luminaire may be required
2.45.3	Reading and colour vision tests with vision charts	500	16	0.70	90	

For more information see SLL Lighting Guide 2: *Hospitals and health care buildings* (SLL, 2008).

Table 2.46 Health care premises – Ear examination rooms

RefNo.	Type of area, task or activity	$\bar{E}_\mathrm{m}$ /lx	UGR_L	U_o	R_a	Specific requirements
2.46.1	*General lighting*	*500*	*19*	*0.60*	*90*	
2.46.2	*Ear examination*	*1000*	*–*	*–*	*90*	Examination luminaire may be required

For more information see SLL Lighting Guide 2: *Hospitals and health care buildings* (SLL, 2008).

Table 2.47 Health care premises – Scanner rooms

RefNo.	Type of area, task or activity	$\bar{E}_\mathrm{m}$ /lx	UGR_L	U_o	R_a	Specific requirements
2.47.1	*General lighting*	*300*	*19*	*0.60*	*80*	
2.47.2	*Scanners with image enhancers and television systems*	*50*	*19*	*–*	*80*	*DSE work, see 2.1.9*

For more information see SLL Lighting Guide 2: *Hospitals and health care buildings* (SLL, 2008).

Table 2.48 Health care premises – Delivery rooms

RefNo.	Type of area, task or activity	$\bar{E}_\mathrm{m}$ /lx	UGR_L	U_o	R_a	Specific requirements
2.48.1	*General lighting*	*300*	*19*	*0.60*	*80*	
2.48.2	*Examination and treatment*	*1000*	*19*	*0.70*	*80*	Examination luminaire may be required

Table 2.49 Health care premises – Treatment rooms (general)

Ref No.	Type of area, task or activity	$\bar{E}_{m}$ /lx	UGR_{L}	U_{o}	R_{a}	Specific requirements
2.49.1	Dialysis	500	19	0.60	80	Lighting should be controllable
2.49.2	Dermatology	500	19	0.60	90	
2.49.3	Endoscopy rooms	300	19	0.60	80	
2.49.4	Plaster rooms	500	19	0.60	80	
2.49.5	Medical baths	300	19	0.60	80	
2.49.6	Massage and radiotherapy	300	19	0.60	80	

For more information see SLL Lighting Guide 2: *Hospitals and health care buildings* (SLL, 2008).

Table 2.50 Health care premises – Operating areas

Ref No.	Type of area, task or activity	$\bar{E}_{m}$ /lx	UGR_{L}	U_{o}	R_{a}	Specific requirements
2.50.1	Pre-op and recovery rooms	500	19	0.60	90	
2.50.2	Operating theatre	1000	19	0.60	90	
2.50.3	Operating cavity			–		$\bar{E}_{m}$: 10 000 lx to 100 000 lx

For more information see SLL Lighting Guide 2: *Hospitals and health care buildings* (SLL, 2008).

Table 2.51 Health care premises – Intensive care unit

Ref No.	Type of area, task or activity	$\bar{E}_m$ / lx	UGR_L	U_o	R_a	Specific requirements
2.51.1	General lighting	100	19	0.60	90	Illuminance at floor level
2.51.2	Simple examinations	300	19	0.60	90	Illuminance at bed level
2.51.3	Examination and treatment	1000	19	0.70	90	Illuminance at bed level
2.51.4	Night watch	20	19	–	90	Illuminance at floor level

For more information see SLL Lighting Guide 2: *Hospitals and health care buildings* (SLL, 2008).

Table 2.52 Health care premises – Dentists

Ref No.	Type of area, task or activity	$\bar{E}_m$ / lx	UGR_L	U_o	R_a	Specific requirements
2.52.1	General lighting	500	19	0.60	90	Lighting should be glare-free for the patient
2.52.2	At the patient	1000	–	0.70	90	
2.52.3	Operating cavity	–	–	–	–	Specific requirements are given in BS EN ISO 9680 (ISO, 2007)
2.52.4	White teeth matching	–	–	–	–	Specific requirements are given in BS EN ISO 9680 (ISO, 2007)

For more information see SLL Lighting Guide 2: *Hospitals and health care buildings* (SLL, 2008).

Table 2.53 Health care premises – Laboratories and pharmacies

Ref No.	Type of area, task or activity	$\bar{E}_m$ / lx	UGR_L	U_o	R_a	Specific requirements
2.53.1	*General lighting*	*500*	*19*	*0.60*	*80*	
2.53.2	*Colour inspection*	*1000*	*19*	*0.70*	*90*	*$6000\ K \leq T_{CP} \leq 6500\ K$*

For more information see SLL Lighting Guide 2: *Hospitals and health care buildings* (SLL, 2008).

Table 2.54 Health care premises – Decontamination rooms

Ref No.	Type of area, task or activity	$\bar{E}_m$ / lx	UGR_L	U_o	R_a	Specific requirements
2.54.1	*Sterilisation rooms*	*300*	*22*	*0.60*	*80*	Luminaires may be subject to high humidity and temperatures as well as an aggressive cleaning regime
2.54.2	*Disinfection rooms*	*300*	*22*	*0.60*	*80*	Luminaires may be subject to high humidity and temperatures as well as an aggressive cleaning regime

For more information see SLL Lighting Guide 2: *Hospitals and health care buildings* (SLL, 2008).

Table 2.55 Health care premises – Autopsy rooms and mortuaries

Ref No.	Type of area, task or activity	$\bar{E}_m$ / lx	UGR_L	U_o	R_a	Specific requirements
2.55.1	*General lighting*	*500*	*19*	*0.60*	*90*	Luminaires may be subject an aggressive cleaning regime
2.55.2	*Autopsy table and dissecting table*	*5000*	*–*	*–*	*90*	*Values higher than 5000 lx may be required*

For more information see SLL Lighting Guide 2: *Hospitals and health care buildings* (SLL, 2008).

Table 2.56 Transportation areas – Airports

Ref No.	Type of area, task or activity	$\bar{E}_m$ / lx	UGR_L	U_o	R_a	Specific requirements
2.56.1	*Arrival and departure halls, baggage claim areas*	200	22	0.40	80	
2.56.2	*Connecting areas*	150	22	0.40	80	
2.56.3	*Information desks, check-in desks*	500	19	0.70	80	*DSE work, see 2.1.9*
2.56.4	*Customs and passport control desks*	500	19	0.70	80	*Facial recognition has to be provided*
2.56.5	*Waiting areas*	200	22	0.40	80	
2.56.6	*Luggage store rooms*	200	25	0.40	80	
2.56.7	*Security check areas*	300	19	0.60	80	*DSE work, see 2.1.9*
2.56.8	*Air traffic control tower*	500	16	0.60	80	*1. Lighting should be dimmable* *2. DSE work, see 2.1.9* *3. Glare from daylight shall be avoided* *4. Reflections in windows, especially at night shall be avoided*
2.56.9	*Testing and repair hangars*	500	22	0.60	80	
2.56.10	*Engine test areas*	500	22	0.60	80	
2.56.11	*Measuring areas in hangars*	500	22	0.60	80	

See Table 3.6 in section 3.2.3 for information on the requirements for outdoor areas of airports.

Table 2.57 Transportation areas – Railway installations

Ref No.	Type of area, task or activity	$\bar{E}_m$ /lx	UGR_L	U_o	R_a	Specific requirements
2.57.1	Fully enclosed platforms, small number of passengers	100	–	0.40	40	1. Special attention to the edge of the platform 2. Avoid glare for vehicle drivers 3. Illuminance at floor level
2.57.2	Fully enclosed platforms, large number of passengers	200	–	0.50	60	1. Special attention to the edge of the platform 2. Avoid glare for vehicle drivers 3. Illuminance at floor level
2.57.3	Passenger subways (underpasses), small number of passengers	50	28	0.50	40	Illuminance at floor level
2.57.4	Passenger subways (underpasses), large number of passengers	100	28	0.50	40	Illuminance at floor level
2.57.5	Ticket hall and concourse	200	28	0.50	40	
2.57.6	Ticket and luggage offices and counters	300	19	0.50	80	
2.57.7	Waiting rooms	200	22	0.40	80	
2.57.8	Entrance halls, station halls	200	–	0.40	80	
2.57.9	Switch and plant rooms	200	28	0.40	60	Safety colours must be recognisable
2.57.10	Access tunnels	50	–	0.40	20	Illuminance at floor level
2.57.11	Maintenance and servicing sheds	300	22	0.50	60	

See Table 3.16 in section 3.2.3 for information on the requirements for outdoor areas associated with railways and tramways.

2.2.3   Verification procedures

2.2.3.1 General

Specified design criteria which are included in this section of the SLL Code shall be verified by the following procedures.

In lighting design, calculations and measurements, certain assumptions including degree of accuracy have been made. These shall be declared.

Reference should be made to Chapter 15 of this Code for measurement procedures and properties of light measuring equipment.

The installation and the environment shall be checked against the design assumptions.

2.2.3.2 Illuminances

When verifying conformity to the illuminance requirements the measurement points shall coincide with any design points or grids used. Verification shall be made to the criteria of the relevant surfaces.

For subsequent measurements, the same measurement points shall be used.

Verification of illuminances that relate to specific tasks shall be measured in the plane of the task.

Note: When verifying illuminances, account should be taken of the calibration of the light meters used, the conformity of the lamps and luminaires to the published photometric data, and of the design assumptions made about surface reflectance, etc, compared with the real values.

The average illuminance and uniformity shall be calculated and shall be not less than the values specified.

2.2.3.3 Unified glare rating

Authenticated UGR data produced by the tabular method shall be provided for the luminaire scheme by the manufacturer of the luminaire. The spacing shall be declared for the UGR-tables provided.

Details of the use of glare tables are given in section 12.2.8, and section 12.3.5 gives details of how the tables are generated.

2.2.3.4 Colour rendering and colour appearance

Authenticated colour rendering index R_a and correlated colour temperatures T_{cp} data shall be provided for the lamps in the scheme by the manufacturer of the lamps. The lamps shall be checked against the design specifications.

2.2.3.5 Luminaire luminance

The average luminance of the luminous parts of the luminaire shall be measured and/or calculated in the C-plane (azimuth) at intervals of 15° starting at 0° and the γ-plane (elevation) for angles of 65°, 70°, 75°, 80° and 85°. Normally the manufacturer of the luminaire shall provide these data based on maximum (lamp/ luminaire) output (see also sections 12.2.7 and 12.3.4).

Where controlled values of luminaire luminance are required, *the values shall not exceed the limits specified in* Table 2.4.

2.2.3.6 Maintenance schedule

The maintenance schedule shall be provided and should be according to section 2.1.10.

Chapter 3: Outdoor workplaces

This chapter of the code is based on BS EN 12464-2: 2007: *Light and lighting – Lighting of work places – Part 2: Outdoor work places* (BSI, 2007c). All of the text in this chapter that is in italics and the tables of lighting requirements have been taken directly from the standard. The other material in this chapter is advice on how best to apply the standard whilst at the same time following what the Society of Light and Lighting regards to be good lighting practice. It should be noted that this chapter does not include all of BS EN 12464-2 (BSI, 2007c); the scope, normative references, terms and definitions together with the original bibliography and some calculations have been omitted as they are either not appropriate to this document or they are covered elsewhere in this *Code*. This section is broadly similar to the section on outdoor workplaces in the 2009 edition of the *SLL Code*; however, the format has been changed and some extra advice and notes have been added.

3.1 Lighting design criteria

3.1.1 Luminous environment

For good lighting practice it is essential that, in addition to the required illuminance, qualitative and quantitative needs are satisfied.

Lighting requirements are determined by the satisfaction of three basic human needs:

- *visual comfort, where the workers have a feeling of well-being; in an indirect way also contributing to a high productivity level*

- *visual performance, where the workers are able to perform their visual tasks, even under difficult circumstances and during longer periods*

- *safety.*

Main parameters determining the luminous environment are:

- *luminance distribution*

- *illuminance*

- *glare*

- *directionality of light*

- *colour rendering and colour appearance of the light*

- *flicker.*

Values for illuminance, glare rating and colour rendering are given in section 3.2.

The following sections give information on the above topics together with recommendations on the control of spill light.

3.1.2　Luminance distribution

The luminance distribution in the field of view controls the adaptation level of the eyes, which affects task visibility.

A well balanced luminance distribution is needed to increase:

● *visual acuity (sharpness of vision)*

● *contrast sensitivity (discrimination of small relative luminance differences)*

● *efficiency of the ocular functions (such as accommodation, convergence, pupillary contraction, eye movements, etc).*

The luminance distribution in the field of view also affects visual comfort. Sudden changes in luminance should be avoided.

3.1.3　Illuminance

The illuminance and its distribution on the task area and the surrounding area have a great impact on how quickly, safely and comfortably a person perceives and carries out the visual task.

All values of illuminances specified in the schedule (section 3.2) are maintained illuminances.

3.1.3.1 Illuminance on the task area

The illuminance values given in the schedule (section 3.2) are maintained illuminances over the task area on the reference surface, which may be horizontal, vertical or inclined. The average illuminance for each task shall not fall below the value given in the schedule, regardless of the age and condition of the installation.

Note:　The values are valid for normal visual conditions and take into account the following factors:

● *psycho-physiological aspects such as visual comfort and well-being*

● *requirements for visual tasks*

● *visual ergonomics*

● *practical experience*

● *safety*

● *economy.*

The value of illuminance may be adjusted by at least one step in the scale of illuminances (see below), if the visual conditions differ from the normal assumptions.

A factor of approximately 1.5 represents the smallest significant difference in subjective effect of illuminance. The recommended scale of illuminance (in lx) is:

5–10–15–20–30–50–75–100–150–200–300–500–750–1000–1500–2000

The required maintained illuminance should be increased, when:

● *visual work is critical*

● *visual task or worker is moving*

● *errors are costly to rectify*

● *accuracy or higher productivity is of great importance*

● *the visual capacity of the worker is below normal*

● *task details are of unusually small size or low contrast*

● *the task is undertaken for an unusually long time.*

The required maintained illuminance may be decreased when:

● *task details are of an unusually large size or high contrast*

● *the task is undertaken for an unusually short time or on only rare occasions.*

3.1.3.2 Illuminance of surroundings

The maintained illuminance of surrounding areas shall be related to the maintained illuminance of the task area and should provide a well-balanced luminance distribution in the field of view. Large spatial variations in illuminances around the task area may lead to visual stress and discomfort. The illuminance of the surrounding areas may be lower than the task illuminance but shall be not less than the values given in Table 3.1.

Table 3.1 Relationship of illuminances of surrounding area to task area

Task illuminance lx	Illuminance of surrounding areas lx
≥500	100
300	75
200	50
150	30
$50 \leq \bar{E}_\mathrm{m} \leq 100$	20
<50	No specification

In addition to the task illuminance the lighting shall provide adequate adaption luminance in accordance with section 3.1.2.

Most areas that have a recommended task illuminance of less than 50 lx do not need to consider light on the surrounding area, however, in situations where people may be working in these areas for a long time, it may be necessary to consider providing light in the surrounding areas.

3.1.3.3 Illuminance grid
Grid systems shall be created to indicate the points at which the illuminance values are calculated and verified.

Grid cells approximating to a square are preferred; the ratio of length to width of a grid cell shall be kept between 0.5 and 2. (The same method of grid specification is used in other parts of this *Code* and there is a worked example of how to apply the formulae given in section 15.2.2.) *The maximum grid size shall be:*

$$p = 0.2 \times 5^{\log d} \tag{3.1}$$

where $p \leq 10\ m$

d is the longer dimension of the calculation area (m), however if the ratio of the longer to the shorter side is 2 or more then d becomes the shorter dimension of the area, and p is the maximum grid cell size (m).

3.1.3.4 Uniformity and diversity
The task area shall be illuminated as uniformly as possible. The illuminance uniformity of the task area shall be not less than the values given in section 3.2. *The uniformity of the surroundings shall not be less than 0.10.*

In some cases, e.g. railways, illuminance diversity is also an important quality criterion.

Note: See Chapter 19 – *Glossary* for definitions of the terms Uniformity and Diversity.

3.1.4   Glare
Glare is the sensation produced by bright areas within the field of view and may be experienced as either discomfort glare or disability glare. Glare caused by reflections in specular surfaces is usually known as veiling reflections or reflected glare. It is important to limit the glare to the users to avoid errors, fatigue and accidents.

Note: *Special care is needed to avoid glare when the direction of view is above horizontal.*

3.1.4.1 Glare rating
The glare directly from the luminaires of an outdoor lighting installation shall be determined using the CIE Glare Rating (GR) method (CIE, 1994). The method is covered in detail in section 14.5 of this Code.

Note: *GR should be computed at grid positions as defined in* section 3.1.3.3, *at 45° interval radially about the grid points with 0° direction parallel to the long side of the task area.*

All assumptions made in the determination of GR shall be stated in the scheme documentation. The GR value of the lighting installation shall not exceed the GR_L value given in Section 3.2.

3.1.4.2 Veiling reflections and reflected glare
High brightness reflections in the visual task may alter task visibility, usually detrimentally. Veiling reflections and reflected glare may be prevented or reduced by the following measures:

- *appropriate arrangement of luminaires and work places*

- *surface finish (e.g. matt surfaces)*

- luminance restriction of luminaires

- increased luminous area of the luminaire.

Note: Veiling glare is often significantly worse when the surfaces being illuminated are wet.

3.1.5   Obtrusive light

To safeguard and enhance the night time environment it is necessary to control obtrusive light (also known as light pollution), which can present physiological and ecological problems to surroundings and people.

The limits of obtrusive light for exterior lighting installations, to minimise problems for people, flora and fauna, are given in Table 3.2 and for road users Table 3.3.

Table 3.2 Maximum obtrusive light permitted for exterior lighting installations

Environmental zone	Light on properties		Luminaire intensity		Upward light	Luminance	
	E_v		I		ULR	L_b	L_s
	lx		cd		%	cd·m^{-2}	cd·m^{-2}
	Pre curfew $^{(a)}$	Post curfew	Pre curfew	Post curfew		Building	Signs
E1	2	0 $^{(b)}$	2500	0	0	0	50
E2	5	1	7500	500	0.05	5	400
E3	10	2	10 000	1000	0.15	10	800
E4	25	5	25 000	2500	0.25	25	1000

$^{(a)}$ In case no curfew regulations are available, the higher values shall not be exceeded and the lower values should be taken as preferable limits.
$^{(b)}$ If the luminaire is for public (road) lighting, then this value may be up to 1 lx.

Where:
E1   represents intrinsically dark areas, such as national parks or protected sites
E2   represents low district brightness areas, such as industrial or residential rural areas
E3   represents medium district brightness areas, such as industrial or residential suburbs
E4   represents high district brightness areas, such as town centres and commercial areas
E_v is the maximum value of vertical illuminance on properties in lx
I is the light intensity of each source in the potentially obtrusive direction in cd
ULR is the proportion of the flux of the luminaire(s) that is emitted above the horizontal, when the luminaire(s) is (are) mounted in its (their) installed position and attitude, and given in %
L_b is the maximum average luminance of the facade of a building in cd·m^{-2}
L_s is the maximum average luminance of signs in cd·m^{-2}.

Notes:

- **Curfew** – the time after which stricter requirements will apply; this is often a condition of use of the lighting applied by the local authority.
- **Light on properties** – the values in Table 3.2 are maxima and need to take account of existing light trespass at the point of measurement.
- **Luminaire intensity** – this applies to each luminaire in any potentially obtrusive direction.
- **Luminance** – this is the maximum luminance of any illuminated or self-luminous surface as seen from any potentially obtrusive direction.

Table 3.3 Maximum values of threshold increment and veiling luminance from non-road lighting installations

Light technical parameter	Road lighting classes [a]			
	No road lighting	ME5	ME4/ME3	ME2/ME1
Threshold increment (TI) [b] [c]	15% based on adaptation luminance of 0.1 cd·m^{-2}	15% based on adaptation luminance of 1 cd·m^{-2}	15% based on adaptation luminance of 2 cd·m^{-2}	15% based on adaptation luminance of 5 cd·m^{-2}
Veiling luminance (L$_v$) [d]	0.04	0.23	0.4	0.84

[a] Road lighting classes as given in BS EN 13201-2 (BSI, 2003a). See also section 4.2 of this Code.
[b] TI calculation as given in BS EN 13201-3. See also section 14.4 of this Code.
[c] Limits apply where users of transport systems are subject to a reduction in the ability to see essential information. Values given are for relevant positions and for viewing directions in the path of travel.
[d] Veiling luminance may be used when assessing the impact on roads for which the average value of luminance is not known.

It is possible that in some circumstances, obtrusive light may be considered a statutory nuisance as set out in section 102 of Chapter 16 of the Clean Neighbourhoods and Environment Act 2005 (HMSO, 2005). The act does make it clear under what conditions lighting becomes statutory nuisance but it does list a number of situations that are exempt. Exemptions include airports, harbour premises, railway premises, tramway premises, bus stations, public service vehicle operating centres, goods vehicle operating centres, lighthouses and prisons.

3.1.6   Directional lighting
Directional lighting may be used to highlight objects, reveal texture and improve the appearance of people. This is described by the term 'modelling'. Directional lighting of a visual task may also affect its visibility.

3.1.6.1 Modelling
Modelling is the balance between diffuse and directional light. It is a valid criterion of lighting quality in virtually all applications. The people and objects should be lit so that form and texture are revealed clearly and pleasingly. This occurs when the light comes predominantly from one direction; the shadows so essential to good modelling are then formed without confusion. The lighting should not be too directional or it will produce harsh shadows.

Note: In areas where it is common to have temporary objects, such as container depots and truck parks, it is important to provide light from multiple high mounted points to ensure that there are no large areas of shadow caused when the area is in use.

3.1.6.2 Directional lighting of visual tasks

Lighting from a specific direction may reveal details within a visual task, increasing their visibility and making the task easier to perform. Veiling reflections and reflected glare should be avoided, see 3.1.4.2.

3.1.7   Colour aspects

The colour qualities of a near-white lamp are characterised by two attributes:

● *the colour appearance of the lamp itself*

● *its colour rendering capabilities, which affect the colour appearance of objects and persons illuminated by the lamp.*

These two attributes shall be considered separately.

3.1.7.1 Colour appearance

The 'colour appearance' of a lamp refers to the apparent colour (chromaticity) of the light emitted. It is quantified by its correlated colour temperature (T_{CP}).

Colour appearance may also be described as in Table 3.4.

Note: Colour appearance does not uniquely specify the colour appearance of a light source. It is possible for two sources with the same colour temperature to have different appearances, one looking slightly purple and the other looking greenish. See section 16.2 for details on how to characterise the colour properties of a light source.

Table 3.4 Lamp colour appearance groups

Colour appearance	Correlated colour temperature T_{CP}
Warm	Below 3300 K
Intermediate	3300 to 5300 K
Cool	Above 5300 K

The choice of colour appearance is a matter of psychology, aesthetics and what is considered to be natural.

3.1.7.2 Colour rendering

It is important for visual performance and the feeling of comfort and well being that colours in the environment, of objects and of human skin are rendered naturally, correctly and in a way that makes people look attractive and healthy.

To provide an objective indication of the colour rendering properties of a light source, the general colour rendering index R_a has been introduced. The maximum value of R_a is 100. This figure decreases with decreasing colour rendering quality.

Safety colours shall always be recognisable as such and therefore light sources shall have colour rendering indices ≥ 20 (see also BS ISO 3864-1 (ISO, 2009)).

The minimum values of colour rendering index for distinct areas, tasks or activities are given in section 3.2.

3.1.8 Flicker and stroboscopic effects

Flicker causes distraction and may give rise to physiological effects such as headaches.

Stroboscopic effects can lead to dangerous situations by changing the perceived motion of rotating or reciprocating machinery.

Lighting systems should be designed to avoid flicker and stroboscopic effects.

Note: This can usually be achieved by technical measures adjusted to the chosen lamp type (i.e. operating discharge lamps at high frequencies).

3.1.9 Maintenance factor (MF)

The lighting scheme should be designed with a maintenance factor calculated for the selected lighting equipment, space environment and specified maintenance schedule, as defined in CIE 154:2003 (CIE, 2003).

The recommended illuminance for each task is given as maintained illuminance. The maintenance factor depends on the maintenance characteristics of the lamp and control gear, the luminaire, the environment and the maintenance programme.

The designer shall:

- *state the maintenance factor and list all assumptions made in the derivation of the value*

- *specify lighting equipment suitable for the application environment*

- *prepare a comprehensive maintenance schedule to include frequency of lamp replacement, luminaire cleaning intervals and cleaning method.*

Chapter 18 of this *Code* gives a comprehensive method for the evaluation of maintenance factor; it is based on CIE 154: 2003 and other similar documents.

In selecting lighting equipment and a maintenance programme, the lighting designer should seek to keep the overall maintenance as high as possible without imposing too onerous a maintenance schedule on the site owner.

3.1.10 Energy considerations

A lighting installation should meet the lighting requirements of a particular area without waste of energy. However, it is important not to compromise the visual aspects of a lighting installation simply to reduce energy consumption. This requires the consideration of appropriate lighting systems, equipment and controls.

It is also important to ensure that lighting is only provided when necessary; if some activities are only carried out for part of the night then the lighting should be dimmed or turned off when not needed.

3.1.11 Sustainability

Consideration should be given to the sustainability of the lighting installation. The selected lighting equipment shall be fit for the purpose.

It is important that the luminaires used may have to stand up to potentially onerous conditions from a mechanical point of view. This includes having the necessary strength to withstand wind loading and vandal attack as well as being sealed to prevent the ingress of dust and moisture. These matters are discussed in Chapter 4 of the *SLL Lighting Handbook* (SLL, 2009); details of the tests necessary for luminaires are given in BS EN 60598-2-3 (BSI, 2003b) and BS EN 60598-2-5 (BSI, 1998).

3.1.12 Emergency lighting

Emergency lighting should be provided to operate in the event of failure of the supply to the normal lighting system and conform to the relevant standards. In general, the emergency lighting must allow people to get to a place of safety, usually off the site in question.

3.2 Schedule of lighting requirements

The lighting requirements for various areas, tasks and activities are given in the tables of section 3.2.3 (see also BS EN 12193: 2007 (BSI, 2007b)).

Lighting recommendations with respect to safety and health of workers at work are given in section 3.2.4.

3.2.1 Composition of Tables 3.5 to 3.19

● *Column 1* lists the reference number for each area, task or activity.

● *Column 2* lists those areas, tasks or activities for which specific requirements are given. If the particular area, task or activity is not listed, the values given for a similar, comparable situation should be adopted.

● *Column 3* gives the maintained illuminance $\bar{E}_m$ on the reference surface (see 3.1.3) for the area, task or activity given in column 2.
Note: Lighting control may be required to achieve adequate flexibility for the variety of tasks performed.

● *Column 4* gives the minimum illuminance uniformity U_o on the reference surface (see 3.1.3) for the area, task or activity given in column 2.

● *Column 5* gives the Glare Rating limits (GR_L) where these are applicable to the situations listed in column 2 (see 3.1.4).

● *Column 6* gives the minimum colour rendering indices (R_a) (see 3.1.7.2) for the situation listed in column 2.

● *Column 7* contains advice and footnotes for exceptions and special applications for the situations listed in column 2.

3.2.2   Schedule of areas, tasks and activities

Table 3.5 *General circulation areas at outdoor work places*

Table 3.6 *Airports*

Table 3.7 *Building sites*

Table 3.8 *Canals, locks and harbours*

Table 3.9 *Farms*

Table 3.10 *Fuel filling stations*

Table 3.11 *Industrial sites and storage areas*

Table 3.12 *Offshore gas and oil structures*

Table 3.13 *Parking areas*

Table 3.14 *Petrochemical and other hazardous industries*

Table 3.15 *Power, electricity, gas and heat plants*

Table 3.16 *Railways and tramways*

Table 3.17 *Saw mills*

Table 3.18 *Shipyards and docks*

Table 3.19 *Water and sewage plants*

3.2.3 Lighting requirements for areas, tasks and activities

Table 3.5 General circulation areas at outdoor work places

Ref No.	Type of area, task or activity	$\bar{E}_m$ / lx	U_o	GR_L	R_a	Remarks
3.5.1	Walkways exclusively for pedestrians	5	0.25	50	20	Where there are other hazards present higher values of $\bar{E}_m$ are required. For example in water and sewerage treatment works 20 lx, on building sites 20–50 lx and in petroleum and chemical works 50 lx
3.5.2	Traffic areas for slowly moving vehicles (max. 10 km/h), e.g. bicycles, trucks and excavators	10	0.40	50	20	
3.5.3	Regular vehicle traffic (max. 40 km/h)	20	0.40	45	20	At shipyards and in docks, GR_L may be 50
3.5.4	Pedestrian passages, vehicle turning, loading and unloading points	50	0.40	50	20	For reading labels and signs: $\bar{E}_m$ 50 lx

Table 3.6 Airports

Ref No.	Type of area, task or activity	$\bar{E}_m$ / lx	U_o	GR_L	R_a	Remarks
	In all areas					1. Direct light in the direction of the control tower and landing aircraft shall be avoided 2. Direct light emitted above horizontal from floodlights should be restricted to the minimum
3.6.1	Hangar apron	20	0.10	55	20	
3.6.2	Terminal apron	30	0.20	50	40	
3.6.3	Loading areas	50	0.20	50	40	For reading labels: $\bar{E}_m$ 50 lx
3.6.4	Fuel depot	50	0.20	50	40	
3.6.5	Aircraft maintenance stands	200	0.50	45	60	

See Table 2.56 in Section 2.2.2 for information on the requirements for indoor areas of airports.

Table 3.7 Building sites

Ref No.	Type of area, task or activity	$\bar{E}_{m}$ / lx	U_o	GR_L	R_a	Remarks
3.7.1	Clearance, excavation and loading	20	0.25	55	20	
3.7.2	Construction areas, drain pipes mounting, transport, auxiliary and storage tasks	50	0.40	50	20	
3.7.3	Framework element mounting, light reinforcement work, wooden mould and framework mounting, electric piping and cabling	100	0.40	45	40	
3.7.4	Element jointing, demanding electrical, machine and pipe mountings	200	0.50	45	40	

Table 3.8 Canals, locks and harbours

Ref No.	Type of area, task or activity	$\bar{E}_{m}$ / lx	U_o	GR_L	R_a	Remarks
3.8.1	Waiting quays at canals and locks	10	0.25	50	20	
3.8.2	Gangways and passages exclusively for pedestrians	10	0.25	50	20	
3.8.3	Lock control and ballasting areas	20	0.25	55	20	
3.8.4	Cargo handling, loading and unloading	30	0.25	55	20	For reading labels: $\bar{E}_{m}$ 50 lx
3.8.5	Passenger areas in passenger harbours	50	0.40	50	20	
3.8.6	Coupling of hoses, pipes and ropes	50	0.40	50	20	
3.8.7	Dangerous part of walkways and driveways	50	0.40	45	20	

Table 3.9 Farms

Ref No.	Type of area, task or activity	$\bar{E}_{m}$ / lx	U_o	GR_L	R_a	Remarks
3.9.1	Farm yard	20	0.10	55	20	
3.9.2	Equipment shed (open)	50	0.20	55	20	
3.9.3	Animals sorting pen	50	0.20	50	40	

See Table 2.10 in Section 2.2.2 for information on the requirements for indoor area agricultural tasks.

Table 3.10 Fuel filling stations

Ref No.	Type of area, task or activity	$\bar{E}_m$ / lx	U_o	GR_L	R_a	Remarks
3.10.1	Vehicle parking and storage areas	5	0.25	50	20	
3.10.2	Entry and exit driveways: dark environment (i.e. rural areas and suburbs)	20	0.40	45	20	
3.10.3	Entry and exit driveways: light environment (i.e. cities)	50	0.40	45	20	
3.10.4	Air pressure and water checking points and other service areas	150	0.40	45	20	
3.10.5	Meter reading area	150	0.40	45	20	

Table 3.11 Industrial sites and storage areas

Ref No.	Type of area, task or activity	$\bar{E}_m$ / lx	U_o	GR_L	R_a	Remarks
3.11.1	Short term handling of large units and raw materials, loading and unloading of solid bulk goods	20	0.25	55	20	
3.11.2	Continuous handling of large units and raw materials, loading and unloading of freight, lifting and descending location for cranes, open loading platforms	50	0.40	50	20	
3.11.3	Reading of addresses, covered loading platforms, use of tools, ordinary reinforcement and casting tasks in concrete plants	100	0.50	45	20	
3.11.4	Demanding electrical, machine and piping installations, inspection	200	0.50	45	60	Use local lighting

See Tables 2.8 and 2.9 in Section 2.2.2 for information on the requirements for internal storage areas.

Table 3.12 Offshore gas and oil structures

Ref No.	Type of area, task or activity	$\bar{E}_m$ / lx	U_o	GR_L	R_a	Remarks
3.12.1	Sea surface below the rig	30	0.25	50	20	
3.12.2	Ladders, stairs, walkways	100	0.25	45	20	On treads
3.12.3	Boat landing areas /transport areas	100	0.25	50	20	
3.12.4	Helideck	100	0.40	45	20	1. Direct light in the direction of the control tower and landing aircraft shall be avoided 2. Direct light emitted above horizontal from floodlights should be restricted to the minimum
3.12.5	Derrick	100	0.50	45	40	
3.12.6	Treatment areas	100	0.50	45	40	
3.12.7	Pipe rack area/deck	150	0.50	45	40	
3.12.8	Test station, shale shaker, wellhead	200	0.50	45	40	
3.12.9	Pumping areas	200	0.50	45	20	
3.12.10	Life boat areas	200	0.40	50	20	
3.12.11	Drill floor and monkey board	300	0.50	40	40	Special attention to string entry is needed
3.12.12	Mud room, sampling	300	0.50	40	40	
3.12.13	Crude oil pumps	300	0.50	45	40	
3.12.14	Plant areas	300	0.50	40	40	
3.12.15	Rotary table	500	0.50	40	40	

Table 3.13 Parking areas

Ref No.	Type of area, task or activity	$\bar{E}_m$ / lx	U_o	GR_L	R_a	Remarks
3.13.1	*Light traffic, e.g. parking areas of shops, terraced and apartment houses; cycle parks*	5	0.25	55	20	
3.13.2	*Medium traffic, e.g. parking areas of department stores, office buildings, plants, sports and multipurpose building complexes*	10	0.25	50	20	
3.13.3	*Heavy traffic, e.g. parking areas of schools, churches, major shopping centres, major sports and multipurpose building complexes*	20	0.25	50	20	

Note: All the illuminance values listed in the table above should be regarded as absolute minima to be used in areas where there are no extra factors such as fear of crime that might require the lighting level to be increased.

See Table 2.38 in Section 2.2.2 for information on the requirements for indoor car parks.

Table 3.14 Petrochemical and other hazardous industries

Ref No.	Type of area, task or activity	$\bar{E}_m$ / lx	U_o	GR_L	R_a	Remarks
3.14.1	*Handling of servicing tools, utilisation of manually regulated valves, starting and stopping motors, lighting of burners*	20	0.25	55	20	
3.14.2	*Filling and emptying of container trucks and wagons with risk free substances, inspection of leakage, piping and packing*	50	0.40	50	20	
3.14.3	*Filling and emptying of container trucks and wagons with dangerous substances, replacements of pump packing, general service work, reading of instruments*	100	0.40	45	40	
3.14.4	*Fuel loading and unloading sites*	100	0.40	45	20	
3.14.5	*Repair of machines and electric devices*	200	0.50	45	60	*Use local lighting*

Table 3.15 Power, electricity, gas and heat plants

Ref No.	Type of area, task or activity	$\bar{E}_{m}$ / lx	U_o	GR_L	R_a	Remarks
3.15.1	Pedestrian movements within electrically safe areas	5	0.25	50	20	
3.15.2	Handling of servicing tools, coal	20	0.25	55	20	
3.15.3	Overall inspection	50	0.40	50	20	
3.15.4	General servicing work and reading of instruments	100	0.40	45	40	
3.15.5	Wind tunnels: servicing and maintenance	100	0.40	45	40	
3.15.6	Repair of electric devices	200	0.50	45	60	Use local lighting

See Table 2.24 in Section 2.2.2 for information on the requirements for power stations.

Table 3.16 Railways and tramways

Ref No.	Type of area, task or activity	$\bar{E}_{m}$ / lx	U_o	GR_L	R_a	Remarks
	Railway areas including light railways, tramways, monorails, miniature rails, metro, etc					Avoid glare for vehicle drivers
3.16.1	Tracks in passenger station areas, including stabling	10	0.25	50	20	$U_d \geq 1/8$
3.16.2	Railway yards: flat marshalling, retarder and classification yards	10	0.40	50	20	$U_d \geq 1/5$
3.16.3	Hump areas	10	0.40	45	20	$U_d \geq 1/5$
3.16.4	Freight track, short duration operations	10	0.25	50	20	$U_d \geq 1/8$
3.16.5	Open platforms, rural and local trains, small number of passengers	15	0.25	50	20	1. Special attention to the edge of the platform 2. $U_d \geq 1/8$
3.16.6	Walkways	20	0.40	50	20	
3.16.7	Level crossings	20	0.40	45	20	
3.16.8	Open platforms, suburban and regional trains with large number of passengers or inter-city services with small number of passengers	20	0.40	45	20	1. Special attention to the edge of the platform 2. $U_d \geq 1/5$

Table 3.16 Continued

Ref No.	Type of area, task or activity	$\bar{E}_m$ / lx	U_o	GR_L	R_a	Remarks
3.16.9	Freight track, continuous operation	20	0.40	50	20	$U_d \geq 1/5$
3.16.10	Open platforms in freight areas	20	0.40	50	20	$U_d \geq 1/5$
3.16.11	Servicing trains and locomotives	20	0.40	50	40	$U_d \geq 1/5$
3.16.12	Railway yards handling areas	30	0.40	50	20	$U_d \geq 1/5$
3.16.13	Coupling area	30	0.40	45	20	$U_d \geq 1/5$
3.16.14	Stairs, small and medium-size stations	50	0.40	45	40	
3.16.15	Open platforms, inter-city services	50	0.40	45	20	1. Special attention to the edge of the platform 2. $U_d \geq 1/5$
3.16.16	Covered platforms, suburban or regional trains or inter-city services with small number of passengers	50	0.40	45	40	1. Special attention to the edge of the platform 2. $U_d \geq 1/5$
3.16.17	Covered platforms in freight areas, short duration operations	50	0.40	45	20	$U_d \geq 1/5$
3.16.18	Covered platforms, inter-city services	100	0.50	45	40	1. Special attention to the edge of the platform 2. $U_d \geq 1/3$
3.16.19	Stairs, large stations	100	0.50	45	40	0
3.16.20	Covered platforms in freight areas, continuous operation	100	0.50	45	40	$U_d \geq 1/5$
3.16.21	Inspection pit	100	0.50	40	40	Use low-glare local lighting

Note: Diversity (U_d) is the ratio of minimum illuminance to maximum illuminance.

See Table 2.57 in Section 2.2.2 for information on the requirements for the indoor parts of railway stations.

Table 3.17 Saw mills

Ref No.	Type of area, task or activity	$\bar{E}_m$ / lx	U_o	GR_L	R_a	Remarks
3.17.1	Timber handling on land and in water, sawdust and chip conveyors	20	0.25	55	20	
3.17.2	Sorting of timber on land or in water, timber unloading points and sawn timber loading points, mechanical lifting to timber conveyor, stacking	50	0.40	50	20	
3.17.3	Reading of addresses and markings of sawn timber	100	0.40	45	40	
3.17.4	Grading and packaging	200	0.50	45	40	
3.17.5	Feeding into stripping and chopping machines	300	0.50	45	40	

See Table 2.29 in Section 2.2.2 for information on the requirements for wood working and processing.

Table 3.18 Shipyards and docks

Ref No.	Type of area, task or activity	$\bar{E}_m$ / lx	U_o	GR_L	R_a	Remarks
3.18.1	General lighting of shipyard area, storage areas for prefabricated goods	20	0.25	55	40	
3.18.2	Short term handling of large units	20	0.25	55	20	
3.18.3	Cleaning of ship hull	50	0.25	50	20	
3.18.4	Painting and welding of ship hull	100	0.40	45	60	
3.18.5	Mounting of electrical and mechanical components	200	0.50	45	60	

Table 3.19 Water and sewage plants

Ref No.	Type of area, task or activity	$\bar{E}_m$ / lx	U_o	GR_L	R_a	Remarks
3.19.1	Handling of service tools, utilisation of manually operated valves, starting and stopping of motors, piping packing and raking plants	50	0.40	45	20	
3.19.2	Handling of chemicals, inspection of leakage, changing of pumps, general servicing work, reading of instruments	100	0.40	45	40	
3.19.3	Repair of motors and electric devices	200	0.50	45	60	

3.2.4 Lighting requirements for safety and security

The values in Table 3.20 provide guidance on the minimum lighting requirements to ensure safety and security. They are intended for use in areas when none of the tasks listed in the tables of section 3.2.3 are being carried out.

Table 3.20 Lighting requirements for safety and security

Risk level	$\bar{E}_m$ / lx	U_o	GR_L	R_a	Remarks
Very low risks, i.e. • Storage areas with occasional traffic in industrial yards • Coal fields in power plants • Timber storage, sawdust and wood chip fields in saw mills • Occasionally used service passages and stairs, waste water cleaning and aeration tanks, filter and sludge digestion tanks in water and sewage plants	5	0.25	55	20	
Low risks, i.e. • General lighting in harbours • Areas of risk free process and occasionally used platforms and stairs in petrochemical and other hazardous industries • Sawn timber storage areas in saw mills	10	0.40	50	20	In harbours, U_o may be 0.25
Medium risks, i.e. • Vehicle storage areas and container terminals with frequent traffic in harbours, industrial yards and storage areas • Vehicle storage areas and conveyors in petrochemical and other hazardous industries • Oil stores in power plants • General lighting and storage areas for prefabricated goods in shipyards and docks • Regularly used stairs, basins and filters of clean water plants in water and sewage plants	20	0.40	50	20	In shipyards and docks, U_o may be 0.25

Risk level	$\bar{E}_m$ / lx	U_o	GR_L	R_a	Remarks
High risks, i.e. • Element mould, timber and steel storage, building foundation hole and working areas on sides of the hole at building sites • Fire, explosion, poison and radiation risk areas in harbours, industrial yards and storage areas • Oil stores, cooling towers, boilers compressors, pumping plants, valves, manifolds, operating platforms, regularly used stairs, crossing points of conveyors, electric switch-yards in petrochemical and other hazardous industries • Switch yards in power plants • Crossing points of conveyors, fire risk areas in saw mills	50	0.40	45	20	At building sites and in saw mills, GR_L may be 50

3.3 Verification procedures

Verification of the lighting installation shall be by measurement, calculation or inspection of data (see Chapters 15, 14 and 12*).*

3.3.1 Illuminance
Verification of illuminances and uniformities that relate to specific tasks shall be measured in the plane of the task and the measurement points chosen shall coincide with the design points or grid used.

Note: When verifying illuminance, account should be taken of the calibration of the light meters used, the conformity of the lamps and luminaires to the published photometric data, and of the design assumptions made about surface reflectances, etc, compared with the real values.

The average illuminance and uniformity shall be not less than the values given in section 3.2 *and* Table 3.1, respectively.

3.3.2 Glare rating
Verification shall be by inspection of the design data and parameters provided for the scheme. All assumptions shall be declared.

3.3.3 Colour Rendering Index
Authenticated R_a data shall be provided for the lamps in the scheme by the manufacturer of the lamps. The lamps shall conform to the requirements.

The lamps shall be as specified in the design.

3.3.4 Obtrusive light
Calculated values for E_v, I, ULR, L_b, L_s and TI shall be provided by the scheme designer.

Verification of E_v, L_b, and L_s shall be made by measurement taking into account all design assumptions.

Chapter 4: Road lighting

There are a number of standards that are important in road lighting. The recommendations of this section are based on these standards but not all material from the standards is covered. The following bullet points discuss the various standards for road lighting and the way they relate to road lighting is covered in this *Code*.

● **BS EN 13201-2** (BSI, 2003a) – this standard defines the lighting characteristics of a number of lighting classes for roads. The classes that are recommended by BS 5489-1 are included in this section of the *Code*.

● **BS EN 13201-3** (BSI, 2003c) – this standard defines the way that the various luminous characteristics recommended in the tables of BS EN 13201-2 must be calculated. The main calculation methods defined in the standard are given in Chapter 14 *Outdoor lighting calculations*.

● **BS EN 13201-4** (BSI, 2003d) – this standard covers the measurement of road lighting. The key points of this measurement process are covered in section 15.2.4 of this *Code*, however, reference to the standard is recommended before any measurement is carried out.

● **BS 5489-1** (BSI, 2003e) – this standard provides advice on selecting the most appropriate type of lighting for a given section of road. It explains the general principle of street lighting, and gives advice on the location and maintenance of street lighting equipment. Recommendations on the selection of lighting classes taken from this standard are included in this section of the *Code*.

4.1   Classification of roads

Road lighting may be divided into three classes:

● traffic routes where the needs of the driver are dominant

● subsidiary roads where the lighting is primarily intended for pedestrians and cyclists; and urban centres, where the lighting is designed to do what can be done for public safety and security, while also providing an attractive night-time environment

● areas where conflict between streams of traffic or traffic and pedestrians may be a problem.

The selection of lighting classes for traffic routes, subsidiary roads and conflict areas is covered in the following sections.

4.1.1   Traffic routes

The primary function of the lighting of traffic routes is to make other vehicles or obstructions on the road visible. Road lighting does this by producing a difference between the luminance of the vehicle or obstruction and the luminance of its immediate background, the road surface. This difference is achieved by increasing the luminance of the road surface above that of the vehicle so that the vehicle is seen in silhouette against the road surface. The following lighting criteria are used to define the lighting on traffic routes:

● Average road surface luminance: The luminance of the road surface averaged over the carriageway (cd/m^2).

- Overall luminance uniformity (U_o): The ratio of the lowest luminance at any point on the carriageway to the average luminance of the carriageway.

- Longitudinal luminance uniformity (U_l): The ratio of the lowest to the highest luminance found along a line along the centre of a driving lane. For the whole carriageway, this is the lowest longitudinal luminance uniformity found for the driving lanes of the carriageway.

- Threshold increment: A measure of the loss of visibility caused by disability glare from the road lighting luminaires.

- Surround Ratio (SR): average illuminance on strips just outside the edges of the carriageway in proportion to the average illuminance on strips just inside the edges. People and objects adjacent to the carriageway need to be seen by the driver; lighting of the area adjacent to the carriageway should conform to the surround ratio.

For more information on how the above terms are defined and their values calculated, see Chapter 14. The ME classes set values for the above list of parameters and thus provide the lighting criteria for the different types of main road that need to be lit.

Traffic routes are predominantly lit using ME classes, selection of the class is based on the type of road, the average daily traffic flow (ADT), the speed of vehicles, the type of vehicles in the traffic and the frequency of conflict areas and pedestrians. Table 4.1 specifies the different classes and identifies the recommended lighting criteria. Details of the recommended lighting criteria for dry roads are given. These are the lighting criteria usually adopted in the UK.

Table 4.1 Lighting classes for traffic routes

Hierarchy description	Type of road/general description	Detailed description	ADT	Lighting class
Motorway	Limited access	Main carriageway in complex interchange areas	< 40 000	ME1
			> 40 000	ME1
		Main carriageway with interchanges at < 3 km	< 40 000	ME2
			> 40 000	ME1
		Main carriageways with interchanges > 3 km	< 40 000	ME2
			> 40 000	ME2
		Emergency lanes		ME4a
Strategic route	Trunk roads and some main A roads between primary destinations	Single carriageway	< 15 000	ME3a
			> 15 000	ME2
		Dual carriageway	< 15 000	ME3a
			> 15 000	ME2

Hierarchy description	Type of road/general description	Detailed description	ADT	Lighting class
Main distributor	Major urban network and inter-primary links, short to medium distance traffic	Single carriageway	< 15 000	ME3a
			> 15 000	ME2
		Dual carriageway	< 15 000	ME3a
			> 15 000	ME2
Secondary distributor	Classified road (B or C road) and unclassified urban bus route, carrying local traffic with frontage access and frequent junctions	Rural areas (Environmental zones 1 or 2). These roads link larger villages and HGV generators to the strategic and main distributor network	< 7000	ME4a
			7000–15 000	ME3b
			> 15 000	ME3a
		Urban areas (Environmental zone 3). These roads have 30 mph speed limits and very high levels of pedestrian activity with some crossing facilities including zebra crossings. On-street parking is generally unrestricted except for safety reasons	< 7000	ME3c
			7000–15 000	ME3b
			> 15 000	ME2
Link road	Road linking the main and secondary distribution network with frontage access and frequent junctions	Rural areas (Environmental zones 1 or 2). These roads link smaller villages to the distributor network. They are of varying width and not always capable of carrying two-way traffic	Any	ME5
		Urban areas (Environmental zone 3). These roads are residential or industrial interconnecting roads with 30 mph speed limits, random pedestrian movements and uncontrolled parking	Any	ME4b or S2
			Any with high pedestrian or cyclist traffic	S1

Notes:

1. See Table 4.3 in section 4.1.3 for conflict areas.

2. The guidance on lighting class selection for motorways and traffic routes uses average daily traffic (ADT), which is the normal concept in traffic planning, and is usually known. Peak traffic is generally taken to be 10% and 12% of ADT in rural and urban areas, respectively. If hourly flows are known, and the peak hour traffic is significantly greater than 12%, the peak traffic should be taken into account when selecting the lighting class.

3. Traffic flow can vary significantly during the night, and the use of different lighting levels at some periods may be considered. For this purpose, a detailed analysis of traffic flow is carried out, to assess the hourly flow through the night.

4. Where lighting levels are reduced at certain periods, any lower levels selected can use the values from appropriate lower ME classes, but retain the U_o and U_l values of the ME class selected for the peak period. This in practice means that it is only possible to reduce the lighting by dimming each of the lanterns and not by switching off alternate lanterns.

4.1.2 Subsidiary roads

Subsidiary roads consist of access roads and residential roads and associated pedestrian areas, footpaths and cycle tracks. The main function of lighting of subsidiary roads and the areas associated with them is to enable pedestrians and cyclists to orientate themselves and to detect vehicular and other hazards, and to discourage crime against people and property. The lighting in such areas can provide some help to drivers but it is unlikely to be sufficient for revealing objects on the road without the use of headlamps. The main purpose of lighting footpaths and cycle tracks separated from roads is to show the direction the route takes, to enable cyclists and pedestrians to orientate themselves, to detect the presence of other cyclists, pedestrians and hazards, and to discourage crime against people and property.

Illuminance on the horizontal is used as the lighting criterion for subsidiary roads and associated areas. The lighting class to be used is determined by the traffic flow, the environmental zone, the level of crime and the colour rendering of the light source used. In the table below, low traffic flow refers to areas where traffic is typical of a residential road and solely associated with adjoining properties. Normal traffic flow refers to areas where traffic flow is equivalent to a housing estate access road. High traffic flow refers to areas where traffic usage is high and can be associated with local amenities such as clubs, shopping facilities and public houses. The crime rates should be considered relative to the local area. The environmental zones (E1 to E4) are as defined in section 3.1.5 on obtrusive light. The divide in CIE general colour rendering index (CRI) at 60 means that the use of low pressure sodium or high pressure sodium light sources calls for a higher illuminance than fluorescent and metal halide light sources. The S-class may be increased one step where there are traffic calming measures.

Table 4.2 may be used to select the appropriate class of lighting for a given road.

Table 4.2 Lighting classes for subsidiary roads

Crime rate	CRI	Low traffic flow E1 or E2	Normal traffic flow E1 or E2	Normal traffic flow E3 or E4	High traffic flow E1 or E2	High traffic flow E3 or E4
Low	<60	S5	S4	S3	S3	S2
Low	>60	S6	S5	S4	S4	S3
Moderate	<60	S4	S3	S2	–	S1
Moderate	>60	S3	S4	S3	–	S2
High	<60	S2	S2	S1	–	S1
High	>60	S3	S3	S2	–	S2

Note: It is recommended that the actual overall uniformity of illuminance U_o be at least 0.25.

To control glare on subsidiary roads, it is recommended that the luminaires used should meet the requirements of class G1 or higher. See section 4.2.4.

4.1.3   Conflict areas

A conflict area is one in which traffic flows merge or cross, e.g. at intersections or roundabouts, or where vehicles and other road users, are in close proximity, e.g. on a shopping street or at a pedestrian crossing. Lighting for conflict areas is intended for drivers rather than pedestrians. The criteria used to define lighting for conflict areas are based on the illuminance on the road surface rather than road surface luminance. This is because drivers' viewing distances may be less than the 60 m assumed for traffic routes and there are likely to be multiple directions of view. The criteria used for the lighting of conflict areas are:

● Average road surface illuminance: the illuminance of the road surface averaged over the carriageway (lx).

● Overall illuminance uniformity (U_o): the ratio of the lowest illuminance at any point on the carriageway to the average illuminance of the carriageway.

These recommendations can be applied to all parts of the conflict area or only to the carriageway when separate recommendations are used for pedestrians or cyclists. The CE classes are used for conflict areas, the class chosen has to be matched to the lighting of the traffic routes approaching the conflict area. Table 4.3 below shows lighting classes of comparable level.

Table 4.3 Lighting classes of comparable level

ME class (traffic routes)	CE class (conflict areas)	S class (subsidiary roads)
–	CE0	–
ME1	CE1	–
ME2	CE2	–
ME3	CE3	S1
ME4	CE4	S2
ME5	CE5	S3
ME6	–	S4

However, it is common to make the level of the conflict area one step higher than the surrounding road network. When using a CE class on a junction of traffic routes, it is normal to pick the class based on the ME class. Table 4.4 shows the normal class selection used based on the highest ME class of any of the approach roads.

Table 4.4 Lighting classes for conflict areas at junctions of traffic routes

Traffic route lighting class	Conflict area lighting class
ME1	CE0
ME2	CE1
ME3	CE2
ME4	CE3
ME5	CE4

In any conflict area, glare should be at least as well controlled as on the approach roads, as the conflict area situation increases the visual demands on the driver. In order to limit glare, an appropriate installed intensity (G) class should be selected, normally classes G4, G5 and G6 are appropriate. See section 4.2.4 for details of the G classes.

Table 4.5 Lighting classes for town and city centres

Type of traffic	Lighting class			
	Normal traffic flow		High traffic flow	
	E3[a]	E4[a]	E3[a]	E4[a]
Pedestrian only	CE3	CE2	CE2	CE1
Mixed vehicle and pedestrian with separate footways	CE2	CE1	CE1	CE1
Mixed vehicle and pedestrian on same surface	CE2	CE1	CE1	CE1

[a] *The environmental zones (E1 to E4) are as defined in section 3.1.5 on obtrusive light.*

The selection of lighting class for a specific city or town centre road type may be varied up or down from the classes indicated in Table 4.5, taking account of:

- vehicular traffic use

- pedestrian and cyclist use

- on street parking

- amenities such as shops, public houses, etc

- level of crime

- CCTV requirements.

4.2 Lighting classes

This section gives tables of the lighting parameters required for the different lighting classes.

4.2.1 ME classes

ME classes are mainly used on traffic routes; the requirements of the ME classes are given in Table 4.6

Table 4.6 ME series of lighting classes

	Luminance of the road surface of the carriageway for the dry road surface condition			Disability glare	Lighting of surroundings
	L / cd/m²	U_o	U_1	TI / %[a]	SR[b]
	(minimum maintained)	(minimum)	(minimum)	(maximum)	(minimum)
ME1	2	0.4	0.7	10	0.5
ME2	1.5	0.4	0.7	10	0.5
ME3a	1	0.4	0.7	15	0.5
ME3b	1	0.4	0.6	15	0.5
ME3c	1	0.4	0.5	15	0.5
ME4a	0.75	0.4	0.6	15	0.5
ME4b	0.75	0.4	0.5	15	0.5
ME5	0.5	0.35	0.4	15	0.5
ME6	0.3	0.35	0.4	15	No requirement

[a] *An increase of 5 percentage points in threshold increment (TI) can be permitted where low luminance light sources are used (see note 5).*
[b] *This criterion can be applied only where there are no traffic areas with their own requirements adjacent to the carriageway; examples of such areas include cycle paths and service roads.*

Notes:

1. The road surface luminance is the result of the illumination of the road surface, the reflection properties of the road surface and the geometric conditions of observation.

2. The average luminance (L) reflects the general luminance level at which the driver performs. At the low level of lighting used for road lighting, performance improves with luminance in terms of increasing contrast sensitivity, increasing visual acuity and amelioration of glare.

3. The overall uniformity (U_o) measures in a general way the variation of luminances and indicates how well the road surface serves as a background for road markings, objects and other road users.

4. The longitudinal uniformity (U_l) provides a measure of the conspicuity of the repeated pattern of bright and dark patches on the road. It relates to visual conditions on long uninterrupted sections of road.

5. The threshold increment (TI) indicates that, although road lighting improves visual conditions, it also causes disability glare to a degree depending on the type of luminaires, lamps and geometric situation. Low pressure sodium lamps and fluorescent tubes are normally considered to be low luminance lamps. For these lamps, and luminaires providing less or equivalent luminance, footnote *a* of Table 4.6 permits higher values.

6. Lighting confined to the carriageway is inadequate for revealing the immediate surrounds of the road and revealing road users at the kerb. The requirements for the surround ratio (SR) apply only where there are no traffic areas with their own requirements adjacent to the carriageway, including footways, cycle ways or emergency lanes.

4.2.2 S classes

The S classes are mainly used on subsidiary roads and they are defined in Table 4.7.

Table 4.7 The S lighting classes

Class	Horizontal illuminance	
	E_{av} / lx [a]	E_{min} / lx
	(minimum maintained)	(maintained)
S1	15	5
S2	10	3
S3	7.5	1.5
S4	5	1
S5	3	0.6
S6	2	0.6
S7	Performance not determined	Performance not determined

[a] *To provide for uniformity, the actual value of the maintained average illuminance may not exceed 1.5 times the minimum E_{av} value indicated for the class, and ideally should not exceed four times the E_{min} value.*

4.2.3 CE classes

CE classes specify the lighting used in conflict zones and they are defined in Table 4.8.

Table 4.8 The CE lighting classes

Class	Horizontal illuminance	
	E_{av} / lx	U_o
	(minimum maintained)	(minimum)
CE0	50	0.4
CE1	30	0.4
CE2	20	0.4
CE3	15	0.4
CE4	10	0.4
CE5	7.5	0.4

4.2.4 G classes

In some situations, it can be necessary to restrict disability glare from installations where the threshold increment (TI) cannot be calculated.

Table 4.9 gives installed luminous intensity classes G1, G2, G3, G4, G5 and G6 from which a class can be chosen to meet the appropriate requirements for restriction of disability glare and/or the control of obtrusive light.

Table 4.9 Luminous intensity (G) classes

Class	Maximum luminous intensity / cd/klm			Other requirements
	at 70° [a]	at 80° [a]	at 90° [a]	
G1		200	50	None
G2		150	30	None
G3		100	20	None
G4	500	100	10	Luminous intensities above 95° [a] to be zero
G5	350	100	10	Luminous intensities above 95° [a] to be zero
G6	350	100	0	Luminous intensities above 90° [a] to be zero

[a] *Any direction forming the specified angle from the downward vertical, with the luminaire installed for use.*

Note: The values of luminous intensity in Table 4.9 are all normalised intensity values for the luminaires used and are thus expressed in terms of candelas per kilo lumen.

Chapter 5: Daylight

Daylighting gives to a building a unique variety and interest. An interior which looks gloomy, or which does not have a view to the outside when this could reasonably be expected, will be considered unsatisfactory by its users. Unfortunately the introduction of daylight into buildings without consideration of its impact on the users can have negative consequences. A short walk around any city will reveal numerous well-glazed office buildings where the blinds on many windows are permanently closed. Such behaviour demonstrates the existence of a failed daylighting design for at least some people within the building. Nonetheless, unless there is a good reason why there should be no daylight in the building, daylighting should always be encouraged.

This section of the code is based on some of the key recommendations of BS 8206-2 (BSI, 2008) on daylight. However, the recommendations here are of a general nature and reference should be made to Chapter 7 of the *SLL Lighting Handbook* (SLL, 2009) for more practical advice and SLL Lighting Guide 10 (SLL, 1999) for more details on the application of daylight. Some of the calculation methods defined in BS 8206-2 are covered in Chapter 17 of this *Code*.

5.1 Daylight and health

Most of the health benefits that are derived from daylight can also be obtained with electric lighting (using lamps with an appropriate spectrum). However, as most of the benefits come from long term exposure to relatively high levels of light, it is usually not practical to provide the necessary light from electrical sources.

5.1.1 Regulation of the circadian system

The role of the circadian system (which controls daily and seasonal body rhythms) is to link the functions of the body (e.g. the sleep/wake cycle, and changes in core body temperature and in hormone secretion) with the external day/night cycle. Disruption to this system (from lack of light, for example) can cause problems such as depression and poor sleep quality which could lead to more serious problems. Therefore, it is important that occupants of buildings, particularly those of limited mobility in, for example, hospitals and nursing homes, and people who might be unable to go outside much, are given access to high levels of daylight, particularly in the mornings, to assist the entrainment of circadian rhythms. Therefore, buildings used by such people should have spaces with high levels of daylight, such as conservatories, which are readily accessible to them.

5.1.2 Mood

Mood can be modified by lighting. The dynamic nature of daylight is strongly favoured by building occupants. Adequate access to daylight can have a positive impact on mood, especially in situations where people are static for long periods of time, for example, in a school or a hospital ward.

5.1.3 Seasonal affective disorder (SAD)

A small percentage of people suffer a seasonal mood disorder known as seasonal affective disorder (SAD) with a further number suffering a mild form known as sub-syndromal SAD (S-SAD). Symptoms include depression, lack of energy, increased need for sleep and increased appetite and weight gain, occurring in the winter, when there is little daylight with symptoms lessening in the summer when there is more daylight. Such symptoms can be reduced by exposure to daylight.

5.1.4   Ultraviolet (UV) radiation

The ultraviolet (UV) radiation in sunlight can be damaging to the skin. However, with people now spending many daylight hours inside buildings, there is the danger of vitamin D deficiency caused by insufficient exposure to UV radiation. A vitamin D deficiency leads to rickets in children and softening of the bones in adults.

Exposure to sunlight, even through glass, can kill many types of viruses and bacteria and so can be of great value in winter when there is a high incidence of respiratory infections.

5.2   Windows and view

Unless an activity requires the exclusion of daylight, a view out-of-doors should be provided irrespective of its quality.

All occupants of a building should have the opportunity for the refreshment and relaxation afforded by a change of scene and focus. Even a limited view to the outside can be valuable. If an external view cannot be provided, occupants should have an internal view possessing some of the qualities of a view out-of-doors, for example, into an atrium.

5.2.1   Analysis of view

In planning the position of windows, the following factors are important.

● Most people like a view of a natural scene: trees, grass, plants and open space.

● In densely built-up areas, a view of the natural scene may not be available. When only buildings, sky and street can be seen, it is especially desirable that the view is dynamic, i.e. including the activities of people outside and the changing weather, however, a static view is usually better than none.

● A specific close view may be essential, particularly for security and supervision of the space around buildings.

Most unrestricted views have three 'layers', as follows:

● upper (distant), being the sky and its boundary with the natural or man-made scene

● middle, being the natural or man-made objects themselves

● lower (close), being the nearby ground.

Views which incorporate all three 'layers' are the most completely satisfying. The role of the size and shape of windows in determining the elements of a view is illustrated in Figure 5.1.

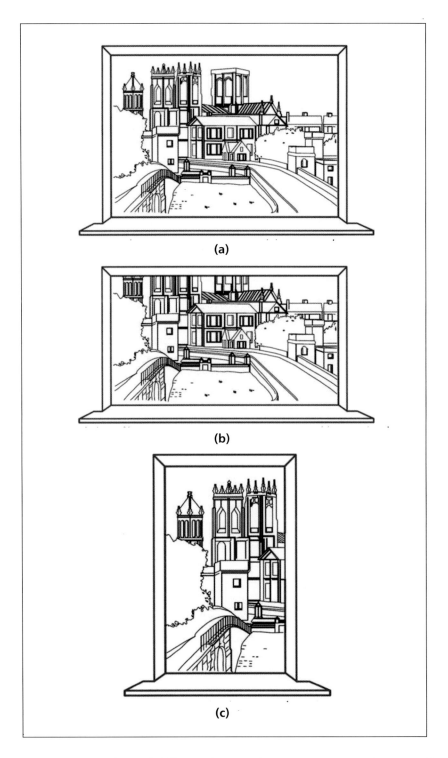

Figure 5.1 Views from windows of different shapes and sizes. (a) Large window providing a view containing all three layers; (b) Smaller window of horizontal proportions providing a view with some sky; (c) Smaller window of vertical proportions showing all three layers but with a restricted view

5.2.2 Size and proportion of windows

The size and proportion of windows should depend on the type of view, the size of the internal space, and the position and mobility of occupants. A variety of window shapes and sizes is illustrated in Figure 5.1. Some circumstances may suggest a tall window which allows occupants anywhere to enjoy the full vertical span of the view. A narrow horizontal window will only offer a similar prospect to those close to it; a narrow vertical window is also restrictive yet will admit a deeper penetration of daylight. For a given area of window, the more exaggerated the horizontal or vertical proportions, the more restricted will be the position of occupants who can experience the views. A view of the immediate foreground will be experienced normally only by those close to the window.

Unless a view of the sky is to be deliberately excluded (and the penetration of daylight severely limited), window heads should be above standing eye height. Sills, normally, should be below the eye level of people seated. Special consideration should be given to window heights in buildings such as nurseries, schools, hospitals and care homes, especially if the windows may be opened.

The most limited views occur in a deep room when windows are confined to one wall only. Table 5.1 gives guidance on minimum window area for a satisfactory view when fenestration is restricted to one wall; higher proportions are recommended. The table gives total glazed area of the room as a percentage of the internal window wall area. When there are windows in two or more walls, the total area of glazing should not be less than the area that would be recommended if the windows were restricted to any one wall. The openings should be distributed to give views from all occupied areas of the room.

Table 5.1 Minimum glazed area for view when windows are restricted to one wall

Depth of room from outside wall (max) / m	Glazed area as percentage of window wall as seen from inside (min) / %
<8	20
≥8 ≤11	25
>11 ≤14	30
>14	35

Note: *Windows which are primarily designed for view may not provide adequate task illumination.*

When windows are confined to one wall only, it is recommended that the total width of the windows should be at least 35 per cent of the length of the wall.

5.3 Daylight and room brightness

The value of daylight goes beyond the illumination of tasks. A daylit room varies in brightness with time, colours are rendered well and architectural form and surface texture can be enhanced by the direction of illumination. Above all, windows give information to the people in a building about their surroundings. Weather and the time of day can be inferred from the changing light.

The user's perception of the character of a daylit interior (often described in terms such as 'bright and well-lit', or 'gloomy') is related to the brightness of all the visible surfaces. This overall luminance depends on the quantity of light admitted and the reflectance of interior surfaces. The reflected light within the room can be as important as the direct illumination.

Sunlight and skylight are both important in general room lighting, but they differ greatly in their qualities. Skylight provides a diffuse illuminance as the source area is the entire sky, whereas, sunlight comes from a single point and thus the light gives areas of high luminance and hard shadows. The criteria for each should be satisfied. Sunlight gives patches of high illuminance and strong contrasts; adequate skylight ensures that there is not excessive contrast between one area of the room and another, or between the interior and the view outside.

If the total glazed area cannot be made large enough for adequate general daylight, supplementary electric lighting is needed to enhance the general room brightness in addition to any need there may be for task illumination.

5.3.1   Sunlight
Sunlight should be admitted unless it is likely to cause thermal or visual discomfort to the users, or deterioration of materials.

Provided that the entry of sunlight is properly controlled, it is generally welcome in most buildings in the UK. Sunlight may be controlled in a number of different ways, each of which block sunlight from reaching areas where it is not wanted, see SLL Lighting Guide 10 (SLL, 1999) for more information. Dissatisfaction can arise as much from the permanent exclusion of sunlight as from its excess. However, uncontrolled sunlight is unacceptable in most types of building. Good control is particularly important in working interiors and other rooms where the occupants are unable to move around freely. Generally, sunlight should not fall on visual tasks or directly on people at work. It should, on the other hand, be used to enhance the overall brightness of interiors with patches of high illuminance.

Considerations of sunlight should influence the form of the building from the early stages of design, because incorrect decisions about the orientation of rooms or the geometrical shape of the building may preclude the admission of sunlight or cause excessive overshadowing of surroundings. The orientation of windows should take into account the periods of occupancy and any preferences for sunlight at particular times of day. It is also necessary to consider the thermal performance of the building when considering sunlight. Whilst in the winter sunlight can make an important contribution to heating a room, in the summer, it may well cause excessive heat gain.

5.3.2   Skylight
The general illumination from skylight should be such that there is not excessive contrast between the interior and the view outside.

The interior of a room will appear gloomy not only if the total quantity of light entering is too small but also if its distribution is poor. In addition, high contrast between the surfaces surrounding windows (or rooflights) and the sky can cause glare.

The average daylight factor, the ratio of internal illuminance due to daylight to unobstructed external illuminance (see section 17.1) is used as the measure of general illumination from skylight. It is considered good practice to ensure that rooms in dwellings and in most other buildings have a predominantly daylit appearance. In order to achieve this, the average daylight factor should be at least 2 per cent.

In dwellings there are recommendations for minimum average daylight factors in different rooms; these are given in Table 5.2.

Table 5.2 Minimum average daylight factor

Room type	Minimum average daylight factor / %
Bedrooms	1.0
Living rooms	1.5
Kitchens	2.0

Levels of daylight greater than those listed in Table 5.2 may be beneficial in many situations and if a daylight factor of 5 per cent is achieved in a space then it is commonly found that electric lighting is not needed during the day time.

The uniformity of daylight is important and SLL Lighting Guide 10 (SLL, 1999) gives methods to determine whether the uniformity of daylight in a space will be considered unsatisfactory.

5.3.3   Contrast between the interior and the view outside

Glare from windows can arise from excessive contrast between the luminance of the visible sky and the luminance of the interior surfaces within the field of view. The window walls, the window reveals, and the interior surfaces adjacent to rooflights should be of high reflectance (white or light-coloured). Walls generally should not be glossy.

In addition, glare from the sky and bright external surfaces can be reduced by:

● providing additional illumination on the window wall from other windows

● reducing the luminance of the sky as seen from the interior with translucent blinds, curtains or tinted/solar-control glazing; if adequate illumination can be provided by other sources, it should be noted that some translucent blinds may give a perception of glare if sunlight falls on them

● splaying window reveals, to give a larger area of intermediate brightness between the exterior view and the window wall.

The aim should be to achieve a subtle gradation of luminance from the darker parts of the room to the visible sky.

Glare from direct sunlight, or from sunlight reflected in glossy external surfaces, should be controlled with shading devices.

The use of tinted glazing will reduce the amount of daylight entering and can affect colour perception. External colours might appear distorted, especially when the view outside is seen simultaneously through different types of glass. The perception of internal colours can be altered, unknown to the viewer, when the main source of light is a window of tinted glazing. Care should be taken in the use of tinted glazing materials when safety or task performance requires good colour recognition. Some heavily tinted glazings can affect the view out of a building. It has been found that if the transmittance of the glass falls below 25 per cent, a significant proportion of the people using the building find the view out unacceptable.

5.4   Daylight for task lighting

When there are visual tasks to be carried out, the principles of lighting design using daylight are the same as those for electric lighting: it is necessary both to achieve a given level of illumination and to take account of the circumstances that determine its quality.

Daylight has the following characteristics as a task illuminant.

- A constant illuminance on the task cannot be maintained. When the sky becomes brighter, the interior illuminance increases; and, although control is possible with louvres, blinds and other methods, fluctuations cannot be avoided. Conversely, in poor weather and at the ends of the working day, daylighting needs to be supplemented with electric lighting.

- The direction of light from windows, which act as large diffuse light sources to the side of a worker, gives good three-dimensional modelling. Rooflights, which give a greater downward component, have a modelling effect similar to that from large ceiling-mounted luminaires.

- The spectral distribution of daylight varies significantly during the course of a day, but the colour rendering is usually considered to be excellent.

- Daylight when it is conducted into a space via a light pipe may provide sufficient illuminance for movement through the space but is generally not sufficient to provide enough light for more complex tasks.

Details of the lighting requirements of various tasks in indoor workplaces are given in Chapter 2. The quantity of illumination is not the sole criterion of good task lighting. There are two aspects of task daylighting which need particular attention: glare and specular reflection, see section 5.6.

5.4.1   Glare

Windows may fill a greater part of a worker's field of view than electric light fittings. Distraction, a poor luminance balance between task and background, and discomfort glare can all occur if the visual task is viewed directly against the bright sky. Although a view outside should be provided, it is usually better if the glazing is at the side of workers, rather than directly facing them.

There is no standard procedure for calculating discomfort glare from skylight. Sky luminance can be very high, and the size of the apparent source is large; so by the criteria adopted for electric lighting most windows cause glare. It should be reduced by ensuring that the sky is not in the immediate field of view with the task.

Highly reflective sunlit external surfaces are more likely to add vitality to a scene than constitute an objectionable glare source. This stimulus will be welcomed in all but the most demanding visual situations. However, glare from the sun, viewed directly or specularly reflected, can be unacceptable in a working environment. If the sun or its mirrored image is likely to lie within 45° of the direction of view, then shading devices should be used, see section 5.6. Low transmittance glazing is unlikely to attenuate the beam sufficiently to eliminate glare; diffusing glazing materials, in scattering the beam, may cause the window or rooflight itself to become an unacceptably bright source of light.

5.4.2 Specular reflection

The visibility of tasks can be seriously impaired by bright reflections of the sky in glossy surfaces. With windows, troublesome reflections occur predominantly in vertical surfaces. With rooflights, horizontal task areas are the most seriously affected. However, openings of either type can affect surfaces of all orientations if the geometry is incorrect. Special attention should be given to the avoidance of reflections of windows in display screens, whiteboards, and pictures in galleries, and it is preferable that these surfaces do not face a window directly.

5.5 Electric lighting used in conjunction with daylight

Electric lighting has two distinct functions in a daylit building, which are:

- to enhance the overall appearance of the room, by improving the distribution of illuminance and by reducing the luminance contrast between the interior and the view outside

- to achieve satisfactory illuminance on visual tasks.

These two functions correspond with the recommendations about room brightness and task illumination in daylight, described in section 5.3 and section 5.4 as well as sections of Chapter 2.

5.5.1 Balance of daylight and electric light

Unless the purpose of the windows is only to provide a view, daylight should appear to the users to be dominant in the interior. This is normally achieved when the average daylight factor is 2 per cent or more, even though the horizontal illuminance from electric lighting may be greater than the daylight illuminance in places.

The design of electric lighting should be such that occupants are aware of the natural gradation of daylight across interior surfaces and of changes in the light outside. In spaces that have a significant amount of daylight during the day, consideration should be given to an automatic control system that dims the electric lighting when there is sufficient daylight.

5.5.2 Modelling

The sideways component of light from windows is important in the enhancement of modelling. It is apparent in the articulation of mouldings and in the highlights and shadows of three-dimensional features. The electric lighting should be designed with the daylighting to achieve optimum modelling, reinforcing the directionality where the natural illumination is too diffuse, and providing infill lighting where windows alone would give harsh modelling. See also section 2.1.6.3.

5.5.3 Contrast between exterior and interior

When the general level of inter-reflected light is low, or the surfaces surrounding a window or rooflight are of low reflectance, there will be a high luminance contrast with the view outside. The brighter the view, the higher should be the luminance of the room surfaces which frame the view. Often the best solution is to increase the reflectance of the surfaces surrounding a window rather than use electric lighting.

5.5.4 Colour appearance of lamps

The sky varies in colour with time and position in the sky. These variations are considerable and no electric lamp matches continuously the colour appearance of daylight. For instance, the appearance of a lamp with a colour temperature close to that of light from a clear sky at midday may seem excessively blue as evening approaches. Sunlight reflected into a room from vegetation or brightly coloured surfaces outside can have a noticeable hue and can affect the colour appearance of lamps.

Apparent discrepancies between the colour of electric light and of daylight may be reduced by:

● using lamps of cool or intermediate class correlated colour temperature (see Table 2.3)

● screening lamps from the view of occupants (see section 2.1.5.3).

5.5.5 Changes of lighting at dusk

An interior with some supplementary lighting yet which is primarily daylit will change in character when, late in the day, the electric lighting becomes predominant. As dusk approaches, additional electric illumination is often needed, both to increase task illuminance near the windows and to improve the general brightness of the room, but not for reducing sky glare.

5.6 Sunlight shading

It is essential that the admission of sunlight be controlled in all work spaces and other interiors where the thermal or visual consequences might lead to personal discomfort or cause materials to undergo unacceptable deterioration. In general, the best control of sunlight penetration is achieved by careful planning of the orientation and disposition of rooms and their windows.

All fenestration in positions where sunlight could cause discomfort or damage should be provided with shading. For some interiors, it is acceptable if sunlight is restricted during the warmer months by shading the apertures with elements such as balconies or overhanging roofs, or by fixed louvres or screens. It may be possible to arrange fixed shading devices or install daylight redirecting systems so that daylight is redistributed to better effect, but fixed devices generally reduce the skylight admitted and glazed areas may need to be increased. The effectiveness of fixed shading devices will depend on window orientation. Internal shading devices, such as blinds, may provide a means to control glare caused by low winter sun, but will not provide an effective means of reducing unwanted thermal gain at other times of the year.

For a full discussion of shading and glazing options, see SLL Lighting Guide 10 (SLL, 1999).

Chapter 6: Energy

The objective of any lighting installation is to meet all of the lighting needs of the people using the area being lit whilst consuming a minimum of energy. This point is critical and any attempt to save energy by skimping on the lighting is doomed to failure if a space is poorly lit. In such a situation, people using the space are likely to perform less well and thus the productivity of the whole space is compromised; this reduces the energy effectiveness of the space.

This chapter reviews the steps necessary to ensure that lighting is as efficient as possible and then looks at the requirements of regulations and standards to see how far they support the objective of minimising energy use.

Section 6.1 gives some simple advice on energy efficient lighting. Section 6.2 discusses some of the standards and regulations that cover lighting energy. In particular, it covers the building regulations (see section 6.2.1) which must be applied in almost all lighting schemes. The European standard BS EN 15193 is discussed in section 6.2.2; this is important as the standard defines the metric that best describes the energy used by lighting and gives methods by which it can be calculated.

6.1   Simple guidance for energy efficient lighting

The basis of energy efficient lighting is to provide the right amount of light, in the right place, at the right time with the right lighting equipment. The following sections look at these four points in more detail.

6.1.1   The right amount of light

It is always necessary to provide the correct amount of light for a particular task or activity. Guidance on the choice of lighting level is given in Chapters 2, 3 and 4 of this *Code*. It is often the case that the task at a given place may change with time as may the preference of the person using that space. Thus it is often a good idea to provide lighting that can be adjusted by the user. This has the double benefit of saving energy by not overlighting some tasks and improving user satisfaction by giving some degree of local control.

6.1.2   Light in the right place

The lighting requirements of Chapters 2 and 3 give values for the required illuminances for given tasks. The area over which most tasks occur is relatively small and so there is in general no reason to provide lighting suitable for the task outside the area where it will be carried out. For indoor work spaces, the situation is slightly more complex as for a given task area, it is also necessary to light the immediate surrounding area and the background area. The situation is further complicated by the need to provide light on the walls and ceiling, together with providing a given level of cylindrical illuminance. The situation is complicated even more by the fact that multiple tasks may be done in the same area.

Consider the situation in an office; assume the task is *writing, typing, reading, data processing*; the required illuminance value (from Table 2.30) is 500 lx over the task area. In general, the task area will be on the desks in the office; the area immediately surrounding (a band 0.5 m wide) the desks will need to be lit to 300 lx (see Table 2.1) and the rest of the area may be used as a circulation space so a level of 100 lx may be appropriate. In addition, as visual communication in offices is very important, there is an additional requirement to provide a cylindrical illuminance of 150 lx. Furthermore, it is also necessary to provide some light on the wall and ceilings (see section 2.1.2.3). Faced with this long list of requirements, it may be tempting for the lighting designer to just put 500 lx everywhere on a plane parallel to the floor at desk height safe in the knowledge that all requirements will be met or exceeded. However, this approach is highly wasteful and may well result in an unnecessary increase in energy consumption of over 50 per cent. The correct approach is to provide a basic level of lighting to the space to meet 100 lx for circulation and 150 lx for visual communication, then add some localised lighting for the desk-based tasks and rely on spill light to provide the illuminance for the immediate surrounding area. This task/ambient approach may make the lighting design process more complex but it is necessary to ensure energy efficiency.

6.1.3 Light at the right time

Clearly there is no point providing electric lighting when it is not needed. It is therefore important for lights to be switched off or dimmed when no one is using them or there is enough daylight available. It is often the best solution to provide some form of automatic control system to ensure light is only provided when necessary, however, care is need to ensure such systems are accepted by the users of the lit space. Lighting control systems may provide a wide range of functions; the ones that are important from an energy point of view are discussed below.

Inside buildings, users usually appreciate some form of manual control of the lighting; however, in shared spaces with a single lighting system there is the potential for conflict as different people may have different preferred levels of light; this may well lead to the control not being fully used and thus a default setting of the lights full on may be adopted. Thus the solution of localised lighting topping up a general ambient lighting scheme, as discussed in section 6.1.2, may well be a good solution, as individual users will have some level of control over their lighting.

In general to improve energy savings, the control system should be configured to be turned on by manual control and to be turned off automatically. In such situations the lights may be switched on or off by users and even dimmed up and down, but when there is no one in the room, the lights are switched off automatically and when there is enough daylight available, the lights are automatically dimmed.

In some areas, such as hotel corridors and loading docks in warehouses, there may be problems with manual on and auto off switching. In such areas, there are a number of options such as time clock switching, turning on the lighting only when the area is in use. Presence/absence detection is when the lighting comes on automatically when someone enters the space and goes off after they leave, but in some applications, for example hotel corridors, the short delay between the person entering the space and lights coming on may not be acceptable, so it may be necessary to have a low level of lighting on at all times and have full lighting come on when the space is in use.

6.1.4 The right lighting equipment

Lighting equipment comprises lamps, luminaires and control systems. The choices of lighting equipment must be first driven by the lighting needs of a given installation, for example, there is no point selecting a very efficient lamp for a job if the colour of the light that it produces is not suitable for the application in which it is to be used. There is a vast range of lamps and luminaires available so there will nearly always be a range of equipment that is able to do a given lighting job. All other factors being equal, the most efficient equipment should be selected.

It is also important to consider the controllability of lighting equipment; a dimmable luminaire might be marginally less efficient at full output than a luminaire that may only be switched on and off, however, when the full output is not needed, the dimming luminaire may provide the amount of light needed and use less energy than a luminaire that is not dimmable.

Consideration of maintenance is also important. Some lighting equipment, for example, up-lighters, may get dirty very quickly and thus it is important for them to be easily accessible so that they can be cleaned frequently. Occasionally it may be necessary to mount luminaires in places that are relatively inaccessible; here it will be necessary to select lamps and luminaires where the luminous output reduces slowly with time. Chapter 18 discusses how maintenance factors may be calculated; it is important to ensure that the maintenance factor for a scheme is as high as possible, however, the maintenance regime necessary for a given maintenance factor must be realistically achievable by the owner of the lighting scheme; for example, no factory owner is going to shut down a plant so that they can access their high bay lighting units to clean and re-lamp them every 6 months.

As the light produced by a lighting system drops with time, it is usual practice to provide too much light when the scheme is new or has just been maintained. This over lighting is wasteful. The situation may be improved by using a control system that dims the lights initially and slowly increases the amount of energy they use so that the system maintains a constant illuminance on the required task. Such systems are often called 'constant illuminance systems', and are often used in conjunction with daylight dimming systems.

To ensure the best performance from a lighting system, the users and the owners must understand how to use and maintain it. For simple systems which are just switched on and off, this is generally not too much of a problem. However, with more complex control systems, the user interface to the lighting control system must be easy to understand and use. Moreover, the building owner needs to have enough information on the lighting installation for the maintenance of the equipment and to permit modifications to the lighting as the use of the installation evolves during its life.

6.2 Energy regulations, and standards

There are two sets of regulations and standards for energy use by indoor lighting relevant to the UK. As yet, there are no comprehensive published guides to energy use in outdoor lighting but building regulations cover some aspects of outdoor lighting.

Perhaps the set of regulations that has the largest impact is Part L of the building regulations; its impact is because it has the force of law behind it, not because it is a good regulation. BS EN 15193: 2007: *Energy performance of buildings. Energy requirements for lighting* (BSI, 2007a) provides a standardised method of calculating the energy used by lighting in terms of kilowatt hours per square metre per year; this metric follows actual energy consumption and thus is a good way to discuss energy use. The standard also gives a series of benchmark values for different buildings; however, this part of the standard is not particularly useful as the target values given are quite high and very easy to achieve.

There are a number of schemes, voluntary and mandatory, to encourage energy saving and good practice; these are discussed in section 6.2.5.

6.2.1   Building regulations

The building regulations apply to fixed lighting within a building, thus light fittings that are plugged in may not be included within their scope. The exact definition of fixed lighting is not given in the building regulations. Whilst it is fairly clear that permanently wired luminaires screwed to the structure of the buildings are fixed, there are a whole range of lighting installations that may not be included. A good example of a type of installation where there is uncertainty is track lighting.

The application of the building regulations in the UK is complex as there are a number of parts to the documents and the regulations only cover England and Wales although there are parallel regulations in Scotland and Northern Ireland. The following sections summarise the documents to be used.

England and Wales
- *Approved Document L1A – Conservation of fuel and power (New dwellings)*

- *Approved Document L1B – Conservation of fuel and power (Existing dwellings)*

- *Approved Document L2A – Conservation of fuel and power (New buildings other than dwellings)*

- *Approved Document L2B – Conservation of fuel and power (Existing buildings other than dwellings).*

The lighting requirements can be found in the sections indicated and the requirements are limited to fixed internal or external lighting but not including emergency escape lighting or specialist process and temporary or plug in lighting. The lighting requirements appear in the compliance documents which have the same legal standing as the Regulations. There are two compliance documents:

- *Domestic Buildings Services Compliance Guide: 2010 Edition*

- *Non-Domestic Buildings Services Compliance Guide: 2010 Edition*

Scotland
- *Domestic Technical Handbook 2010*

- *Non-domestic Technical Handbook 2010*

Northern Ireland
- *Department of Finance and Personnel Technical Booklet F1: 2006 (domestic)*

- *Department of Finance and Personnel Technical Booklet F2: 2006 (non-domestic)*

As the requirements in Northern Ireland have not changed in since 2006, they are little behind the regulations in the rest of the UK and are thus less onerous. This section does not cover the NI regulations; however, compliance with the regulations mentioned in this section should in most cases exceed the NI requirements.

The building regulations are divided between dwellings and other types of buildings. They also treat refurbishment differently to new build; however, most of the requirements for lighting are the same.

6.2.2　Dwellings

The Building Regulations mention adequate daylight levels in domestic premises and limiting solar gain in section 4.27 of Approved Document L1A. The Building Regulations do not specify minimum daylight requirements; however, they mention reducing window area, for thermal reasons. This produces conflicting impacts on predicted CO_2 emissions: reduced solar gain but increased use of electric lighting. As a general guide, if the window area is much less than 20 per cent of the total floor area, some parts of the dwelling may experience poor levels of daylight, resulting in increased use of electric lighting.

For electric lighting in and around new and existing refurbished buildings, the requirements are given in Table 6.1.

Table 6.1 Recommended standards for fixed internal and external lighting for dwellings

Lighting	New and replacement systems	Supplementary information
Fixed internal lighting	a. In the areas affected by the building work, provide low energy light fittings (fixed lights or lighting units) that number not less than three per four of all the light fittings in the main dwelling spaces of those areas (excluding infrequently accessed spaces used for storage, such as cupboards and wardrobes) b. Low energy light fittings should have lamps with the luminous efficacy greater than 45 lamp lumens per circuit watt and total output greater than 400 lamp lumens c. Light fittings whose supplied power is less than five circuit watts are excluded from the overall count of the total number of light fittings	Light fittings may be either: • dedicated fittings which will have separate control gear and will take only low energy lamps (e.g. pin-based fluorescent or compact fluorescent lamps); or • standard fittings supplied with low energy lamps with integrated control gear (e.g. bayonet or Edison screw base compact fluorescent lamps) Light fittings with GLS tungsten filament lamps or tungsten halogen lamps would not meet the standard The energy saving trust publication GI 020, *Low energy domestic lighting*, gives guidance on identifying suitable locations for fixed energy efficient lighting
Fixed external lighting	Where fixed external lighting is installed, provide light fittings with the following characteristics: a. either: 　i. the lamp capacity not greater than 100 lamp watts per light fitting; and 　ii. all lamps automatically controlled so as to switch off after the area lit by the fitting becomes unoccupied; and 　iii. all lamps automatically controlled so as to switch off when daylight is sufficient b. or 　i. lamp efficacy greater than 45 lumen per circuit watt; and 　ii. all lamps automatically controlled so as to switch off when daylight is sufficient; and 　iii. light fittings controllable manually by occupants	

In the 2010 Scottish *Domestic Technical Handbook* (DTH), the following additional guidance is given:

- All fixed light fittings and lamps provided to corridors, stairs and other circulation areas should be low energy type, controls to such lighting to enable safe use of the areas in question are identified in DTH Section 4.6.

- The dwelling should have an electric lighting system providing at least one lighting point to every circulation space, kitchen, bathroom, toilet and other space having floor area of 2 m² or more.

- Any lighting point serving a stair should have the controlling switch at, or in immediate vicinity of, the stair landing on each story. DTH Section 4.6.2 recommends that common areas should have artificial lighting capable of providing a uniform lighting level, at floor level, of not less than 100 lx on stair flights and landings and 50 lx elsewhere within circulation areas. Lighting should not present sources of glare and should avoid creation of areas of strong shadow that may cause confusion or mis-step. A means of automatic control should be provided to ensure that lighting is operable during the hours of darkness.

These additional requirements in Scotland are important, as they stop some switching strategies such as the *landlord's switch*; they also force some minimum level of lighting.

6.2.3   Non-domestic buildings

For new build, there is a requirement that the building energy rating (BER) be calculated using a calculation tool that follows the rule of the national calculation method (NCM; see http://www.ncm.bre.co.uk). This calculation is done for all aspects of the building and daylight and electric lighting are just minor elements in this calculation. One of the most common tools used to do this calculation is called SBEM (Simplified Building Energy Model; see http://www.bre.co.uk/page.jsp?id=706).

Because these tools have to cover all aspects of building design, they do not cover lighting particularly well. In general, they do not consider actual lighting designs but make crude assumptions about the amount of light needed and the efficiency of a given lighting technology in a room and make assumptions about given lighting levels being provided throughout entire spaces.

There are also restrictions on the energy efficiency of the lighting equipment used for both new build and refurbished existing buildings. Table 6.2 gives the recommended minimum lighting efficacy as taken from Table 44 of *Non-Domestic Buildings Services Compliance Guide* (2010 edition).

Table 6.2 Recommended minimum lighting efficacy

Lighting	Lighting efficacy
General lighting in office, industrial and storage areas	The average initial efficacy should be not less than 55 luminaire lumens per circuit-watt In calculating the average luminaire lumens per circuit-watt, the circuit-watts for each luminaire may first be multiplied by the control factors in Table 6.3
General lighting in other types of space	The average initial efficacy should be not less than 55 lamp lumens per circuit-watt
Display lighting	The average initial efficacy should be not less than 22 lamp lumens per circuit-watt

If lighting control systems are used in office, industrial and storage areas, then it is possible to use less efficient light fittings; the amount that the efficacy can be reduced is given in Table 6.3 which shows the control factors and minimum permitted average luminaire lumens per circuit-watt. The table is based on Table 45 of *Non-Domestic Buildings Services Compliance Guide* (2010 edition).

Table 6.3 Control factors and minimum permitted average luminaire lumens per circuit-watt

Light output control	Control factor	Minimum efficacy / luminaire lumens per circuit-watt
a. The luminaire is in a daylit space and its light output is controlled by photoelectric switching or dimming control, with or without override	0.90	49.5
b. The luminaire is in a space that is likely to be unoccupied for a significant number of operating hours, and where a sensor switches off the lighting in the absence of occupants but switching on is done manually except where this would be unsafe	0.90	49.5
c. Circumstances a. and b. combined	0.85	46.8
d. None of the above	1.00	55

Note that, as the above table applies to office, industrial and storage areas, there are a number of areas when automatic switching could be applied where it is not possible to use lamps with a luminous efficacy of less than 55 lumens per circuit-watt.

6.2.4   BS EN 15193

BS EN 15193: 2007 (BSI, 2007a) was developed to help support the European Directive on the Energy Performance of Buildings 2002/91/EC (EC (2002)). It is one of a set of standards that cover all aspects of energy consumption in buildings.

The standard introduces the concept of the Lighting Energy Numeric Indicator (LENI) which is used to express the total amount of energy used by a lighting system per square metre per year. The standard then goes on to describe how this value may be calculated or measured. Figure 6.1 gives a schematic of how the process works.

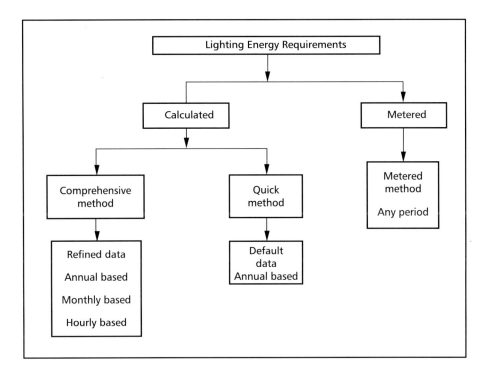

Figure 6.1 Methods for determining energy use in BS EN 15193: 2007

The quick calculation method makes many simplifications and assumptions but the standard is set up so that, for a given situation, the quick method will nearly always result in a higher estimation of energy use. The section on metering gives general advice on how to use metering to measure energy consumed by lighting.

The basic process of calculation depends on the following variables:

A **Useful area**
floor area inside the outer walls excluding non-habitable cellars and un-illuminated spaces, measured in m^2

F_C **Constant illuminance factor**
factor relating to the usage of the total installed power when constant illuminance control is in operation in the room or zone

F_D **Daylight dependency factor**
factor relating the usage of the total installed lighting power to daylight availability in the room or zone

F_O **Occupancy dependency factor**
factor relating the usage of the total installed lighting power to occupancy period in the room or zone

P_{em} **Total installed charging power of the emergency lighting luminaires in the room or zone**
input charging power of all emergency lighting luminaires in the room or zone, measured in watts

P_n **Total installed lighting power in the room or zone**
power of all luminaires in the room or zone, measured in watts

P_{pc} **Total installed parasitic power of the controls in the room or zone**
input power of all control systems in luminaires in the room or zone when the lamps are not operating, measured in watts

t_D **Daylight time usage**
operating hours during the daylight time, measured in hours

t_{em} **Emergency lighting charge time**
operating hours during which the emergency lighting batteries are being charged, measured in hours

t_N **Non-daylight time usage**
operating hours during the non-daylight time, measured in hours

t_y **Standard year time**
time taken for one standard year to pass, taken as 8760 h

$W_{L,t}$ **Energy consumption used for illumination**
energy consumed in period t, by the luminaire when the lamps are operating to fulfil the illumination function and purpose in the building, measured in kW h

$W_{P,t}$ **Luminaire parasitic energy consumption**
parasitic energy consumed in period t, by the charging circuit of emergency lighting luminaire and by the standby control system controlling the luminaires when the lamps are not operating, measured in kW h

W_t **Total lighting energy used**
energy consumed in period t, by the sum of the luminaires when the lamps are operating, plus the parasitic loads when the lamps are not operating, in a room or zone, measured in kW h

The total energy required to provide lighting for a period may be estimated using equation 6.1:

$$W_t = W_{L,t} + W_{P,t} \quad [\text{kW·h}] \tag{6.1}$$

$W_{L,t}$ may be estimated using equation 6.2:

$$W_{L,t} = \frac{\sum (P_n \times F_C) \times \left[(t_D \times F_O \times F_D) + (t_N \times F_O) \right]}{1000} \quad [\text{kW·h}] \tag{6.2}$$

$W_{P,t}$ may be estimated using equation 6.3:

$$W_{P,t} = \frac{\sum \left\{ P_{pc} \times \left[t_y - (t_D + t_N) \right] \right\} + (P_{em} \times t_{em})}{1000} \quad [\text{kW·h}] \tag{6.3}$$

When the time period in equation 6.1 is set to one year, then W, the annual estimate of energy use is equal to W_t and the Lighting Energy Numeric Indicator may be calculated using equation 6.4:

$$LENI = \frac{W}{A} \quad \left[\text{kW·h·m}^{-2}\text{·yr}^{-1}\right] \tag{6.4}$$

Using the quick method of calculation, the time is always one year and default values for t_D, t_N, F_c, F_D, F_O, P_{em} and P_{pc} are taken from Table 6.4. Moreover, for the quick calculation, equations 6.2 and 6.3 simplify to give equations 6.5 and 6.6:

$$W_{L,t} = \frac{(P_n \times F_C) \times \left[(t_D \times F_O \times F_D) + (t_N \times F_O)\right]}{1000} \quad \left[\text{kW·h}\right] \tag{6.5}$$

$$W_{P,t} = \frac{\left\{P_{pc} \times \left[t_y - (t_d + t_n)\right]\right\} + (P_{em} \times t_e)}{1000} \quad \left[\text{kW·h}\right] \tag{6.6}$$

Table 6.4 Factors for the quick method calculation

	P_{em} per sq. m	P_{pc} per sq. m	t_D	t_N	F_C		F_O		F_D	
	(kW·h· m^{-2}·yr^{-1})	(kW·h· m^{-2}·yr^{-1})	(h)	(h)	No CTE*	With CTE*	Manual	Auto	Manual	Auto
Office	1	5	2250	250	1	0.9	1	0.9	1	0.9
Education	1	5	1800	200	1	0.9	1	0.9	1	0.8
Hospital	1	5	3000	2000	1	0.9	0.9	0.8	1	0.8
Hotel	1	5	3000	2000	1	0.9	0.7	0.7	1	1
Restaurant	1	5	1250	1250	1	0.9	1	1	1	–
Sports	1	5	2000	2000	1	0.9	1	1	1	0.9
Retail	1	5	3000	2000	1	0.9	1	1	1	–
Industry	1	5	2500	1500	1	0.9	1	1	1	0.9

*CTE = constant illuminance control.

For the comprehensive calculation, it is necessary to apply equations 6.2 and 6.3 doing a much more detailed analysis of the building in question to determine the factors used and if needed, summing the results over different time periods. The process is quite complex. The calculation process is given in the European standard and it has been implemented in commonly available lighting software.

The metering method of calculating the LENI value for a given installation involves measuring the amount of energy used over the course of a year and dividing by the area being illuminated.

6.2.5   Schemes to support energy efficient lighting

To support good practice, there are a number of schemes to recognise good performance.

Display Energy Certificates (DEC) (http://www.communities.gov.uk/documents/planningandbuilding/pdf/20.pdf) are required in buildings occupied by public authorities and institutions that provide services to the public, have a floor area of over 1000 m² and are frequently visited by the public; they cover all energy use in the building of which lighting is one part. The energy rating on the certificate is based on energy consumed over the course of a year so for the lighting element, it is effectively the same as the metered method of BS EN 15193: 2007.

BREEAM (Building Research Establishment's Environmental Assessment Method) (http://www.breeam.org) is an assessment of a number of features of a building. In each area of assessment, there are a number of points available, then based on the percentage of the points available, a building is assessed into the categories given in Table 6.5.

Table 6.5 BREEAM rating benchmarks

BREEAM rating	Percentage score
Outstanding	≥85
Excellent	≥70
Very good	≥55
Good	≥45
Pass	≥30
Unclassified	<30

In the BREEAM assessment system, there are quite a few points available for lighting, including daylighting, however, the assessment is not purely about minimising energy consumption but also covers meeting the correct light requirements.

There is also a similar system to BREEAM set up in the USA called LEED (Leadership in Energy and Environmental Design) (http://www.usgbc.org/DisplayPage.aspx?CMSPageID=1988) and it is run by the US Green Building council. Whilst LEED is less used in the UK than BREEAM as both are voluntary schemes, it is possible that clients can ask for a LEED assessment on UK buildings.

Chapter 7: Construction (Design and Management) Regulations

The Construction (Design and Management) Regulations (HMSO, 2007), often referred to as the CDM regulations, place a number of duties upon people involved in construction, with the aim of improving health and safety during construction and maintenance.

Note: This chapter attempts to broadly set out the requirements of the Construction (Design and Management) Regulations 2007 (HMSO, 2007) and the effects particularly relevant to the lighting. It does not address every detail of the Regulations and the accompanying Approved Code of Practice (ACoP) L144: Managing health and safety in construction (HSE, 2007). Therefore, this guidance must be used with caution, and reference made to the Approved Code of Practice, as well as the regulations themselves.

For simplicity in reading, this chapter together with the regulations and ACoP has been split into four sections that cover sections 1 to 4 of the regulations. Section 5 of the regulations is not covered here as it does not directly impact upon lighting.

7.1   Introduction

The CDM regulations apply equally across all of the UK and in a few other places such as the Channel Islands. In the introduction, many terms are defined; the following are selected abridged definitions.

Client
A person who in the course or furtherance of a business seeks or accepts the services of another to carry out the project for him, or carry out the project himself.

CDM co-ordinator
The person appointed as the CDM co-ordinator if the project is *notifiable*. This applies to projects that are likely to last more than 30 days or require more than 500 person days to complete. The duties of the CDM co-ordinator are given in section 7.3.4.

Construction site
Includes any place where construction work is being carried out or to which the construction workers have access.

Construction work
The carrying out of building, civil engineering and construction. The term includes alteration, fitting out, commissioning, repair and maintenance.

Contractor
Any person (potentially this could also be a client) who, in the course or furtherance of a business, carries out or manages construction work.

Design
Includes drawings, design details, specification and bill of quantities (including specification of equipment) relating to a structure, and calculations prepared for the purpose of a design.

Designer

Any person who prepares or modifies a design or instructs someone else under his control to do so.

Hazard

Something (e.g. an object, a property of a substance, a phenomenon or an activity) that can cause adverse effects. For example:

- Water on a staircase is a hazard, because you could slip on it, fall and hurt yourself.

- Loud noise is a hazard because it can cause hearing loss.

- Breathing in asbestos dust is a hazard because it can cause cancer.

Place of work

Any place which is used by any person at work for the purposes of construction work or for the purposes of any associated activity.

Principal contractor

The person appointed to perform certain specified duties as set out in section 7.3.

Risk

A risk is the likelihood that a hazard will actually cause its adverse effects, together with a measure of the effect. It is a two-part concept and you have to have both parts to make sense of it. Likelihoods can be expressed as probabilities (e.g. 'one in a thousand'), frequencies (e.g. '1000 cases per year') or in a qualitative way (e.g. 'negligible', 'significant', etc). The effect can be described in many different ways. For example:

- The annual risk of a worker in Great Britain experiencing a fatal accident [effect] at work [hazard] is less than one in 100 000 [likelihood].

- About 1500 workers each year [likelihood] in Great Britain suffer a non-fatal major injury [effect] from contact with moving machinery [hazard].

- The lifetime risk of an employee developing asthma [effect] from exposure to substance X [hazard] is significant [likelihood].

7.2   General management duties

Competence

The regulations state that no one should appoint or engage a CDM co-ordinator, designer, principal contractor or contractor unless they have taken reasonable steps to ensure the person being appointed or engaged is competent. Moreover, no one should accept such an appointment unless they are competent. The need for competence also extends to construction workers where workers must not be instructed to carry out any duties unless they are competent to perform them or they are under the direct supervision of a competent person.

This need for competence can give rise to problems in establishing whether a person has the necessary skills to do the job. This need to establish competency has given rise to many *competency schemes* organised by a whole variety of bodies which register people as competent after they have met certain criteria such as attending training and maintaining certain records of work. There are no such schemes currently dedicated to lighting, however, there are some within the electrical industry that impinge on lighting; for example, the Highway Electrical Registration Scheme (see http://www.highwayelectrical.org.uk/HERS/) covers the area of street lighting. Such schemes need to be used with care as lighting is such a diverse discipline that competence in one area does not necessarily mean the ability to work in another. Thus someone employing a person to work on a lighting project has to use judgement to select a person who is able to carry out the work.

The key question in selecting a person for a job is whether or not they are able to carry out their duties listed below in this section and, if the project is notifiable, the additional duties listed in section 7.3 as well. In making a decision, there are a number of factors that may guide a person who wishes to employ a lighting person; these factors include the education and experience of the person and if they belong to any of the professional bodies for the lighting industry. There are currently four such bodies active in the UK, and they are:

- The Society of Light and Lighting (SLL; http://www.sll.org.uk)

- The Institution of Lighting Professionals (ILP; http://www.theilp.org.uk)

- The International Association of Lighting Designers (IALD; http://www.iald.org/home.asp)

- The Professional Lighting Designers Association (PLDA; http://www.pld-a.org)

Full membership of one of these bodies means that the person has been reviewed by fellow professionals and they have been assessed to be suitably qualified and experienced to design lighting. Moreover, all of the bodies have codes of conduct or rules which if broken could lead to them losing their membership status; thus any member who took on work for which they were not competent would risk losing their membership.

Co-operation
Everybody connected with a project or who has particular duties placed on them by the CDM regulations should seek co-operation with and in turn provide co-operation to permit all parties to perform their required duties. In addition, everybody has a duty to report anything they are aware of that may endanger the health and safety of themselves or anyone else working on the project.

Co-ordination
Everybody (clients, designers, contractors and principal contractors) working on a construction project must co-ordinate their activities with each other, as far as reasonably possible, to ensure the health and safety of people carrying out the construction work and those affected by the work.

7.2.1 Duties of clients
There are two sets of duties of clients; one associated with the management of the project, the other associated with information on the project.

Duties in relation to arrangements for managing projects
Every client shall take reasonable steps to ensure that the arrangements made for managing the project (including the allocation of sufficient time and other resources) by persons with a duty under these Regulations (including the client himself) are suitable to ensure that:

- the construction work can be carried out so far as is reasonably practicable without risk to the health and safety of any person

- any structure designed for use as a workplace has been designed taking account of the provisions of the Workplace (Health, Safety and Welfare) Regulations 1992 (HMSO, 1992) which relate to the design of, and materials used in, the structure.

The client needs to take reasonable steps that the above measures are maintained and reviewed throughout the project and that all persons involved in the construction are aware of the current arrangements.

Duties in relation to information
The client must ensure that all designers and contractors involved in the projects must have access to the pre-construction information, listed below, that is relevant to their part of the project:

- any information about or affecting the site or the construction work

- any information concerning the proposed use of the structure as a workplace

- the minimum amount of time before the construction phase which will be allowed to the contractors appointed by the client for planning and preparation for construction work

- any information in any existing health and safety file.

7.2.2 Duties of designers
No designer should start work on a project if the client is unaware of his duties under the CDM regulations.

Every designer must take reasonable steps when preparing or modifying a design to avoid foreseeable risks to the health and safety of any person:

- carrying out construction work

- liable to be affected by such construction work

- cleaning any window or any transparent or translucent wall, ceiling or roof in or on a structure

- maintaining the permanent fixtures and fittings of a structure

- using a structure designed as a workplace.

In addressing the above issues, the designer should eliminate risks that give rise to hazards and reduce the risks from any remaining hazards.

The designer must provide his design with sufficient information to assist the client, other designers and contractors in the construction and maintenance of the building.

7.2.3 Duties of contractors
No contractor should start work on a project if the client is unaware of his duties under the CDM regulations. The contractor must plan, manage and monitor construction work carried

out by him or under his control to ensure that it is done without risks to health and safety. Every sub-contractor needs to be informed of the time available for planning before commencement of work.

Every contractor must provide any construction workers under his control with information and training, including:

- suitable site induction, where not provided by any principal contractor

- information on the risks to their health and safety; identified by his risk assessment under regulation 3 of the Management of Health and Safety at Work Regulations 1999 (HMSO, 1999), or arising out of the conduct by another contractor of his undertaking and of which he is or ought reasonably to be aware

- information about measures which have been identified by the contractor in consequence of the risk assessment as being needed to take to meet statutory health and safety requirements

- any site rules

- the procedures to be followed in the event of serious and imminent danger to such workers

- the identity of the persons nominated to implement those procedures.

The above requirements are in line with those in Regulation 13.2.b of Health and Safety at Work Regulations 1999 (see http://www.legislation.gov.uk/uksi/1999/3242/regulation/13/made).

Work should not start on a construction site unless reasonable steps have been taken to prevent unauthorised persons gaining access to the site.

7.3 Additional duties if the project is notifiable

A project becomes notifiable if it is likely to last more than 30 days or require more than 500 person days to complete. When a project becomes notifiable, there are some additional duties placed upon the client, contractors and designers. There are also two other roles that come into play; those of principal contractor and CDM co-ordinator.

7.3.1 Additional duties of the client
The client must appoint a CDM co-ordinator and a principal contractor as soon as practical and these posts must be filled for the duration of the construction project; however, it is permitted for the people filling these roles to change during the life of a project. The client must also pass all information listed in section 7.2.1 to the CDM co-ordinator. The client must also ensure that the principal contractor has prepared a construction phase plan (as described in 7.3.5) before permitting construction to start.

7.3.2 Additional duties of designers
When a project is notifiable, designers should not commence work, apart from initial design work, on a project unless a CDM co-ordinator has been appointed. Furthermore, the designers should provide the CDM co-ordinator with sufficient information to allow the performance of his duties as described in section 7.3.4.

7.3.3   Additional duties of contractors

When a project is notifiable, a contractor must not start work on a project unless both the CDM co-ordinator and the principal contractor have been appointed by the client and the project notification has been sent by the CDM co-ordinator to the Health and Safety Executive.

Contractors must provide the principal contractor with information that

- might affect the health and safety of construction workers and others

- might justify a review of the construction phase plan

- has been identified for inclusion in the health and safety file.

Contractors must:

- in complying with their duties given in section 7.2.3 take all reasonable steps to ensure that the construction work is carried out in accordance with the construction phase plan

- take appropriate action to ensure health and safety where it is not possible to comply with the construction phase plan in any particular case

- notify the principal contractor of any significant finding which requires the construction phase plan to be altered or added to.

7.3.4   Duties of the CDM co-ordinator

The main duties of the CDM co-ordinator are listed below

- give advice and assistance to the client on undertaking the measures he needs to take to comply with the CDM regulations during the project and in particular advising on the required management duties

- ensure that arrangements are made and implemented for the co-ordination of health and safety measures during planning and preparation for the construction phase of the project

- liaise with the principal contractor on the health and safety file, construction phase plan and any design development that affect planning and management of the construction work

- collect the pre-construction information and disseminate it to the designers and contractors

- ensure that designers perform their duties (see sections 7.2.2 and 7.3.2)

- ensure co-operation between the designers and the principal contractor

- prepare and maintain the health and safety file containing information which may be needed to ensure any subsequent work on the building may be carried out safely

- pass the health and safety file to the client at the end of the project.

The CDM co-ordinator has the additional duty of notifying the project. This notification is usually to the Health and Safety Executive (http://www.hse.gov.uk), but for certain transport related projects may be to the Office of Rail Regulation (http://www.rail-reg.gov.uk).

7.3.5 Duties of the principal contractor

The principal contractor has three sets of duties: first, general site requirements, next to the construction phase plan and lastly, co-operation and consultation with the workers.

General duties cover planning and managing the construction plan so the construction is without health and safety risks. It covers:

- co-operation and co-ordination between persons concerned in the project

- liaison with the CDM co-ordinator to ensure he is able to perform his duties

- ensure that welfare facilities are provided throughout the construction phase

- if needed, draw up health and safety rules for the construction site

- give directions to the contractors to ensure they are able to carry out their duties

- ensure that every contractor is informed of the minimum amount of time which will be allowed to him for planning and preparation before he begins construction work and ensure he has sufficient time to enable him to prepare properly for that work

- where necessary, consult a contractor before finalising such part of the construction phase plan as is relevant to the work to be performed by him

- identify to each contractor the information relating to the contractor's activity which is likely to be required by the CDM co-ordinator for inclusion in the health and safety file

- take reasonable steps to prevent access by unauthorised persons to the construction site.

The principal contractor is responsible for the construction phase plan and it is his duty to prepare the plan before construction work starts and from time to time, to review and revise the plan. The objective of the plan is to ensure that the project is planned, managed and monitored so that construction work may proceed in a way that is, as far as possible, free from risks to health and safety.

The principal contractor has duties that require him to consult and co-operate with the workforce. He must maintain arrangements that promote communication and co-operation with the workforce to ensure the safety and welfare of the workers, and must consult with the workers or their representatives on any issue that may affect their health, safety or welfare. The principal contractor must ensure that the workers or their representatives can inspect and copy any information relating to the planning or management of the project. There is a restricted set of reasons why the principal contractor may not show information to the workforce.

7.4 Duties relating to health and safety on construction sites

The CDM regulations provide a lot of detailed information about measures necessary to ensure that construction sites are safe places of work. Some of the advice is general relating to all sites, some advice is particular to specific elements of work. In this section, only the parts that impact on lighting are discussed.

7.4.1   Electricity distribution

When there is a risk from power cables, one of the following should be used:

- they shall be directed away from the area of risk

- the power shall be isolated and, where necessary, earthed.

If neither of these is reasonably practical, suitable warning notices and barriers suitable for excluding work equipment should be placed around the area. Other procedures that provide an equal level of safety may be used.

7.4.2   Emergency routes and exits

It is necessary to provide emergency routes with emergency lighting; see SLL Lighting Guide 12 (SLL, 2004) for more information.

7.4.3   Lighting

The CDM regulations require: *Every place of work and approach thereto and every traffic route shall be provided with suitable and sufficient lighting, which shall be, so far as is reasonably practicable, by natural light.* This leaves open the question as to what is deemed to be *suitable and sufficient.* A good starting guide to the lighting on construction sites is given in Table 3.7.

There are requirements for the colour quality of such lighting and these are usually met by the colour rendering requirements listed in Table 3.7.

There is also a requirement that *suitable and sufficient secondary lighting shall be provided in any place where there would be a risk to the health or safety of any person in the event of failure of primary artificial lighting;* this type of lighting is discussed in SLL Lighting Guide 12 (SLL, 2004) where it is called *High Risk Area* lighting.

Chapter 8: Basic energy and light

Light is a form of energy that may be transmitted through space without the need of any material or substance to help propagate it. Such energy transfer is known as radiation. There are various forms of radiation, some such as nuclear α and β radiation is a stream of particles, whilst light and other forms of electromagnetic radiation are propagated by waves. Light is generally considered to be electromagnetic radiation with a wavelength in the range 380 to 780 nm, where a nano-metre (nm) is 10^{-9} m. However, in certain respects, light behaves like a stream of particles, called photons. This dual nature of light is quite complex and an in-depth knowledge of quantum mechanics is needed before attempting a more detailed description. Thus, in this chapter, most of the description of light will look at the wave properties of light and one section will look at the quantum properties where light behaves like a stream of particles. This dual wave particle nature is true for all electromagnetic radiation. The key point that separates light from the rest of the electromagnetic spectrum is the fact that the human eye is sensitive to it. Figure 8.1 show how light relates to the rest of the electromagnetic spectrum.

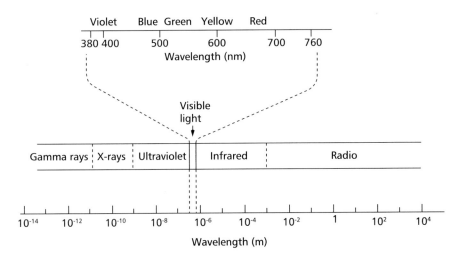

Figure 8.1 Light as part of the electromagnetic spectrum

8.1 Properties of electromagnetic waves

Like all electromagnetic radiation, light travels through a vacuum in straight lines[1] at a speed of approximately 3×10^8 m·s⁻¹. In any medium such as glass, air or water, the speed will be lower by a factor known as the refractive index of the medium. As with any wave, electromagnetic radiation may be characterised by its wavelength (λ) and its frequency (v). Frequency, wavelength and velocity (v) are related by equation 8.1

$$v = v \times \lambda \tag{8.1}$$

---

[1] This is true for all practical terrestrial purposes, however, under certain circumstances light may bend. This topic is covered in the general theory of relativity, see Einstein, Albert (1916) 'Die Grundlage der allgemeinen Relativitätstheorie', *Annalen der Physik* 49.

As light travels from one medium to another, thus changing speed, the frequency stays the same but the wavelength changes. At the boundary between two different media having different refractive indices (see Figure 8.2), light is split into two paths, and the directions of these new rays are discussed below.

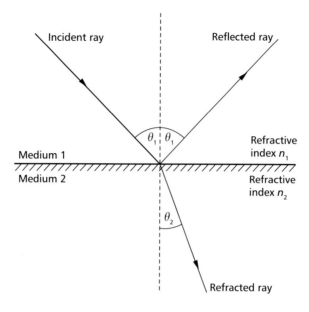

Figure 8.2 Reflection and refraction at a boundary between two media

Laws of specular reflection

At a surface that is smooth when compared with the wavelength of the incident radiation, specular reflection may occur. The following relations hold between the incident ray and the reflected ray:

- The incident ray, the reflected ray and the normal to the surface at the point of reflection all lie in the same plane.

- The incident ray and the reflected ray make equal angles with the normal to the surface and are on opposite sides of it.

The proportion of light reflected at the surface is a function of the angle between the rays and the normal to the surface and the difference in refractive index between the two media. The amount of reflected light may be calculated using Frensel's equations.[2]

Laws of refraction

Light incident on a smooth surface suffers a change of direction when passing into the second medium that is governed by the following rules:

- The incident ray, the refracted ray and the normal to the surface at the point of reflection all lie in the same plane.

---

[2] Frensel's equations may be found in almost any optical text book, for example, *Optics* by Hecht and Zajac, published by Addison Wesley (1974) (ISBN-10 0201028352).

- Snell's law relates the angle of the incident ray to the normal to the angle of the refracted ray to the normal; this is expressed in equation 8.2:

$$n_1 \sin\theta_1 = n_2 \sin\theta_2 \qquad (8.2)$$

where n_1 is the refractive index of the first material, n_2 is the refractive index of the second material, θ_1 is the angle between the normal to the surface and the incident ray and θ_2 is the angle between the normal to the surface and the refracted ray; θ_1 and θ_2 lie on opposite sides of the normal.

These laws of refraction apply to most substances such as glass, liquids and plastics, however, for certain crystals and other materials under strain, these rules are not followed exactly.

Total reflection
When light passes from a high refractive index medium to a low one, such as from glass to air, a refracted ray is only possible if the incident angle (θ_1) is smaller than a value that will give a refraction angle (θ_2) of less than 90°; this critical angle (θ_c) is given by equation 8.3

$$\theta_c = \sin^{-1}\left({n_2}/{n_1}\right) \qquad (8.3)$$

If the angle of the incident ray is greater than the critical angle, then all of the light is reflected; this is known as total internal reflection. This effect has a practical application in a number of devices, the best known of which is fibre optics.

Dispersion
In general, the refractive index of a substance is a function of frequency of the light. This means that different colours of light may be refracted in slightly different directions. For a range of common materials, the refractive index for blue light is higher than that for red. This means that the change in direction for blue light is greater.

Absorption and scattering
When light passes through a perfect vacuum, there is no loss of energy from the beam, although a diverging beam will spread out with distance. However, in other media, it may be possible for the light to interact with the material it is passing through and this causes losses to the beam by absorption and scattering.

Absorption is caused by the energy in the light being converted into other forms of energy by interaction with the material. If the material is homogeneous, then for a given wavelength of light in a parallel beam, the rate of loss follows an exponential curve as given by equation 8.4:

$$i = i_0\, e^{-\alpha x} \qquad (8.4)$$

where i_0 is the initial intensity of the beam, i the intensity of the beam after passing through distance x of the medium and α is the linear absorption coefficient.

Scattering is caused by multiple random reflections and refractions in non homogeneous materials. Examples of where obvious scattering occurs include fog, cloud and smoke in air and suspended particles in water. Scattering may be wavelength selective due to refraction in the scattering media; a good example of this is the atmosphere which scatters blue light more than red light; this gives rise to the blue sky and the red sunset due to scattering by water droplets and dust in the atmosphere.

Diffuse reflection and transmission

When light hits a surface which has irregularities that are greater than the wavelength of the light, then the light is no longer reflected or refracted in a single direction but it is spread out in all directions. The degree of scattering is a function of the surface properties and, in general, any pattern of reflected or transmitted light may be created. If the surface is such that the light is completely scattered then the surface is known as a perfect or Lambertian diffuser. A key property of a perfect diffuser is that it has the same luminance in all directions no matter which direction it is illuminated from. With a perfect diffuser, the luminance of the surface (L) is a function of the illuminance (E) and the reflectance (σ). The relationship is given in equation 8.5:

$$L = \frac{E\,\sigma}{\pi} \tag{8.5}$$

Polarisation

Electromagnetic waves may be thought of as waves of an oscillating electric field in a plane normal to the direction of propagation of the wave. For any given wave, there is a plane that contains the direction of propagation and the oscillation of the electric field. Most light sources emit light where the orientation of the planes of polarisation is random; such light is described as unpolarised. It is possible to select light with a single orientation of the plane of polarisation from unpolarised light using a number of methods, the most used of which are polarising crystals. The resultant light is said to be polarised. The way light of different polarisations is reflected in some specular reflections is different and so it is sometimes possible to use polarisation to remove reflected glare.

Interference and diffraction

The wave nature of light means that it is possible for wave related phenomena to occur, however, due to the short wavelength of light it is only in special circumstances that such phenomena are apparent.

Interference may be seen when a screen is illuminated by two coherent sources of light. This is typically achieved by passing the light through two narrow slits or reflecting light off a multi element reflector. To make the pattern of interference as large as possible, it is necessary to make the slits as close together as possible. There are a number of devices that make use of this principle including the diffraction gratings in monochrometer and dichroic filters, and the iridescence on the reflectors in some luminaires is also caused by interference.

Diffraction is the bending of light round the edge of obstacles. The diffraction effects are generally too small to see but are occasionally visible when using a gobo projector over a very long distance. Diffraction is an important problem when designing optical instruments.

Quantum phenomena

The wave properties satisfactorily explain why light propagates; however, they do not explain the processes by which light is radiated and absorbed. It has been shown that when light is emitted or absorbed, the energy gain or loss is always a discrete amount. This finite amount of energy is known as a quantum. It may therefore be assumed that light is made up of a series of particles, known as photons, which have wave like properties but whenever energy is exchanged, they behave as particles. The energy carried by each photon (Q) is given by equation 8.6:

$$Q = h\,v \tag{8.6}$$

where v is the frequency of the radiation and h is Plank's constant (6.626068×10^{-34} m²·kg·s⁻¹).

8.2   Evaluating energy as light

The eye is not equally sensitive to all wavelengths of light and thus we need to apply a weighting function to the energy at each wavelength so that we can evaluate electromagnetic radiation as light. Because the human visual response varies at different light levels and from person to person, photometry requires the definition of representative standard observers. The CIE system of physical photometry specifies procedures for the quantitative evaluation of optical radiation in terms of the spectral luminous efficiency functions of two such standard observers. One, $V(\lambda)$, represents photopic vision and the other, $V'(\lambda)$, scotopic vision. Photopic vision occurs at luminances over 3 cd·m^{-2} and scotopic vision happens at luminances below 0.001 cd·m^{-2}. At luminances between these two values, a more complex set of visual responses occurs known as mesopic vision. To make matters even more complex, the visual sensitivity varies across the field of vision, the peripheral field having a larger response at the blue end of the spectrum than the central part of the visual field. Thus, there are a plethora of possible visual response functions. However, the one that is used the vast majority of the time is $V(\lambda)$ for photopic vision and when using such terms as lumens, lux and candelas, it is always the photopic spectral efficiency function that has been used to calculate the amount of light present. When using photometric quantities based on any other spectral sensitivity function, it is normal to preface the unit being used by the name of the function used. Thus, occasionally, you may see terms such as scotopic candelas.

$V(\lambda)$, for photopic vision and $V'(\lambda)$, for scotopic vision are the only two spectral sensitivity functions defined by an international standard (ISO, 2005).

The luminous flux associated with electromagnetic radiation may be calculated using equation 8.7 for the photopic condition and equation 8.8 for the scotopic condition.

$$\Phi_v = K_m \int_0^\infty \Phi_{e,\lambda} \cdot V(\lambda) \, d\lambda \tag{8.7}$$

$$\Phi'_v = K'_m \int_0^\infty \Phi_{e,\lambda} \cdot V'(\lambda) \, d\lambda \tag{8.8}$$

where Φ_v and Φ'_v are the photopic and scotopic luminous flux, respectively, K_m and K'_m are constants with values of 683.002 and 1700.005, respectively and $V(\lambda)$ and $V'(\lambda)$ are values obtained from the values given in Tables 8.1 and 8.2, respectively.

Table 8.1 Values of the spectral luminous efficiency function for photopic vision $V(\lambda)$

Wavelength / nm	Spectral luminous efficiency	Wavelength / nm	Spectral luminous efficiency	Wavelength / nm	Spectral luminous efficiency
360	0.0000039170	401	0.0004337147	442	0.0256102400
361	0.0000043936	402	0.0004730240	443	0.0269585700
362	0.0000049296	403	0.0005178760	444	0.0283512500
363	0.0000055321	404	0.0005722187	445	0.0298000000
364	0.0000062082	405	0.0006400000	446	0.0313108300
365	0.0000069650	406	0.0007245600	447	0.0328836800
366	0.0000078132	407	0.0008255000	448	0.0345211200
367	0.0000087673	408	0.0009411600	449	0.0362257100
368	0.0000098398	409	0.0010698800	450	0.0380000000
369	0.0000110432	410	0.0012100000	451	0.0398466700
370	0.0000123900	411	0.0013620910	452	0.0417680000
371	0.0000138864	412	0.0015307520	453	0.0437660000
372	0.0000155573	413	0.0017203680	454	0.0458426700
373	0.0000174430	414	0.0019353230	455	0.0480000000
374	0.0000195838	415	0.0021800000	456	0.0502436800
375	0.0000220200	416	0.0024548000	457	0.0525730400
376	0.0000248397	417	0.0027640000	458	0.0549805600
377	0.0000280413	418	0.0031178000	459	0.0574587200
378	0.0000315310	419	0.0035264000	460	0.0600000000
379	0.0000352152	420	0.0040000000	461	0.0626019700
380	0.0000390000	421	0.0045462400	462	0.0652775200
381	0.0000428264	422	0.0051593200	463	0.0680420800
382	0.0000469146	423	0.0058292800	464	0.0709110900
383	0.0000515896	424	0.0065461600	465	0.0739000000
384	0.0000571764	425	0.0073000000	466	0.0770160000
385	0.0000640000	426	0.0080865070	467	0.0802664000
386	0.0000723442	427	0.0089087200	468	0.0836668000
387	0.0000822122	428	0.0097676800	469	0.0872328000
388	0.0000935082	429	0.0106644300	470	0.0909800000
389	0.0001061361	430	0.0116000000	471	0.0949175500
390	0.0001200000	431	0.0125731700	472	0.0990458400
391	0.0001349840	432	0.0135827200	473	0.1033674000
392	0.0001514920	433	0.0146296800	474	0.1078846000
393	0.0001702080	434	0.0157150900	475	0.1126000000
394	0.0001918160	435	0.0168400000	476	0.1175320000
395	0.0002170000	436	0.0180073600	477	0.1226744000
396	0.0002469067	437	0.0192144800	478	0.1279928000
397	0.0002812400	438	0.0204539200	479	0.1334528000
398	0.0003185200	439	0.0217182400	480	0.1390200000
399	0.0003572667	440	0.0230000000	481	0.1446764000
400	0.0003960000	441	0.0242946100	482	0.1504693000

Table 8.1 Continued

Wavelength / nm	Spectral luminous efficiency	Wavelength / nm	Spectral luminous efficiency	Wavelength / nm	Spectral luminous efficiency
483	0.1564619000	524	0.7778368000	565	0.9786000000
484	0.1627177000	525	0.7932000000	566	0.9740837000
485	0.1693000000	526	0.8081104000	567	0.9691712000
486	0.1762431000	527	0.8224962000	568	0.9638568000
487	0.1835581000	528	0.8363068000	569	0.9581349000
488	0.1912735000	529	0.8494916000	570	0.9520000000
489	0.1994180000	530	0.8620000000	571	0.9454504000
490	0.2080200000	531	0.8738108000	572	0.9384992000
491	0.2171199000	532	0.8849624000	573	0.9311628000
492	0.2267345000	533	0.8954936000	574	0.9234576000
493	0.2368571000	534	0.9054432000	575	0.9154000000
494	0.2474812000	535	0.9148501000	576	0.9070064000
495	0.2586000000	536	0.9237348000	577	0.8982772000
496	0.2701849000	537	0.9320924000	578	0.8892048000
497	0.2822939000	538	0.9399226000	579	0.8797816000
498	0.2950505000	539	0.9472252000	580	0.8700000000
499	0.3085780000	540	0.9540000000	581	0.8598613000
500	0.3230000000	541	0.9602561000	582	0.8493920000
501	0.3384021000	542	0.9660074000	583	0.8386220000
502	0.3546858000	543	0.9712606000	584	0.8275813000
503	0.3716986000	544	0.9760225000	585	0.8163000000
504	0.3892875000	545	0.9803000000	586	0.8047947000
505	0.4073000000	546	0.9840924000	587	0.7930820000
506	0.4256299000	547	0.9874182000	588	0.7811920000
507	0.4443096000	548	0.9903128000	589	0.7691547000
508	0.4633944000	549	0.9928116000	590	0.7570000000
509	0.4829395000	550	0.9949501000	591	0.7447541000
510	0.5030000000	551	0.9967108000	592	0.7324224000
511	0.5235693000	552	0.9980983000	593	0.7200036000
512	0.5445120000	553	0.9991120000	594	0.7074965000
513	0.5656900000	554	0.9997482000	595	0.6949000000
514	0.5869653000	555	1.0000000000	596	0.6822192000
515	0.6082000000	556	0.9998567000	597	0.6694716000
516	0.6293456000	557	0.9993046000	598	0.6566744000
517	0.6503068000	558	0.9983255000	599	0.6438448000
518	0.6708752000	559	0.9968987000	600	0.6310000000
519	0.6908424000	560	0.9950000000	601	0.6181555000
520	0.7100000000	561	0.9926005000	602	0.6053144000
521	0.7281852000	562	0.9897426000	603	0.5924756000
522	0.7454636000	563	0.9864444000	604	0.5796379000
523	0.7619694000	564	0.9827241000	605	0.5668000000

Table 8.1 Continued

Wavelength / nm	Spectral luminous efficiency	Wavelength / nm	Spectral luminous efficiency	Wavelength / nm	Spectral luminous efficiency
606	0.5539611000	647	0.1250248000	688	0.0095333110
607	0.5411372000	648	0.1187792000	689	0.0088461570
608	0.5283528000	649	0.1127691000	690	0.0082100000
609	0.5156323000	650	0.1070000000	691	0.0076237810
610	0.5030000000	651	0.1014762000	692	0.0070854240
611	0.4904688000	652	0.0961886400	693	0.0065914760
612	0.4780304000	653	0.0911229600	694	0.0061384850
613	0.4656776000	654	0.0862648500	695	0.0057230000
614	0.4534032000	655	0.0816000000	696	0.0053430590
615	0.4412000000	656	0.0771206400	697	0.0049957960
616	0.4290800000	657	0.0728255200	698	0.0046764040
617	0.4170360000	658	0.0687100800	699	0.0043800750
618	0.4050320000	659	0.0647697600	700	0.0041020000
619	0.3930320000	660	0.0610000000	701	0.0038384530
620	0.3810000000	661	0.0573962100	702	0.0035890990
621	0.3689184000	662	0.0539550400	703	0.0033542190
622	0.3568272000	663	0.0506737600	704	0.0031340930
623	0.3447768000	664	0.0475496500	705	0.0029290000
624	0.3328176000	665	0.0445800000	706	0.0027381390
625	0.3210000000	666	0.0417587200	707	0.0025598760
626	0.3093381000	667	0.0390849600	708	0.0023932440
627	0.2978504000	668	0.0365638400	709	0.0022372750
628	0.2865936000	669	0.0342004800	710	0.0020910000
629	0.2756245000	670	0.0320000000	711	0.0019535870
630	0.2650000000	671	0.0299626100	712	0.0018245800
631	0.2547632000	672	0.0280766400	713	0.0017035800
632	0.2448896000	673	0.0263293600	714	0.0015901870
633	0.2353344000	674	0.0247080500	715	0.0014840000
634	0.2260528000	675	0.0232000000	716	0.0013844960
635	0.2170000000	676	0.0218007700	717	0.0012912680
636	0.2081616000	677	0.0205011200	718	0.0012040920
637	0.1995488000	678	0.0192810800	719	0.0011227440
638	0.1911552000	679	0.0181206900	720	0.0010470000
639	0.1829744000	680	0.0170000000	721	0.0009765896
640	0.1750000000	681	0.0159037900	722	0.0009111088
641	0.1672235000	682	0.0148371800	723	0.0008501332
642	0.1596464000	683	0.0138106800	724	0.0007932384
643	0.1522776000	684	0.0128347800	725	0.0007400000
644	0.1451259000	685	0.0119200000	726	0.0006900827
645	0.1382000000	686	0.0110683100	727	0.0006433100
646	0.1315003000	687	0.0102733900	728	0.0005994960

Table 8.1 Continued

Wavelength / nm	Spectral luminous efficiency	Wavelength / nm	Spectral luminous efficiency	Wavelength / nm	Spectral luminous efficiency
729	0.0005584547	763	0.0000487184	797	0.0000045697
730	0.0005200000	764	0.0000454475	798	0.0000042602
731	0.0004839136	765	0.0000424000	799	0.0000039717
732	0.0004500528	766	0.0000395610	800	0.0000037029
733	0.0004183452	767	0.0000369151	801	0.0000034522
734	0.0003887184	768	0.0000344487	802	0.0000032183
735	0.0003611000	769	0.0000321482	803	0.0000030003
736	0.0003353835	770	0.0000300000	804	0.0000027971
737	0.0003114404	771	0.0000279913	805	0.0000026078
738	0.0002891656	772	0.0000261136	806	0.0000024312
739	0.0002684539	773	0.0000243602	807	0.0000022665
740	0.0002492000	774	0.0000227246	808	0.0000021130
741	0.0002313019	775	0.0000212000	809	0.0000019699
742	0.0002146856	776	0.0000197786	810	0.0000018366
743	0.0001992884	777	0.0000184529	811	0.0000017122
744	0.0001850475	778	0.0000172169	812	0.0000015962
745	0.0001719000	779	0.0000160646	813	0.0000014881
746	0.0001597781	780	0.0000149900	814	0.0000013873
747	0.0001486044	781	0.0000139873	815	0.0000012934
748	0.0001383016	782	0.0000130516	816	0.0000012058
749	0.0001287925	783	0.0000121782	817	0.0000011241
750	0.0001200000	784	0.0000113625	818	0.0000010480
751	0.0001118595	785	0.0000106000	819	0.0000009771
752	0.0001043224	786	0.0000098859	820	0.0000009109
753	0.0000973356	787	0.0000092173	821	0.0000008493
754	0.0000908459	788	0.0000085924	822	0.0000007917
755	0.0000848000	789	0.0000080091	823	0.0000007381
756	0.0000791467	790	0.0000074657	824	0.0000006881
757	0.0000738580	791	0.0000069596	825	0.0000006415
758	0.0000689160	792	0.0000064880	826	0.0000005981
759	0.0000643027	793	0.0000060487	827	0.0000005576
760	0.0000600000	794	0.0000056394	828	0.0000005198
761	0.0000559819	795	0.0000052578	829	0.0000004846
762	0.0000522256	796	0.0000049018	830	0.0000004518

Table 8.2 Values of the spectral luminous efficiency function for scotopic vision $V'(\lambda)$

Wavelength / nm	Spectral luminous efficiency	Wavelength / nm	Spectral luminous efficiency	Wavelength / nm	Spectral luminous efficiency
380	0.0005890000	421	0.1052000000	462	0.5880000000
381	0.0006650000	422	0.1141000000	463	0.5990000000
382	0.0007520000	423	0.1235000000	464	0.6100000000
383	0.0008540000	424	0.1334000000	465	0.6200000000
384	0.0009720000	425	0.1436000000	466	0.6310000000
385	0.0011080000	426	0.1541000000	467	0.6420000000
386	0.0012680000	427	0.1651000000	468	0.6530000000
387	0.0014530000	428	0.1764000000	469	0.6640000000
388	0.0016680000	429	0.1879000000	470	0.6760000000
389	0.0019180000	430	0.1998000000	471	0.6870000000
390	0.0022090000	431	0.2119000000	472	0.6990000000
391	0.0025470000	432	0.2243000000	473	0.7100000000
392	0.0029390000	433	0.2369000000	474	0.7220000000
393	0.0033940000	434	0.2496000000	475	0.7340000000
394	0.0039210000	435	0.2625000000	476	0.7450000000
395	0.0045300000	436	0.2755000000	477	0.7570000000
396	0.0052400000	437	0.2886000000	478	0.7690000000
397	0.0060500000	438	0.3017000000	479	0.7810000000
398	0.0069800000	439	0.3149000000	480	0.7930000000
399	0.0080600000	440	0.3281000000	481	0.8050000000
400	0.0092900000	441	0.3412000000	482	0.8170000000
401	0.0107000000	442	0.3543000000	483	0.8280000000
402	0.0123100000	443	0.3673000000	484	0.8400000000
403	0.0141300000	444	0.3803000000	485	0.8510000000
404	0.0161900000	445	0.3931000000	486	0.8620000000
405	0.0185200000	446	0.4060000000	487	0.8730000000
406	0.0211300000	447	0.4180000000	488	0.8840000000
407	0.0240500000	448	0.4310000000	489	0.8940000000
408	0.0273000000	449	0.4430000000	490	0.9040000000
409	0.0308900000	450	0.4550000000	491	0.9140000000
410	0.0348400000	451	0.4670000000	492	0.9230000000
411	0.0391600000	452	0.4790000000	493	0.9320000000
412	0.0439000000	453	0.4900000000	494	0.9410000000
413	0.0490000000	454	0.5020000000	495	0.9490000000
414	0.0545000000	455	0.5130000000	496	0.9570000000
415	0.0604000000	456	0.5240000000	497	0.9640000000
416	0.0668000000	457	0.5350000000	498	0.9700000000
417	0.0736000000	458	0.5460000000	499	0.9760000000
418	0.0808000000	459	0.5570000000	500	0.9820000000
419	0.0885000000	460	0.5670000000	501	0.9860000000
420	0.0966000000	461	0.5780000000	502	0.9900000000

Table 8.2 Continued

Wavelength / nm	Spectral luminous efficiency	Wavelength / nm	Spectral luminous efficiency	Wavelength / nm	Spectral luminous efficiency
503	0.9940000000	544	0.5810000000	585	0.0899000000
504	0.9970000000	545	0.5640000000	586	0.0845000000
505	0.9980000000	546	0.5480000000	587	0.0793000000
506	1.0000000000	547	0.5310000000	588	0.0745000000
507	1.0000000000	548	0.5140000000	589	0.0699000000
508	1.0000000000	549	0.4970000000	590	0.0655000000
509	0.9980000000	550	0.4810000000	591	0.0613000000
510	0.9970000000	551	0.4650000000	592	0.0574000000
511	0.9940000000	552	0.4480000000	593	0.0537000000
512	0.9900000000	553	0.4330000000	594	0.0502000000
513	0.9860000000	554	0.4170000000	595	0.0469000000
514	0.9810000000	555	0.4020000000	596	0.0438000000
515	0.9750000000	556	0.3864000000	597	0.0409000000
516	0.9680000000	557	0.3715000000	598	0.0381600000
517	0.9610000000	558	0.3569000000	599	0.0355800000
518	0.9530000000	559	0.3427000000	600	0.0331500000
519	0.9440000000	560	0.3288000000	601	0.0308700000
520	0.9350000000	561	0.3151000000	602	0.0287400000
521	0.9250000000	562	0.3018000000	603	0.0267400000
522	0.9150000000	563	0.2888000000	604	0.0248700000
523	0.9040000000	564	0.2762000000	605	0.0231200000
524	0.8920000000	565	0.2639000000	606	0.0214700000
525	0.8800000000	566	0.2519000000	607	0.0199400000
526	0.8670000000	567	0.2403000000	608	0.0185100000
527	0.8540000000	568	0.2291000000	609	0.0171800000
528	0.8400000000	569	0.2182000000	610	0.0159300000
529	0.8260000000	570	0.2076000000	611	0.0147700000
530	0.8110000000	571	0.1974000000	612	0.0136900000
531	0.7960000000	572	0.1876000000	613	0.0126900000
532	0.7810000000	573	0.1782000000	614	0.0117500000
533	0.7650000000	574	0.1690000000	615	0.0108800000
534	0.7490000000	575	0.1602000000	616	0.0100700000
535	0.7330000000	576	0.1517000000	617	0.0093200000
536	0.7170000000	577	0.1436000000	618	0.0086200000
537	0.7000000000	578	0.1358000000	619	0.0079700000
538	0.6830000000	579	0.1284000000	620	0.0073700000
539	0.6670000000	580	0.1212000000	621	0.0068200000
540	0.6500000000	581	0.1143000000	622	0.0063000000
541	0.6330000000	582	0.1078000000	623	0.0058200000
542	0.6160000000	583	0.1015000000	624	0.0053800000
543	0.5990000000	584	0.0956000000	625	0.0049700000

Table 8.2 Continued

Wavelength / nm	Spectral luminous efficiency	Wavelength / nm	Spectral luminous efficiency	Wavelength / nm	Spectral luminous efficiency
626	0.0045900000	667	0.0001848000	708	0.0000104300
627	0.0042400000	668	0.0001716000	709	0.0000097600
628	0.0039130000	669	0.0001593000	710	0.0000091400
629	0.0036130000	670	0.0001480000	711	0.0000085600
630	0.0033350000	671	0.0001375000	712	0.0000080200
631	0.0030790000	672	0.0001277000	713	0.0000075100
632	0.0028420000	673	0.0001187000	714	0.0000070400
633	0.0026230000	674	0.0001104000	715	0.0000066000
634	0.0024210000	675	0.0001026000	716	0.0000061800
635	0.0022350000	676	0.0000954000	717	0.0000058000
636	0.0020620000	677	0.0000888000	718	0.0000054400
637	0.0019030000	678	0.0000826000	719	0.0000051000
638	0.0017570000	679	0.0000769000	720	0.0000047800
639	0.0016210000	680	0.0000715000	721	0.0000044900
640	0.0014970000	681	0.0000666000	722	0.0000042100
641	0.0013820000	682	0.0000620000	723	0.0000039510
642	0.0012760000	683	0.0000578000	724	0.0000037090
643	0.0011780000	684	0.0000538000	725	0.0000034820
644	0.0010880000	685	0.0000501000	726	0.0000032700
645	0.0010050000	686	0.0000467000	727	0.0000030700
646	0.0009280000	687	0.0000436000	728	0.0000028840
647	0.0008570000	688	0.0000406000	729	0.0000027100
648	0.0007920000	689	0.0000378900	730	0.0000025460
649	0.0007320000	690	0.0000353300	731	0.0000023930
650	0.0006770000	691	0.0000329500	732	0.0000022500
651	0.0006260000	692	0.0000307500	733	0.0000021150
652	0.0005790000	693	0.0000287000	734	0.0000019890
653	0.0005360000	694	0.0000267900	735	0.0000018700
654	0.0004960000	695	0.0000250100	736	0.0000017590
655	0.0004590000	696	0.0000233600	737	0.0000016550
656	0.0004250000	697	0.0000218200	738	0.0000015570
657	0.0003935000	698	0.0000203800	739	0.0000014660
658	0.0003645000	699	0.0000190500	740	0.0000013790
659	0.0003377000	700	0.0000178000	741	0.0000012990
660	0.0003129000	701	0.0000166400	742	0.0000012230
661	0.0002901000	702	0.0000155600	743	0.0000011510
662	0.0002689000	703	0.0000145400	744	0.0000010840
663	0.0002493000	704	0.0000136000	745	0.0000010220
644	0.0002313000	705	0.0000127300	746	0.0000009620
665	0.0002146000	706	0.0000119100	747	0.0000009070
666	0.0001991000	707	0.0000111400	748	0.0000008550

Table 8.2 Continued

Wavelength / nm	Spectral luminous efficiency	Wavelength / nm	Spectral luminous efficiency	Wavelength / nm	Spectral luminous efficiency
749	0.0000008060	760	0.0000004250	771	0.0000002282
750	0.0000007600	761	0.0000004010	772	0.0000002159
751	0.0000007160	762	0.0000003790	773	0.0000002042
752	0.0000006750	763	0.0000003580	774	0.0000001932
753	0.0000006370	764	0.0000003382	775	0.0000001829
754	0.0000006010	765	0.0000003196	776	0.0000001731
755	0.0000005670	766	0.0000003021	777	0.0000001638
756	0.0000005350	767	0.0000002855	778	0.0000001551
757	0.0000005050	768	0.0000002699	779	0.0000001468
758	0.0000004770	769	0.0000002552	780	0.0000001390
759	0.0000004500	770	0.0000002413		

Chapter 9: Luminous flux, intensity, illuminance, luminance and their interrelationships

Luminous flux, intensity, illuminance and luminance are all metrics that we use to describe the amount of light in different situations. They all depend on the basic notion of what a lumen is, but also on the geometry of the way they are applied. The basis of all lighting calculations depends on the way that the basic units of light are defined and their geometric interrelationships. This chapter looks first at the fundamental definitions of the four units of light and then at some of the interrelationships and methods of light calculation.

9.1   Definitions of the units

9.1.1   Flux
The formal definition of flux is:

> *Luminous flux: Quantity derived from radiant flux (radiant power) by evaluating the radiation according to the spectral sensitivity of the human eye (as defined by the CIE standard photometric observer). It is the light power emitted by a source or received by a surface.*

The unit of flux is the lumen.

The lumen is calculated using the radiant energy of a source that has been multiplied by a spectral sensitivity function. Details of this are given in section 8.1.

Flux is often used to compare the outputs of lamps and where some idea of the total amount of light is needed without reference to a particular direction of light flow or a particular surface being illuminated.

9.1.2   Intensity
The formal definition of intensity is:

> *Luminous intensity (of a point source in a given direction) (I): Luminous flux per unit solid angle in the direction in question, i.e. the luminous flux on a small surface, divided by the solid angle that the surface subtends at the source.*

The unit of intensity is the candela which is equal to one lumen per steradian.

This brings in the concept of solid angle and more particularly, unit solid angle. A unit solid angle is called a steradian. A steradian is the measure for a solid angle, enclosing the part of the surface of a sphere, with the centre at its apex, with an area equal to the radius of the sphere squared. There are therefore 4π steradians in a sphere (Figure 9.1).

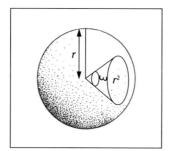

Figure 9.1 The steradian

9.1.3   Illuminance

The formal definition of illuminance is:

> *Illuminance (at a point of a surface) (E): Quotient of the luminous flux, incident on an element of the surface containing the point, by the area of that element.*

The unit of illuminance is the lux which is equivalent to one lumen per square metre.

Note that illuminance is the light falling onto a surface; this surface may be an actual surface but may also be a virtual surface in space. As this surface has an orientation, then illuminance has direction and is thus sometimes treated as a vector. It is also sometimes necessary to consider the amount of light falling onto curved surfaces and methods exist for the calculation of cylindrical, semi-cylindrical, spherical and semi-spherical illuminance.

9.1.4   Luminance

The formal definition of luminance is:

> *Luminance (L): Luminous flux per unit solid angle transmitted by an elementary beam passing through the given point and propagating in the given direction, divided by the area of a section of that beam normal to the direction of the beam and containing the given point.*
>
> *It can also be defined as:*
>
> a. *The luminous intensity of the light emitted or reflected in a given direction from an element of the surface, divided by the area of the element projected in the same direction.*
>
> b. *The illuminance produced by the beam of light on a surface normal to its direction, divided by the solid angle of the source as seen from the illuminated surface.*

The unit of luminance is candelas per square metre.

The definition of luminance brings in the concept of area normal to the direction of propagation; this is sometimes called the projected area of a surface. It is important to understand that area normal to propagation of an object will vary with direction of view. Figure 9.2 shows the various areas for a pyramid.

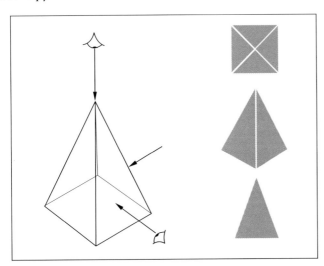

Figure 9.2 Areas normal to the direction of propagation for a pyramid

Luminance is used as it represents a measurable physical quantity that relates to the perceived sensations of brightness and lightness. However, the relationship between luminance and brightness and lightness is very complex but the general rule is that if you increase the luminance of an object or area of a space then if the luminances of all other objects in the field of view are kept the same, the object or area will appear brighter or lighter.

9.2   Interrelationships between the units

Most lighting calculations are based on the relationship between the four units defined in section 9.1.

9.2.1   Flux and intensity

Intensity is flux per unit steradian so provided that the solid angle over which a flux is distributed is known then intensity can be calculated and likewise the flux associated with a given intensity can be calculated if the solid angle associated with it is known. If the distribution of intensity from a light source is regular, then such calculations are simple. If it is more complex, then it is necessary to divide up the beam of light into smaller zones and treat each separately.

Consider a light source of flux (Φ) that radiates light equally in all directions, then the intensity (I) in any direction will be given by equation 9.1:

$$I = \frac{\Phi}{4\pi} \tag{9.1}$$

The other special case is that of a Lambertian radiator. A Lambertian radiator is a perfect diffuser, it has equal luminance in all directions and thus the intensity in any direction (I_θ) is a function of the intensity normal to the surface (I_0); the relationship is given in equation 9.2:

$$I_\theta = I_0 \cos\theta \tag{9.2}$$

where θ is the angle between the normal to the surface and the direction of the intensity in question. This is illustrated in Figure 9.3.

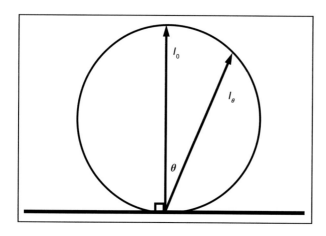

Figure 9.3 A Lambertian light source

For a Lambertian source, it can be shown that the intensity at the normal (I_0) is related to the total flux emitted (Φ) by equation 9.3:

$$I_0 = \frac{\Phi}{\pi} \tag{9.3}$$

For more complex light distributions, say a typical luminaire, it is normal to measure the intensity in a number of locations within an angular web around the light source. Each reading of intensity may be regarded as the best estimate of intensity for the solid angle that is bounded by half of the angular distance to each of the adjacent measurement points. Thus it is possible to calculate the solid angle associated with each intensity value and so the flux associated with it may be calculated. The total flux from the light source may be estimated by summing all of the values obtained in this way. This topic is covered in greater detail in section 12.3.1.

9.2.2 Intensity and illuminance

One of the most used principles in lighting calculation relates intensity and illuminance and is known as the inverse square law. It may be derived in the following way.

Considering the light from a source towards a point as a small beam, the luminous flux (Φ) contained in this beam is given by the average intensity (I) of the beam multiplied by the solid angle (ω) containing the beam. From the definitions in section 9.1, we have equation 9.4:

$$\Phi = I \, \omega \tag{9.4}$$

As the surface of the solid angle is at right angles to the radius of the sphere, the illuminance on a surface at right angles to the direction of the incident light may be derived by equation 9.5:

$$E = \frac{\Phi}{Area} = \frac{I\,\omega}{Area} = \frac{I}{Area} \times \frac{Area}{R^2} = \frac{I}{R^2} \tag{9.5}$$

where $Area$ is the area over which the surface is illuminated and R is the distance from the source to the point of illumination.

Note: The intensity used in this calculation relates to the flux contained in a very small solid angle and the illuminance calculated using the inverse square law will be that at a point, not that over a large area. The section of the sphere forming the solid angle will be very small and can be considered to be a flat surface. It should also be noted that this formula should, strictly speaking be used with point sources. However, in most cases, provided the size of the maximum dimension of the source is less than a fifth of the distance from the source to the point of illumination, the errors involved in any calculations will be very small.

In practice, the direction of light onto a surface is not often normal to the plane and in this circumstance, the area of the plane receiving the flux will be increased as shown in Figure 9.4. The illuminance on plane 'abcd' will be less than that on plane 'ABCD' because the flux on both areas is the same but the area is increased.

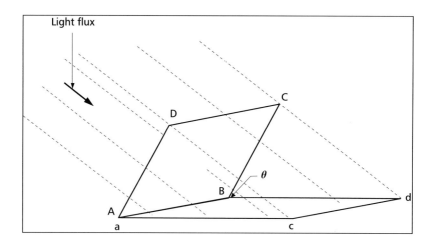

Figure 9.4 Illuminance on an inclined plane

The area of 'ABCD' is equal to the area of 'abcd' multiplied by cos θ, thus the illuminance on an inclined plane may be calculated using equation 9.6

$$E = \frac{I \cos \theta}{D^2} \qquad (9.6)$$

where θ is the angle between the incident light and the normal to the surface.

Chaper 10: Direct lighting

This chapter explores some of the techniques used for predicting the illuminance due to various light sources.

10.1 Illuminance from point sources

In Chapter 9 in the discussion of photometric quantities, it was assumed in the discussion of flux and intensity that the source of light was a point. In practice, all light sources have a finite size and so the formulae start to break down. However, for most practical purposes, it can be shown that if the distance between the source and the point being illuminated is at least five times the maximum dimension of the light source then point source formulae may be used and the errors that arise are generally small enough to be neglected.

10.1.1 Planar illuminance
In Chapter 9, the equation for planar illuminance was derived:

$$E = \frac{I \cos \theta}{D^2} \tag{10.1}$$

where E is the illuminance, I is the intensity from the source to the point where the illuminance is evaluated, θ is the angle between the normal to the plane being illuminated and the line joining the point of illumination to the light source. D is the distance between the point of illumination and the source.

This formula may be developed to make it more useful. Consider the calculation of illuminance across a horizontal plane due to a single light source; see Figure 10.1. The distance D will vary with the point considered on the horizontal plane. The calculation may be simplified so that height H (which is fixed) may be used instead of D.

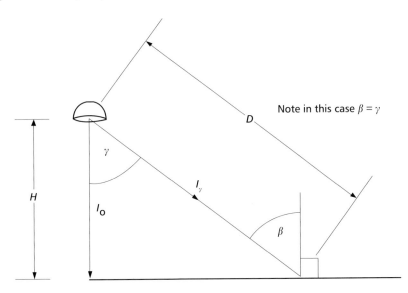

Figure 10.1 Calculation of illuminance across a horizontal plane due to a single light source

In this case, the illuminance at the point is given by equation 10.2:

$$E = \frac{I_\gamma \cos \beta}{D^2} \tag{10.2}$$

where I_γ is the intensity towards the point, β is the angle between the direction of the intensity and the perpendicular to the plane and D is the distance between the source and the illuminated point. Given that angles β and γ are the same, we can rewrite equation 10.2 in the steps set out in equation 10.3:

$$E = \frac{I_\gamma \cos \gamma}{D^2} = \frac{I_\gamma \cos \gamma}{\left(H/\cos \gamma\right)^2} = \frac{I_\gamma \cos^3 \gamma}{H^2} \tag{10.3}$$

10.1.2 Cylindrical illuminance

Cylindrical illuminance is a useful metric as it can be used to give an impression as to how visible an object might be when placed at a certain point in space. It is used as a metric in workplaces to control lighting to ensure that people's heads are visible.

It should be noted that the definition of the term cylindrical illuminance relates to the amount of light falling onto the curved surface of an infinitely small cylinder and does not include light falling onto the ends of the cylinder.

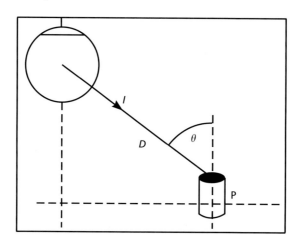

Figure 10.2 Cylindrical illuminance

For cylindrical illuminance, Figure 10.2 gives the geometry of the situation and equation 10.4 gives the cylindrical illuminance (E_{cyl}) at point P:

$$E_{cyl} = \frac{I \sin \theta}{D^2 \pi} \tag{10.4}$$

where E_{cyl} is the cylindrical illuminance at point P, I is the luminous intensity of the source in the direction of the point P, θ is the angle between the direction of light incidence and the direction of the axis of the cylinder and D is the distance between the light source and the point.

10.1.3 Semi-cylindrical illuminance

Semi-cylindrical illuminance has been found to be a good parameter for describing the visibility of people's faces under street lighting conditions. For semi-cylindrical calculations, the situation is slightly more complex than for cylindrical calculations as the direction of view has to be taken into account. Figure 10.3 shows the layout.

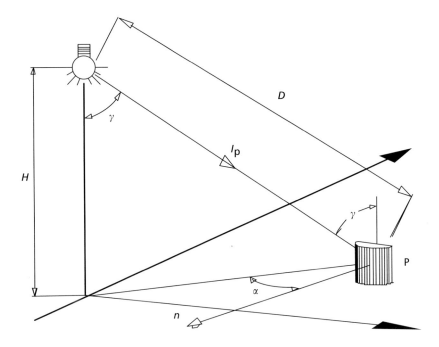

Figure 10.3 Semi-cylindrical illuminance

Semi-cylindrical illuminance may be calculated with equation 10.5:

$$E_{SC} = \frac{I_P}{\pi D^2}(1+\cos\alpha)\sin\gamma \tag{10.5}$$

where E_{SC} is the semi-cylindrical illuminance at point P, I_p is the luminous intensity of the source in the direction of the point P, α is the angle between the direction of light incidence and the direction of observation on which the semi-cylinder stands and γ is the angle between the direction of light incidence and the normal on the semi-cylindrical plane.

10.1.4 Spherical illuminance
Average spherical illuminance at a point is a useful metric to describe the average amount of light falling on a point in space; it is sometimes known as the scalar illuminance and it is used as the basis of a calculation of how strong a modelling effect at that point is.

Average spherical illuminance is quite simple to calculate using equation 10.6:

$$E_{sph} = \frac{I}{4 D^2} \tag{10.6}$$

where E_{sph} is the spherical illuminance, I is the intensity from the source in the direction of the sphere and D is the distance from the source to the sphere.

10.1.5 Hemispherical illuminance
Hemispherical illuminance is used as a metric for assessing the performance of roads using the A-classes as defined in BS EN 13201-2 (BSI, 2003a). Note that the use of A classes is not recommended in the UK by BS 5489-1 (BSI, 2003e), however, it is used in some other European countries.

Hemispherical illuminance is the ratio of luminous flux incident on an infinitely small semi-sphere to the area of that semi-sphere. The orientation of the semi-sphere is defined by the normal to the base. Hemispherical illuminance may be calculated using equation 10.7.

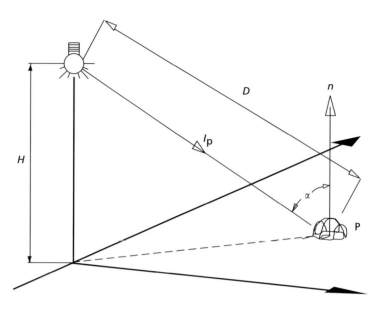

Figure 10.4 Hemispherical illuminance

$$E_{HSP} = \frac{I_P}{4\,D^2}(1+\cos\alpha)$$ (10.7)

where E_{HSP} is the hemispherical illuminance at point P, I_P is the luminous intensity from the source in the direction of point P, D is the distance from the source to point P and α is the angle between the normal to the base of the hemisphere and the incident light.

10.2 Non-point sources

Traditionally, there have been a number of calculation methods that deal with line and area sources, and some of these methods are given below. However, all of these techniques are limited in their application and care has to be taken to ensure they are applied correctly. Since computers have become widely available, the most used and generally applicable technique for dealing with large sources has become recursive source sub division, where sources are split down until each element's maximum dimension is less than one-fifth of the distance from the element to the point being illuminated.

10.2.1 Line source calculations

Line source formulae are used for linear light sources, for example, luminaires using long fluorescent tubes. There are a number of formulae to cover points both on and off the axis of the light source. Some of the formulae make assumptions about the light distribution of the source and so care has to be taken when applying them. The equations 10.8 and 10.9 for illuminance at points below the axis of the light source will work for any photometric distribution. Other equations, not given in this *Code*, will only work where the shape of the photometric distribution follows certain rules.

In equations 10.8 and 10.9, the following symbols are used.

E	Illuminance at point
H	Height of light source above point
L	Length of light source (see Figures 10.5 and 10.6)
I_0	Intensity of light source to the downward vertical
α	Aspect angle (see Figures 10.5 and 10.6)
AF	Aspect factor (see Table 10.1)

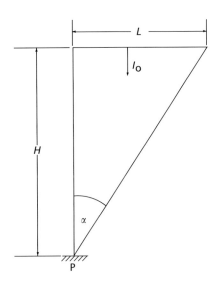

Figure 10.5 Point under the end of a linear source

Equation 10.8 gives the illuminance at a point under the end of a linear source (see Figure 10.5):

$$E = \frac{I_0\, AF(\alpha)}{L\,H} \tag{10.8}$$

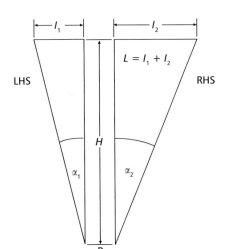

Figure 10.6 Point under a linear source

Equation 10.9 gives the illuminance at a point under a linear source (see Figure 10.6)

$$E = \frac{I_0 \times \left(AF(\alpha_1) + AF(\alpha_2)\right)}{L\,H} \tag{10.9}$$

It is also possible to calculate the illuminance on a point that lies on the axis of a light source but is away from the end, by assuming a virtual light source that extends to the point then doing a calculation by subtracting the result for the virtual source from the combined virtual and real sources.

In order to use equations 10.8 and 10.9, it is necessary to know the aspect factor for the light source. This may be calculated using the process set out in Table 10.1.

Table 10.1 Calculation of aspect factors

a	b	c	d	e	f	g	h
Vertical angle (γ)	Axial intensity $(I_{\gamma,90})$	$(I_{\gamma,90})/$ $(I_{0,90})$	$\cos \gamma$ $(I_{\gamma,90})/$ $(I_{0,90})$	Average	γ / radians	Angular step / radians	Aspect factor
0	310.0	1.000	1.000	–	–	–	–
5	308.8	0.996	0.992	0.996	0.087	0.087	0.087
10	305.3	0.985	0.970	0.981	0.175	0.087	0.173
15	299.4	0.966	0.933	0.951	0.262	0.087	0.256
20	291.3	0.940	0.883	0.908	0.349	0.087	0.335
25	280.9	0.906	0.821	0.852	0.436	0.087	0.409
30	268.5	0.866	0.750	0.786	0.524	0.087	0.478
35	253.9	0.819	0.671	0.711	0.611	0.087	0.540
40	237.5	0.766	0.587	0.629	0.698	0.087	0.595
45	219.2	0.707	0.500	0.543	0.785	0.087	0.642
50	199.3	0.643	0.413	0.457	0.873	0.087	0.682
55	177.8	0.574	0.329	0.371	0.960	0.087	0.714
60	155.0	0.500	0.250	0.289	1.047	0.087	0.740
65	131.0	0.423	0.179	0.214	1.134	0.087	0.758
70	106.0	0.342	0.117	0.148	1.222	0.087	0.771
75	80.2	0.259	0.067	0.092	1.309	0.087	0.779
80	53.8	0.174	0.030	0.049	1.396	0.087	0.783
85	27.0	0.087	0.008	0.019	1.484	0.087	0.785
90	0.0	0.000	0.000	0.004	1.571	0.087	0.785

The calculation process is as follows:

1) From the values of axial intensity in column (b) calculate the ratio of $I_{\gamma,90}/I_{0,90}$ for column (c).
2) Multiply the values in column (c) by the cosine of the respective angles of column (a) to determine the values for column (d).
3) The values of column (e) are the average of the values of $\cos(\gamma)I_{\gamma,90}/I_{0,90}$ for the angle being considered and the preceding angle.
4) Column (f) is the value of the angle of column (a) expressed in radians.
5) Column (g) is the difference in angle between the angle considered and the previous angle, measured in radians.
6) Aspect Factor is shown in column (h); it is derived from the cumulative total of the values of column (e) × column (g).

10.2.2 Area sources

There are a number of formulae available to calculate the illuminance due to an area source at a given point. Most of these formulae have limitations to their applicability. Equation 10.10 has the restriction that the luminance of the source towards the point being illuminated must be the same across the source.

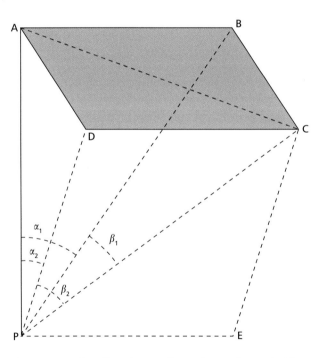

Figure 10.7 Illuminance from an area source

For a rectangle of luminance L, angles as shown in Figure 10.7 and angles β_1 and β_2 expressed in radians, the horizontal illuminance at point P is given by equation 10.10:

$$E = \frac{L}{2}\left(\beta_1 \sin \alpha_1 + \beta_2 \sin \alpha_2\right) \tag{10.10}$$

10.2.3 Recursive source subdivision

Recursive source subdivision is a process whereby a large light source is split up into a number of elements so that each element may be considered to be a point source for a given calculation.

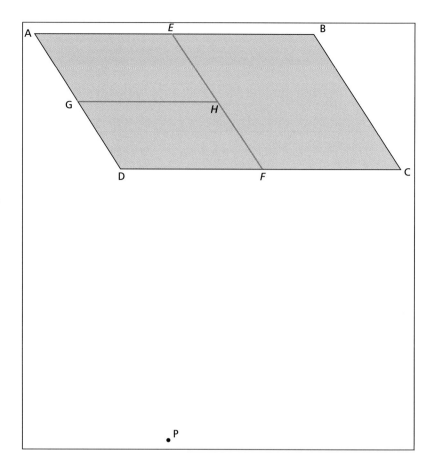

Figure 10.8 Recursive source subdivision

The process is illustrated in Figure 10.8. The source is an area source given by rectangle ABCD; dimension AB is too large for the ratio of the distance between the centre of the rectangle to the maximum dimension of the source to pass the 1 to 5 rule. Thus, ABCD is divided into two rectangles, AEFD and EBCF; each of these two light sources has half the intensity of the original source. Considering source AEFD, it is found that dimension AD is too large for the source to be considered a point source, so it is further subdivided into AEHG and GHFD; each of these sources has half of the intensity of AEFD. This process is repeated until the whole of the original light source is subdivided into elements for which the 1 to 5 rule holds. Then it is possible to calculate the illuminance at point P by summing the illuminance from each of the elements of the original light source.

Chapter 11: Indirect lighting

11.1 Introduction

In any interior space, light may reach any given point either directly from the source or via a number of inter-reflections off the room surfaces. Fully modelling these multiple reflections is highly complex; however, the objective of this chapter is to provide a number of relatively simple techniques that permit the estimation of indirect light in a number of relatively common situations. All of the indirect calculation methods discussed assume that all of the reflecting surfaces reflect light in a way that is diffuse and that the luminance of any surface is the same in all directions and only a function of the illuminance on it and its reflection factor.

11.2 Sumpner's method

Sumpner's method relies on the fact that all of the light entering a space must either leave it or be absorbed on one of the surfaces of the space. Thus, in a room, we are able to calculate the average illuminance onto the room surfaces if we know the total light flux entering the space and the reflective properties of the walls. Equation 11.1 is a mathematical statement of Sumpner's method:

$$E = \frac{\Phi}{A(1 - R_{av})} \qquad (11.1)$$

where: E is the average illuminance on the room surfaces

Φ is the total flux entering the space

A is the total area of the room surfaces

R_{av} is the area weighted average reflectance of the surfaces and thus, $1-R$ is the average absorbance.

As it is quite easy to calculate the direct illuminance onto room surfaces, it is often more useful to be able to look at the average indirect illuminance (E_{ind}) falling on to the surfaces, see equation 11.2:

$$E_{ind} = \frac{\Phi}{A} \times \frac{R_{av}}{1 - R_{av}} \qquad (11.2)$$

11.2.1 Checking the results of lighting calculation software

Sumpner's method may also be used to check the results of a lighting calculation. As all flux entering a space must be absorbed, it is easy to do a quick check. Consider a room 6 m long, 4 m wide and 2.5 m high; see Figure 11.1.

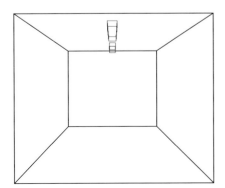

Figure 11.1 Example room

In the room, the reflectance of the ceiling is 0.7, the walls 0.5 and the floors 0.2. The room is illuminated with two luminaires each containing two lamps each with an output of 5200 lumens; the light output ratio of the luminaires is 0.69. From lighting software, the average illuminances have been calculated and are shown in Table 11.1.

Table 11.1 Calculated illuminances

Surface	Average illuminance / lx
Ceiling	132
Floor	417
Left wall	212
Back wall	217
Right wall	212
Front wall	217

Given these data, it is simple to calculate the flux being absorbed on each surface and thus the total flux being absorbed. The necessary calculations are shown in Table 11.2.

Table 11.2 Calculation of absorbed flux

Surface	Area (A)	Reflectance (R)	Absorbance ($1-R$)	Illuminance (E)	Absorbed flux ($A \times (1-R) \times E$)
Ceiling	24	0.7	0.3	132	950
Floor	24	0.2	0.8	417	8006
Left wall	15	0.5	0.5	212	1590
Back wall	10	0.5	0.5	217	1085
Right wall	15	0.5	0.5	212	1590
Front wall	10	0.5	0.5	217	1085
				Total	14306

The total flux coming from the luminaires is equal to the total lamp flux per luminaire (2 × 5200) multiplied by the number of luminaires (2) multiplied by the LOR of each luminaire (0.69). This gives a total flux leaving the luminaires of 14,352 lumens (2 × 5200 × 2 × 0.69). Thus, in this case, it can be seen that the calculation is approximately correct.

11.3 Transfer factors

Transfer factors can be used to assess the amount of light inter-reflected between various surfaces in a room. The method is used as the basis of utilisation factor calculations and has been developed to cope with the room being broken down into three or four sets of surfaces. The three surface case is used for luminaries mounted close to the ceiling, and the room surfaces used are the ceiling, the walls and the floor. The four surface case is for suspended luminaires and the surfaces used are the ceiling, the frieze (the upper walls), the walls and the floor. The underling notion behind form factors may be extended to be used with as many surfaces as necessary to fully describe any complex shape. However, many of the mathematical simplifications that are possible with the transfer factor methods described here are not always possible and so the application of such methods may be numerically complex. However, because such methods provide a method to cope with any spatial shape, they are often used in lighting calculation software.

The description of transfer method requires the use of a large number of equations, and the symbols used in them are:

g_{ij}	The exchange coefficient between two parallel surfaces i and j
A	The width of a surface or room
B	The length of a surface or room
H	The separation between two surfaces or height of a room
T	The ratio of the width of a surface to the separation between the surfaces
U	The ratio of the length of a surface to the separation between the surfaces
$FF(i,j)$	The form factor between surfaces or elements of a room i and j
C	The ceiling surface of a room
W	The wall surface of a room
F	The working plane surface of a room
S	The frieze surface of a room
L	The luminaire plane in a room
M	The luminous exitance of a surface
RI	The Room Index, the ratio of the floor area to half the wall area of a room
$UF(i)$	The utilisation factor for surface i
$DF(i)$	The distribution factor for surface i
$FT[]$	The flux transfer matrix for a room
$R(i)$	The reflectance of surface i
$TF(i,j)$	The transfer factor between surface i and j

11.3.1 Basis of calculation

For any two surfaces, it is possible to consider the flux being transferred between them as a function of the flux leaving the surface per unit area (luminous exitance), the area of the surface and a geometric multiplier, the form factor, which is a function of the geometric relationship between the two surfaces. The form factor represents the fraction of the flux leaving one surface that arrives at another surface. Figure 11.2 diagrammatically represents the situation.

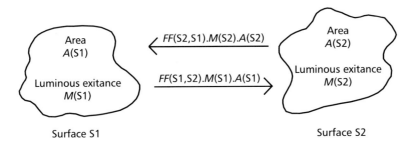

Figure 11.2 Flux transfer

11.3.2 Calculation of form factors

The basis of the calculation of form factors used here is the equation given in CIE 40 (CIE, 1978) for the exchange coefficient, between two parallel surfaces, i and j, of width A, length B and separation H. The exchange coefficients give the fraction of the flux leaving one surface that goes to the other surface.

$$g_{ij} = \frac{2AB}{\pi} \left\{ \begin{array}{l} \dfrac{\left(B^2+H^2\right)^{\frac{1}{2}}}{B} \tan^{-1}\left(\dfrac{A}{\left(B^2+H^2\right)^{\frac{1}{2}}}\right) + \dfrac{\left(A^2+H^2\right)^{\frac{1}{2}}}{A} \tan^{-1}\left(\dfrac{B}{\left(A^2+H^2\right)^{\frac{1}{2}}}\right) \\[2ex] -\dfrac{H}{B}\tan^{-1}\left(\dfrac{A}{H}\right) - \dfrac{H}{A}\tan^{-1}\left(\dfrac{B}{H}\right) + \dfrac{H^2}{2AB}\ln\dfrac{\left(A^2+H^2\right)\left(B^2+H^2\right)}{\left(A^2+B^2+H^2\right)H^2} \end{array} \right\} \quad (11.1)$$

This equation needs to be developed in two ways. First, in the way of working of this method, it is necessary to remove the surface area from the equation so that it gives a form factor rather than an exchange coefficient. Second, to simplify the equation, as area is to be removed from the equation, it is then that the ratios of A to H and B to H can be derived from the value of room index and A, B and H can be replaced by T and U where:

$$T = \frac{A}{H} \qquad \text{and} \qquad U = \frac{B}{H} \qquad (11.2)$$

Thus, the basic equation for the form factor between two parallel planes is given by:

$$FF(i,j) = \frac{2}{\pi} \left\{ \begin{array}{l} \dfrac{\sqrt{T^2+1}}{T} \times \tan^{-1}\left(\dfrac{U}{\sqrt{T^2+1}}\right) + \dfrac{\sqrt{U^2+1}}{U} \times \tan^{-1}\left(\dfrac{T}{\sqrt{U^2+1}}\right) \\[2ex] -\dfrac{\tan^{-1}U}{T} - \dfrac{\tan^{-1}T}{U} + \dfrac{1}{2TU}\ln\dfrac{\left(T^2+1\right)\left(U^2+1\right)}{T^2+U^2+1} \end{array} \right\} \quad (11.3)$$

All of the other form factors may be derived from this equation provided that appropriate values are given for T and U.

There are two sets of calculations of form factors, one for recessed and ceiling mounted luminaires, the three surface case and another for suspended luminaires, the four surface case.

11.3.3 The three surface case

The values of T and U are related to Room Index (RI) as RI is a function of length (L), width (W) and height (H).

$$RI = \left(\frac{LW}{H(L+W)}\right) = \frac{TU}{T+U} \qquad (11.4)$$

For the calculation of standard utilisation factor tables (see section 12.3.3), it is normal to take the ratio of room width to length as 1:1.6. However, in principle, any ratio may be used and a much higher value may be appropriate if designing lighting for corridors, however for the rest of this section, a value of 1:1.6 will be assumed. This is expressed in equation 11.5:

$$U = 1.6T \qquad (11.5)$$

Therefore:

$$RI = \frac{1.6T^2}{2.6T} = \frac{T}{1.625}$$

$$\therefore T = 1.625RI$$

$$\therefore U = 2.6RI \qquad (11.6)$$

Thus with T set to 1.625 RI and U set to 2.6 RI, equation 11.3 may be used to calculate the form factor for the ceiling and the working plane, $FF(C,F)$ and $FF(F,C)$.

The form factors for the floor to walls and ceiling to walls are easy to calculate as the flux that leaves the ceiling and does not get to the floor must go to the walls. Thus:

$$FF(C,W) = 1 - FF(C,F)$$
$$FF(F,W) = 1 - FF(F,C)$$
(11.7)

The form factors for the walls to the ceiling and floor can be derived by using the reversible nature of form factors thus:

$$FF(W,C) = FF(C,W)\frac{Area(C)}{Area(W)}$$
(11.8)

As Room Index is in fact the area of the ceiling divided by half the area of the walls, then the above equation may be rewritten:

$$FF(W,C) = FF(C,W)\frac{RI}{2}$$
(11.9)

Likewise:

$$FF(W,F) = FF(F,W)\frac{RI}{2}$$
(11.10)

The wall to wall transfer factor is similarly the flux that leaves the wall that does not go to the ceiling or the floor. Thus:

$$FF(W,W) = 1 - FF(W,C) - FF(W,F)$$
(11.11)

With these equations, it is then possible to calculate all form factors needed for the evaluation of a standard set of utilisation factors. Table 11.3 gives the calculated form factors.

Table 11.3 Form factors for the three surface cases

RI	$FF(F,W)\ FF(C,W)$	$FF(F,C)\ FF(C,F)$	$FF(W,F)\ FF(W,C)$	$FF(W,W)$
0.60	0.74727	0.25273	0.22418	0.55164
0.80	0.65810	0.34190	0.26324	0.47352
1.00	0.58466	0.41534	0.29233	0.41534
1.25	0.51122	0.48878	0.31951	0.36097
1.50	0.45330	0.54670	0.33998	0.32005
2.00	0.36877	0.63123	0.36877	0.26246
2.50	0.31051	0.68949	0.38814	0.22372
3.00	0.26809	0.73191	0.40213	0.19573
4.00	0.21056	0.78944	0.42113	0.15774
5.00	0.17341	0.82659	0.43352	0.13296

11.3.4 The four surface case

In the case of suspended luminaires, the situation is more complex. To start with, there are three parallel planes to consider, the working plane, the luminaire plane and the ceiling. Again equation 11.3 may be used to calculate the form factors for ceiling and working plane $FF(C,F)$ and $FF(F,C)$; luminaire plane and ceiling $FF(L,C)$ and $FF(C,L)$; and finally, the working plane and the luminaire plane $FF(L,F)$ and $FF(F,L)$. The values of T and U used are given Table 11.4.

Table 11.4 Values of T and U

Form factor calculated	T	U
$FF(C,F)$ and $FF(F,C)$	$RI \times 1.625 \times 3/4$	$RI \times 2.6 \times 3/4$
$FF(F,L)$ and $FF(L,F)$	$RI \times 1.625$	$RI \times 2.6$
$FF(L,C)$ and $FF(C,L)$	$RI \times 1.625 \times 3$	$RI \times 2.6 \times 3$

The other form factors are derived in a similar way to that used in the three surface case. The factor for the ceiling to the frieze is given by

$$FF(C,S) = 1 - FF(C,L) \tag{11.12}$$

The form factor for the frieze to the ceiling may be calculated by using the following equation:

$$FF(S,C) = FF(C,S)\frac{Area(C)}{Area(S)} \tag{11.13}$$

Given that the relationship between the area of the frieze and ceiling is a function of room index, the above equation may be rewritten as:

$$FF(S,C) = FF(C,S)\frac{3RI}{2} \tag{11.14}$$

The form factor for the ceiling to the walls is given by:

$$FF(C,W) = FF(C,L) - FF(C,F) \tag{11.15}$$

The reverse form factor for the walls to the ceiling is given by

$$FF(W,C) = FF(C,W)\frac{Area(C)}{Area(W)} \tag{11.16}$$

The above equation may be rewritten in terms of the room index as follows

$$FF(W,C) = FF(C,W)\frac{RI}{2} \tag{11.17}$$

The frieze to frieze form factor is that fraction of the flux from the frieze that does not go to the ceiling or the luminaire plane. Its calculation is simplified by the fact that the form factors from the frieze to the luminaire plane and the ceiling are the same

$$FF(S,S) = 1 - 2FF(S,C) \tag{11.18}$$

The form factor for the floor to the frieze is the fraction of the flux leaving the floor that reaches the luminaire plane but does not reach the ceiling, so the equation for the factor is given by

$$FF(F,S) = FF(F,L) - FF(F,C) \tag{11.19}$$

The reverse of the above factor, the frieze to floor factor may be evaluated by the following equation

$$FF(S,F) = FF(F,S)\frac{Area(F)}{Area(S)}$$ (11.20)

This may be rewritten in terms of room index

$$FF(S,F) = FF(F,S)\frac{3RI}{2}$$ (11.21)

The floor to wall factor is then given by the fraction of the flux leaving the floor that does not reach the luminaire plane

$$FF(F,W) = 1 - FF(F,L)$$ (11.22)

The reverse of the above factor is calculated by multiplying it by a ratio of the respective areas to get the form factor floor to wall

$$FF(W,F) = FF(F,W)\frac{Area(F)}{Area(W)}$$ (11.23)

This may be rewritten in terms of room index as

$$FF(W,F) = FF(F,W)\frac{RI}{2}$$ (11.24)

The wall to wall form factor is given by the following equation

$$FF(W,W) = 1 - 2FF(W,F)$$ (11.25)

The frieze to wall form factor is calculated on the basis that the fraction of the flux leaving the frieze that ends up on the walls is that flux that does not go to any of the other room surfaces

$$FF(S,W) = 1 - FF(S,C) - FF(S,S) - FF(S,F)$$ (11.26)

The final form factor of the walls to the frieze is the reverse of the above factor and may be calculated

$$FF(W,S) = FF(S,W)\frac{Area(S)}{Area(W)}$$ (11.27)

The area of the frieze is always one-third of the area of the walls so the above equation may be rewritten as

$$FF(W,S) = FF(S,W)\frac{1}{3}$$ (11.28)

With these equations, it is then possible to calculate all form factors needed for the evaluation of a standard set of utilisation factors for suspended luminaires. Table 11.5 gives the calculated form factors.

Table 11.5 Form factors for the four surface case

RI	FF(F,C) FF(C,F)	FF(C,S)	FF(S,C)	FF(C,W)	FF(W,C)	FFS,S)	FF(F,S)
0.60	0.175969	0.398586	0.358727	0.425445	0.127634	0.282545	0.076762
0.80	0.252731	0.320654	0.384785	0.426615	0.170646	0.230430	0.089168
1.00	0.321140	0.268090	0.402135	0.410770	0.205385	0.195730	0.094200
1.25	0.393953	0.222495	0.417178	0.383552	0.239720	0.165644	0.094824
1.50	0.454306	0.190181	0.427908	0.355512	0.266634	0.144185	0.092392
2.00	0.546698	0.147434	0.442302	0.305868	0.305868	0.115396	0.084534
2.50	0.613134	0.120425	0.451594	0.266441	0.333051	0.096812	0.076354
3.00	0.662836	0.101807	0.458133	0.235357	0.353035	0.083735	0.069074
4.00	0.731910	0.077799	0.466791	0.190291	0.380583	0.066417	0.057525
5.00	0.777505	0.062975	0.472312	0.159520	0.398800	0.055377	0.049087

RI	FF(S,F)	FF(F,W)	FF(W,F)	FF(W,W)	FF(S,W)	FF(W,S)
0.60	0.069086	0.747269	0.224181	0.551639	0.289641	0.096547
0.80	0.107001	0.658101	0.263240	0.473519	0.277784	0.092595
1.00	0.141299	0.584660	0.292330	0.415340	0.260836	0.086945
1.25	0.177796	0.511223	0.319514	0.360972	0.239382	0.079794
1.50	0.207881	0.453302	0.339976	0.320047	0.220026	0.073342
2.00	0.253601	0.368768	0.368768	0.262463	0.188701	0.062900
2.50	0.286326	0.310512	0.388140	0.223719	0.165268	0.055089
3.00	0.310834	0.268090	0.402135	0.195730	0.147299	0.049100
4.00	0.345152	0.210565	0.421129	0.157742	0.121639	0.040546
5.00	0.368149	0.173408	0.433521	0.132958	0.104162	0.034721

11.3.5 Derivation of transfer factors

The fraction of the flux intercepted by a room surface, the utilisation factor of that surface, is the sum of the flux directly received and the flux transferred from the other room surfaces. The flux transferred from any room surface to another is the total flux leaving the surface multiplied by the form factor between the two surfaces. The fraction of flux leaving a surface is the utilisation factor of the surface multiplied by the reflection factor of the surface. So for the three surface case, the following equations describe the situation

$$UF(F) = DF(F) + UF(W).R(W).FF(W,F) + UF(C).R(C).FF(C,F)$$
$$UF(W) = UF(F).R(F).FF(F,W) + DF(W) + UF(W).R(W).FF(W,W)$$
$$+ UF(C).R(C).FF(C,W)$$
$$UF(C) = UF(F).R(F).FF(F,C) + UF(W).R(W).FF(W,C) + DF(C)$$

(11.29)

For the four surface case, the equations become

$$UF(F) = DF(F) + UF(W).R(W).FF(W,F) + UF(S).R(S).FF(S,F)$$
$$+ UF(C).R(C).FF(C,F)$$
$$UF(W) = UF(F).R(F).FF(F,W) + DF(W) + UF(W).R(W).FF(W,W)$$
$$+ UF(S).R(S).FF(S,W) + UF(C).R(C).FF(C,W)$$
$$UF(S) = UF(F).R(F).FF(F,S) + UF(W).R(W).FF(W,S) + DF(S)$$
$$+ UF(S).R(S).FF(S,S) + UF(C).R(C).FF(C,S)$$
$$UF(C) = UF(F).R(F).FF(F,C) + UF(W).R(W).FF(W,C)$$
$$+ UF(S).R(S).FF(S,C) + DF(C)$$

$$(11.30)$$

These equations may be arranged into a more logical form with the distribution factors all arranged on the left-hand side.

For the three surface case, the equations become

$$DF(F) = UF(F) - UF(W).R(W).FF(W,F) - UF(C).R(C).FF(C,F)$$
$$DF(W) = -UF(F).R(F).FF(F,W) + UF(W).(1 - R(W).F(W,W))$$
$$- UF(C).R(C).FF(C,W)$$
$$DF(C) = -UF(F).R(F).FF(F,C) - UF(W).R(W).FF(W,C) + UF(C)$$

$$(11.31)$$

For the four surface case, they are

$$DF(F) = UF(F) - UF(W).R(W).FF(W,F) - UF(S).R(S).FF(S,F)$$
$$- UF(C).R(C).FF(C,F)$$
$$DF(W) = -UF(F).R(F).FF(F,W) + UF(W).(1 - R(W).FF(W,W))$$
$$- UF(S).R(S).FF(S,W) - UF(C).R(C).FF(C,W)$$
$$DF(S) = -UF(F).R(F).FF(F,S) - UF(W).R(W).FF(W,S)$$
$$+ UF(S).(1 - R(S).FF(S,S)) - UF(C).R(C).FF(C,S)$$
$$DF(C) = -UF(F).R(F).FF(F,C) - UF(W).R(W).FF(W,C)$$
$$- UF(S).R(S).FF(S,C) + UF(C)$$

$$(11.32)$$

This set of equations enables the distribution factors required to produce a given set of utilisation factors to be found. However, to find the utilisation factors produced by a given set of distribution factors, the equations must be solved simultaneously. The solution of these simultaneous equations will yield the transfer factors. Expressing the above set of equations in matrix form gives:

$$[DF] = [FT][UF]$$

$$(11.33)$$

where $[DF]$ is the array of distribution factors, $[UF]$ is the array of utilisation factors having the same meanings as before and $[FT]$ is the flux transfer matrix. In the three surface case, they are given by equations 11.34 to 11.36 and for the four surface case, they are given by equations 11.37 to 11.39

$$[DF] = \begin{bmatrix} DF(F) \\ DF(W) \\ DF(C) \end{bmatrix}$$

$$(11.34)$$

$$[UF] = \begin{bmatrix} UF(F) \\ UF(W) \\ UF(C) \end{bmatrix} \qquad (11.35)$$

$$[FT] = \begin{bmatrix} 1 & -R(W).FF(W,F) & -R(C).FF(C,F) \\ -R(F).FF(F,W) & 1-R(W).FF(W,W) & -R(C).FF(C,W) \\ -R(F).FF(F,C) & -R(W).FF(W,C) & 1 \end{bmatrix} \qquad (11.36)$$

In the four surface case, the equations are:

$$[DF] = \begin{bmatrix} DF(F) \\ DF(W) \\ DF(S) \\ DF(C) \end{bmatrix} \qquad (11.37)$$

$$[UF] = \begin{bmatrix} UF(F) \\ UF(W) \\ UF(S) \\ UF(C) \end{bmatrix} \qquad (11.38)$$

$$[FT] = \begin{bmatrix} 1 & -R(W).FF(W,F) & -R(S).FF(S,F) & -R(C).FF(C,F) \\ -R(F).FF(F,W) & 1-R(W).FF(W,W) & -R(S).FF(S,W) & R(C).FF(C,W) \\ -R(F).FF(F,S) & -R(W).FF(W,S) & 1-R(S).FF(S,S) & R(C).FF(C,S) \\ -R(F).FF(F,C) & -R(W).FF(W,C) & -R(S).FF(S,C) & 1 \end{bmatrix}$$

$$(11.39)$$

In order to express the utilisation factors in terms of the distribution factors, it is necessary to find a matrix $[TF]$ such that:

$$[TF][FT] = [I] \qquad \text{i.e.} \qquad [TF] = [FT]^{-1} \qquad (11.40)$$

where $[I]$ is the identity matrix

$$[I] = \begin{bmatrix} 1 & 0 & 0 \\ 0 & 1 & 0 \\ 0 & 0 & 1 \end{bmatrix} \quad \text{or} \quad \begin{bmatrix} 1 & 0 & 0 & 0 \\ 0 & 1 & 0 & 0 \\ 0 & 0 & 1 & 0 \\ 0 & 0 & 0 & 1 \end{bmatrix} \qquad (11.41)$$

Thus

$$[TF].[DF] = [UF] \qquad (11.42)$$

In the three surface case, the matrix TF is

$$[TF] = \begin{bmatrix} TF(F,F) & TF(W,F) & TF(C,F) \\ TF(F,W) & TF(W,W) & TF(C,W) \\ TF(F,C) & TF(W,C) & TF(C,C) \end{bmatrix} \qquad (11.43)$$

In the four surface case, it is

$$[TF] = \begin{bmatrix} TF(F,F) & TF(W,F) & TF(S,F) & TF(C,F) \\ TF(F,W) & TF(W,W) & TF(S,W) & TF(C,W) \\ TF(F,S) & TF(W,S) & TF(S,S) & TF(C,S) \\ TF(F,C) & TF(W,C) & TF(S,C) & TF(C,C) \end{bmatrix} \qquad (11.44)$$

The matrix inversion may be achieved using a wide variety of computer programs such as Microsoft® Excel or MathWorks® Matlab.

The above methods may be used to calculate the transfer factors to any of the room surfaces. Tables 11.6 to 11.8 give the transfer factors for the three surface case and Tables 11.9 to 11.12 give the values for the four surface case.

Table 11.6 Transfer factors to the working plane for the three surface case

	Reflectances								
Ceiling	0.8	0.8	0.8	0.7	0.7	0.7	0.5	0.5	0.5
Walls	0.7	0.5	0.3	0.7	0.5	0.3	0.7	0.5	0.3
Floor	0.2	0.2	0.2	0.2	0.2	0.2	0.2	0.2	0.2
k	Transfer factors to working plane								
0.60	TF(C,F)	0.4532	0.3408	0.2708	0.3856	0.2934	0.2349	0.2610	0.2031	0.1650
	TF(W,F)	0.3924	0.2152	0.1048	0.3735	0.2073	0.1017	0.3387	0.1923	0.0957
	TF(F,F)	1.0815	1.0494	1.0293	1.0753	1.0458	1.0271	1.0638	1.0390	1.0226
0.80	TF(C,F)	0.5367	0.4259	0.3522	0.4562	0.3661	0.3051	0.3082	0.2527	0.2137
	TF(W,F)	0.4500	0.2567	0.1282	0.4254	0.2455	0.1236	0.3802	0.2241	0.1144
	TF(F,F)	1.0959	1.0629	1.0410	1.0872	1.0573	1.0371	1.0711	1.0468	1.0297
1.00	TF(C,F)	0.5980	0.4923	0.4184	0.5082	0.4228	0.3620	0.3433	0.2912	0.2529
	TF(W,F)	0.4920	0.2891	0.1473	0.4629	0.2749	0.1411	0.4096	0.2480	0.1291
	TF(F,F)	1.1072	1.0747	1.0520	1.0963	1.0673	1.0466	1.0764	1.0532	1.0361
1.25	TF(C,F)	0.6542	0.5561	0.4843	0.5561	0.4772	0.4185	0.3757	0.3281	0.2916
	TF(W,F)	0.5304	0.3203	0.1665	0.4972	0.3031	0.1586	0.4361	0.2706	0.1435
	TF(F,F)	1.1182	1.0871	1.0644	1.1052	1.0776	1.0571	1.0813	1.0597	1.0432
1.50	TF(C,F)	0.6954	0.6049	0.5361	0.5913	0.5187	0.4628	0.3998	0.3563	0.3219
	TF(W,F)	0.5588	0.3445	0.1818	0.5224	0.3248	0.1725	0.4554	0.2876	0.1547
	TF(F,F)	1.1267	1.0974	1.0751	1.1120	1.0862	1.0662	1.0850	1.0650	1.0492
2.00	TF(C,F)	0.7515	0.6740	0.6120	0.6395	0.5775	0.5274	0.4329	0.3962	0.3657
	TF(W,F)	0.5979	0.3793	0.2047	0.5570	0.3559	0.1931	0.4816	0.3118	0.1711
	TF(F,F)	1.1390	1.1131	1.0924	1.1218	1.0992	1.0808	1.0902	1.0730	1.0588

Table 11.6 Continued

k		Transfer factors to working plane								
2.50	TF(C,F)	0.7877	0.7203	0.6646	0.6706	0.6170	0.5721	0.4545	0.4228	0.3958
	TF(W,F)	0.6234	0.4031	0.2209	0.5796	0.3770	0.2077	0.4987	0.3281	0.1824
	TF(F,F)	1.1473	1.1244	1.1054	1.1285	1.1085	1.0918	1.0936	1.0787	1.0659
3.00	TF(C,F)	0.8129	0.7535	0.7031	0.6924	0.6451	0.6047	0.4696	0.4418	0.4177
	TF(W,F)	0.6414	0.4204	0.2331	0.5955	0.3924	0.2185	0.5107	0.3398	0.1908
	TF(F,F)	1.1534	1.1328	1.1154	1.1333	1.1155	1.1002	1.0961	1.0829	1.0714
4.00	TF(C,F)	0.8456	0.7977	0.7557	0.7207	0.6827	0.6491	0.4894	0.4672	0.4473
	TF(W,F)	0.6651	0.4440	0.2500	0.6165	0.4133	0.2336	0.5265	0.3557	0.2024
	TF(F,F)	1.1615	1.1446	1.1298	1.1398	1.1252	1.1123	1.0994	1.0887	1.0791
5.00	TF(C,F)	0.8659	0.8258	0.7899	0.7383	0.7066	0.6779	0.5018	0.4833	0.4664
	TF(W,F)	0.6801	0.4593	0.2614	0.6298	0.4268	0.2436	0.5365	0.3659	0.2101
	TF(F,F)	1.1667	1.1525	1.1396	1.1439	1.1316	1.1205	1.1016	1.0926	1.0844

Table 11.7 Transfer factors to the walls for the three surface case

	Reflectances								
Ceiling	0.8	0.8	0.8	0.7	0.7	0.7	0.5	0.5	0.5
Walls	0.7	0.5	0.3	0.7	0.5	0.3	0.7	0.5	0.3
Floor	0.2	0.2	0.2	0.2	0.2	0.2	0.2	0.2	0.2

k		Transfer factors to walls								
0.60	TF(C,W)	1.3061	1.0029	0.8139	1.1112	0.8635	0.7061	0.7521	0.5977	0.4959
	TF(W,W)	2.0585	1.5805	1.2827	2.0041	1.5573	1.2734	1.9038	1.5131	1.2554
	TF(F,W)	0.3737	0.2869	0.2328	0.3557	0.2764	0.2260	0.3225	0.2563	0.2127
0.80	TF(C,W)	1.0786	0.8616	0.7172	0.9168	0.7407	0.6213	0.6194	0.5112	0.4351
	TF(W,W)	1.8817	1.5031	1.2513	1.8322	1.4803	1.2417	1.7414	1.4370	1.2232
	TF(F,W)	0.3214	0.2567	0.2137	0.3039	0.2455	0.2059	0.2716	0.2241	0.1907
1.00	TF(C,W)	0.9142	0.7520	0.6387	0.7769	0.6458	0.5526	0.5248	0.4448	0.3860
	TF(W,W)	1.7548	1.4435	1.2260	1.7104	1.4218	1.2165	1.6288	1.3808	1.1983
	TF(F,W)	0.2811	0.2313	0.1964	0.2645	0.2199	0.1882	0.2341	0.1984	0.1722
1.25	TF(C,W)	0.7654	0.6472	0.5605	0.6506	0.5553	0.4843	0.4396	0.3819	0.3375
	TF(W,W)	1.6398	1.3864	1.2008	1.6009	1.3663	1.1917	1.5294	1.3284	1.1742
	TF(F,W)	0.2425	0.2050	0.1776	0.2273	0.1940	0.1692	0.1993	0.1732	0.1530
1.50	TF(C,W)	0.6572	0.5672	0.4989	0.5588	0.4864	0.4306	0.3778	0.3341	0.2995
	TF(W,W)	1.5556	1.3425	1.1807	1.5211	1.3240	1.1721	1.4578	1.2892	1.1555
	TF(F,W)	0.2129	0.1837	0.1616	0.1990	0.1732	0.1533	0.1735	0.1534	0.1375
2.00	TF(C,W)	0.5113	0.4541	0.4084	0.4351	0.3891	0.3520	0.2945	0.2669	0.2441
	TF(W,W)	1.4408	1.2796	1.1509	1.4130	1.2639	1.1432	1.3617	1.2342	1.1285
	TF(F,W)	0.1708	0.1517	0.1364	0.1591	0.1423	0.1288	0.1376	0.1247	0.1140
2.50	TF(C,W)	0.4180	0.3784	0.3456	0.3558	0.3241	0.2975	0.2411	0.2221	0.2058
	TF(W,W)	1.3662	1.2368	1.1298	1.3430	1.2231	1.1229	1.3001	1.1974	1.1098
	TF(F,W)	0.1425	0.1290	0.1178	0.1325	0.1207	0.1108	0.1140	0.1050	0.0973

Table 11.7 Continued

k		Transfer factors to walls								
3.00	TF(C,W)	0.3533	0.3242	0.2996	0.3009	0.2776	0.2577	0.2041	0.1901	0.1780
	TF(W,W)	1.3139	1.2057	1.1141	1.2939	1.1937	1.1079	1.2571	1.1711	1.0961
	TF(F,W)	0.1222	0.1121	0.1036	0.1134	0.1046	0.0971	0.0973	0.0906	0.0848
4.00	TF(C,W)	0.2697	0.2521	0.2366	0.2299	0.2157	0.2032	0.1561	0.1476	0.1400
	TF(W,W)	1.2450	1.1635	1.0921	1.2295	1.1538	1.0870	1.2008	1.1356	1.0772
	TF(F,W)	0.0950	0.0888	0.0833	0.0881	0.0827	0.0779	0.0752	0.0711	0.0675
5.00	TF(C,W)	0.2181	0.2062	0.1956	0.1860	0.1764	0.1678	0.1264	0.1207	0.1155
	TF(W,W)	1.2016	1.1362	1.0775	1.1889	1.1280	1.0731	1.1654	1.1128	1.0648
	TF(F,W)	0.0777	0.0735	0.0697	0.0720	0.0683	0.0650	0.0613	0.0585	0.0560

Table 11.8 Transfer factors to the ceiling for the three surface case

	Reflectances								
Ceiling	0.8	0.8	0.8	0.7	0.7	0.7	0.5	0.5	0.5
Walls	0.7	0.5	0.3	0.7	0.5	0.3	0.7	0.5	0.3
Floor	0.2	0.2	0.2	0.2	0.2	0.2	0.2	0.2	0.2

k		Transfer factors to ceiling								
0.60	TF(C,C)	1.2279	1.1296	1.0684	1.1939	1.1116	1.0594	1.1312	1.0773	1.0417
	TF(W,C)	0.3429	0.1880	0.0916	0.3334	0.1850	0.0908	0.3159	0.1793	0.0893
	TF(F,C)	0.1133	0.0852	0.0677	0.1102	0.0838	0.0671	0.1044	0.0813	0.0660
0.80	TF(C,C)	1.2354	1.1425	1.0807	1.2001	1.1225	1.0699	1.1352	1.0846	1.0490
	TF(W,C)	0.3775	0.2154	0.1076	0.3667	0.2116	0.1065	0.3469	0.2045	0.1044
	TF(F,C)	0.1342	0.1065	0.0881	0.1303	0.1046	0.0872	0.1233	0.1011	0.0855
1.00	TF(C,C)	1.2367	1.1508	1.0908	1.2012	1.1295	1.0785	1.1359	1.0892	1.0549
	TF(W,C)	0.3999	0.2350	0.1198	0.3885	0.2307	0.1184	0.3673	0.2224	0.1158
	TF(F,C)	0.1495	0.1231	0.1046	0.1452	0.1208	0.1034	0.1373	0.1165	0.1012
1.25	TF(C,C)	1.2351	1.1577	1.1011	1.1999	1.1354	1.0873	1.1351	1.0931	1.0609
	TF(W,C)	0.4186	0.2528	0.1314	0.4067	0.2479	0.1297	0.3847	0.2387	0.1266
	TF(F,C)	0.1635	0.1390	0.1211	0.1589	0.1363	0.1196	0.1503	0.1313	0.1166
1.50	TF(C,C)	1.2324	1.1626	1.1095	1.1976	1.1394	1.0945	1.1336	1.0958	1.0657
	TF(W,C)	0.4313	0.2659	0.1403	0.4191	0.2606	0.1384	0.3967	0.2506	0.1348
	TF(F,C)	0.1739	0.1512	0.1340	0.1689	0.1482	0.1322	0.1599	0.1425	0.1287
2.00	TF(C,C)	1.2269	1.1688	1.1224	1.1930	1.1447	1.1055	1.1307	1.0992	1.0732
	TF(W,C)	0.4474	0.2838	0.1532	0.4351	0.2780	0.1509	0.4123	0.2669	0.1464
	TF(F,C)	0.1879	0.1685	0.1530	0.1827	0.1650	0.1507	0.1732	0.1585	0.1463
2.50	TF(C,C)	1.2222	1.1728	1.1319	1.1892	1.1480	1.1135	1.1282	1.1014	1.0785
	TF(W,C)	0.4572	0.2956	0.1620	0.4448	0.2894	0.1594	0.4220	0.2776	0.1544
	TF(F,C)	0.1969	0.1801	0.1661	0.1916	0.1763	0.1635	0.1818	0.1691	0.1583
3.00	TF(C,C)	1.2184	1.1755	1.1391	1.1861	1.1503	1.1196	1.1262	1.1029	1.0826
	TF(W,C)	0.4637	0.3040	0.1685	0.4514	0.2975	0.1656	0.4286	0.2852	0.1602
	TF(F,C)	0.2032	0.1884	0.1758	0.1978	0.1843	0.1728	0.1878	0.1767	0.1671

Table 11.8 Continued

k		Transfer factors to ceiling								
4.00	TF(C,C)	1.2130	1.1790	1.1492	1.1816	1.1532	1.1282	1.1233	1.1049	1.0883
	TF(W,C)	0.4720	0.3151	0.1775	0.4598	0.3082	0.1742	0.4371	0.2953	0.1681
	TF(F,C)	0.2114	0.1994	0.1889	0.2059	0.1951	0.1855	0.1958	0.1869	0.1789
5.00	TF(C,C)	1.2093	1.1812	1.1560	1.1785	1.1551	1.1339	1.1213	1.1061	1.0921
	TF(W,C)	0.4771	0.3222	0.1833	0.4649	0.3151	0.1798	0.4424	0.3017	0.1732
	TF(F,C)	0.2165	0.2065	0.1975	0.2110	0.2019	0.1937	0.2007	0.1933	0.1866

Table 11.9 Transfer factors to the working plane for the four surface case

		Reflectances								
Ceiling		0.8	0.8	0.8	0.7	0.7	0.7	0.5	0.5	0.5
Walls		0.7	0.5	0.3	0.7	0.5	0.3	0.7	0.5	0.3
Floor		0.2	0.2	0.2	0.2	0.2	0.2	0.2	0.2	0.2
k		Transfer factors to working plane								
0.60	TF(C,F)	0.3558	0.2519	0.1929	0.3021	0.2168	0.1674	0.2036	0.1498	0.1176
	TF(S,F)	0.2654	0.1267	0.0547	0.2456	0.1186	0.0515	0.2093	0.1033	0.0453
	TF(W,F)	0.3544	0.1915	0.0931	0.3435	0.1876	0.0917	0.3233	0.1801	0.0890
	TF(F,F)	1.0696	1.0394	1.0215	1.0657	1.0375	1.0204	1.0587	1.0338	1.0181
0.80	TF(C,F)	0.4447	0.3346	0.2671	0.3774	0.2877	0.2315	0.2543	0.1985	0.1623
	TF(S,F)	0.3346	0.1722	0.0788	0.3087	0.1608	0.0741	0.2613	0.1390	0.0649
	TF(W,F)	0.4103	0.2289	0.1133	0.3941	0.2224	0.1108	0.3644	0.2100	0.1060
	TF(F,F)	1.0824	1.0501	1.0298	1.0764	1.0467	1.0276	1.0655	1.0402	1.0233
1.00	TF(C,F)	0.5129	0.4027	0.3310	0.4353	0.3459	0.2867	0.2933	0.2384	0.2007
	TF(S,F)	0.3884	0.2103	0.1000	0.3578	0.1959	0.0939	0.3018	0.1687	0.0820
	TF(W,F)	0.4527	0.2593	0.1303	0.4320	0.2502	0.1267	0.3942	0.2331	0.1196
	TF(F,F)	1.0932	1.0601	1.0384	1.0852	1.0552	1.0350	1.0706	1.0457	1.0284
1.25	TF(C,F)	0.5776	0.4709	0.3978	0.4904	0.4043	0.3442	0.3306	0.2782	0.2404
	TF(S,F)	0.4397	0.2488	0.1224	0.4047	0.2314	0.1147	0.3405	0.1986	0.0998
	TF(W,F)	0.4930	0.2899	0.1481	0.4678	0.2779	0.1431	0.4216	0.2554	0.1333
	TF(F,F)	1.1043	1.0715	1.0488	1.0941	1.0647	1.0439	1.0756	1.0518	1.0345
1.50	TF(C,F)	0.6264	0.5249	0.4525	0.5320	0.4504	0.3911	0.3590	0.3097	0.2727
	TF(S,F)	0.4786	0.2794	0.1409	0.4403	0.2596	0.1319	0.3700	0.2221	0.1143
	TF(W,F)	0.5237	0.3144	0.1629	0.4949	0.2999	0.1565	0.4419	0.2728	0.1443
	TF(F,F)	1.1132	1.0814	1.0585	1.1013	1.0729	1.0522	1.0795	1.0570	1.0400
2.00	TF(C,F)	0.6945	0.6042	0.5357	0.5903	0.5180	0.4623	0.3989	0.3557	0.3215
	TF(S,F)	0.5330	0.3246	0.1693	0.4902	0.3010	0.1580	0.4115	0.2566	0.1363
	TF(W,F)	0.5672	0.3510	0.1859	0.5330	0.3326	0.1773	0.4700	0.2980	0.1608
	TF(F,F)	1.1268	1.0974	1.0751	1.1121	1.0863	1.0663	1.0852	1.0652	1.0493
2.50	TF(C,F)	0.7394	0.6590	0.5955	0.6289	0.5647	0.5133	0.4254	0.3874	0.3561
	TF(S,F)	0.5691	0.3561	0.1898	0.5233	0.3298	0.1768	0.4391	0.2805	0.1520
	TF(W,F)	0.5965	0.3771	0.2030	0.5585	0.3557	0.1926	0.4886	0.3154	0.1727
	TF(F,F)	1.1364	1.1097	1.0885	1.1198	1.0964	1.0776	1.0892	1.0714	1.0567

Table 11.9 Continued

k					Transfer factors to working plane					
3.00	TF(C,F)	0.7711	0.6990	0.6403	0.6562	0.5988	0.5515	0.4444	0.4104	0.3819
	TF(S,F)	0.5946	0.3791	0.2052	0.5469	0.3509	0.1910	0.4588	0.2979	0.1637
	TF(W,F)	0.6175	0.3965	0.2162	0.5768	0.3728	0.2043	0.5018	0.3283	0.1817
	TF(F,F)	1.1435	1.1192	1.0993	1.1255	1.1042	1.0867	1.0921	1.0761	1.0626
4.00	TF(C,F)	0.8127	0.7534	0.7030	0.6922	0.6450	0.6046	0.4694	0.4417	0.4176
	TF(S,F)	0.6284	0.4107	0.2270	0.5781	0.3798	0.2109	0.4850	0.3217	0.1801
	TF(W,F)	0.6457	0.4236	0.2351	0.6013	0.3966	0.2211	0.5193	0.3459	0.1944
	TF(F,F)	1.1534	1.1328	1.1154	1.1333	1.1155	1.1002	1.0962	1.0829	1.0714
5.00	TF(C,F)	0.8389	0.7885	0.7447	0.7148	0.6749	0.6398	0.4853	0.4619	0.4411
	TF(S,F)	0.6498	0.4314	0.2417	0.5979	0.3986	0.2242	0.5017	0.3372	0.1910
	TF(W,F)	0.6637	0.4417	0.2481	0.6170	0.4124	0.2325	0.5305	0.3575	0.2030
	TF(F,F)	1.1598	1.1422	1.1268	1.1384	1.1232	1.1098	1.0988	1.0875	1.0775

Table 11.10 Transfer factors to the walls for the four surface case

		Reflectances								
Ceiling		0.8	0.8	0.8	0.7	0.7	0.7	0.5	0.5	0.5
Walls		0.7	0.5	0.3	0.7	0.5	0.3	0.7	0.5	0.3
Floor		0.2	0.2	0.2	0.2	0.2	0.2	0.2	0.2	0.2
k		Transfer factors to walls								
0.60	TF(C,W)	0.9702	0.6785	0.5111	0.8236	0.5838	0.4434	0.5552	0.4036	0.3115
	TF(S,W)	0.8160	0.4062	0.1830	0.7619	0.3844	0.1746	0.6628	0.3429	0.1583
	TF(W,W)	1.9464	1.5073	1.2448	1.9164	1.4967	1.2411	1.8616	1.4765	1.2340
	TF(F,W)	0.3376	0.2554	0.2068	0.3271	0.2501	0.2038	0.3079	0.2401	0.1978
0.80	TF(C,W)	0.8578	0.6362	0.4996	0.7280	0.5469	0.4330	0.4905	0.3774	0.3036
	TF(S,W)	0.7189	0.3788	0.1775	0.6689	0.3571	0.1687	0.5775	0.3158	0.1516
	TF(W,W)	1.7995	1.4438	1.2185	1.7683	1.4314	1.2139	1.7112	1.4078	1.2049
	TF(F,W)	0.2930	0.2289	0.1888	0.2815	0.2224	0.1847	0.2603	0.2100	0.1766
1.00	TF(C,W)	0.7591	0.5867	0.4745	0.6442	0.5040	0.4109	0.4341	0.3474	0.2876
	TF(S,W)	0.6354	0.3488	0.1682	0.5902	0.3278	0.1594	0.5073	0.2882	0.1424
	TF(W,W)	1.6929	1.3955	1.1982	1.6624	1.3823	1.1929	1.6064	1.3573	1.1828
	TF(F,W)	0.2587	0.2074	0.1737	0.2469	0.2002	0.1689	0.2253	0.1865	0.1595
1.25	TF(C,W)	0.6585	0.5283	0.4390	0.5591	0.4535	0.3798	0.3769	0.3122	0.2653
	TF(S,W)	0.5515	0.3142	0.1557	0.5116	0.2947	0.1472	0.4385	0.2578	0.1307
	TF(W,W)	1.5947	1.3490	1.1780	1.5659	1.3356	1.1724	1.5132	1.3103	1.1616
	TF(F,W)	0.2254	0.1855	0.1580	0.2139	0.1779	0.1526	0.1927	0.1635	0.1422
1.50	TF(C,W)	0.5791	0.4776	0.4050	0.4919	0.4098	0.3501	0.3319	0.2817	0.2441
	TF(S,W)	0.4859	0.2845	0.1439	0.4504	0.2665	0.1358	0.3855	0.2324	0.1201
	TF(W,W)	1.5213	1.3127	1.1619	1.4946	1.2996	1.1562	1.4457	1.2748	1.1452
	TF(F,W)	0.1995	0.1677	0.1448	0.1885	0.1600	0.1391	0.1683	0.1455	0.1282

Table 11.10 Continued

k					Transfer factors to walls				
2.00	TF(C,W)	0.4644	0.3977	0.3472	0.3947	0.3410	0.2997	0.2667	0.2341	0.2084
	TF(S,W)	0.3916	0.2383	0.1242	0.3630	0.2228	0.1169	0.3104	0.1936	0.1028
	TF(W,W)	1.4193	1.2595	1.1375	1.3963	1.2474	1.1319	1.3542	1.2246	1.1212
	TF(F,W)	0.1621	0.1404	0.1240	0.1523	0.1331	0.1182	0.1343	0.1192	0.1072
2.50	TF(C,W)	0.3866	0.3395	0.3023	0.3288	0.2909	0.2606	0.2224	0.1995	0.1808
	TF(S,W)	0.3281	0.2048	0.1089	0.3042	0.1913	0.1024	0.2601	0.1659	0.0898
	TF(W,W)	1.3515	1.2223	1.1198	1.3316	1.2113	1.1145	1.2950	1.1906	1.1044
	TF(F,W)	0.1363	0.1207	0.1083	0.1277	0.1138	0.1027	0.1117	0.1009	0.0921
3.00	TF(C,W)	0.3307	0.2957	0.2671	0.2815	0.2533	0.2301	0.1906	0.1736	0.1593
	TF(S,W)	0.2826	0.1797	0.0970	0.2621	0.1677	0.0910	0.2243	0.1453	0.0797
	TF(W,W)	1.3031	1.1948	1.1063	1.2856	1.1848	1.1013	1.2535	1.1659	1.0919
	TF(F,W)	0.1176	0.1057	0.0961	0.1099	0.0994	0.0908	0.0956	0.0875	0.0808
4.00	TF(C,W)	0.2564	0.2347	0.2163	0.2183	0.2009	0.1860	0.1481	0.1376	0.1285
	TF(S,W)	0.2218	0.1445	0.0797	0.2059	0.1349	0.0747	0.1766	0.1168	0.0652
	TF(W,W)	1.2385	1.1567	1.0870	1.2245	1.1482	1.0827	1.1987	1.1324	1.0745
	TF(F,W)	0.0922	0.0847	0.0784	0.0859	0.0793	0.0737	0.0742	0.0692	0.0648
5.00	TF(C,W)	0.2092	0.1944	0.1816	0.1783	0.1664	0.1560	0.1210	0.1139	0.1076
	TF(S,W)	0.1831	0.1212	0.0677	0.1702	0.1131	0.0634	0.1462	0.0980	0.0554
	TF(W,W)	1.1973	1.1314	1.0739	1.1856	1.1242	1.0701	1.1641	1.1106	1.0629
	TF(F,W)	0.0759	0.0707	0.0661	0.0705	0.0660	0.0620	0.0606	0.0572	0.0541

Table 11.11 Transfer factors to the frieze plane for the four surface case

				Reflectances					
Ceiling		0.8	0.8	0.8	0.7	0.7	0.7	0.5	0.5	0.5
Walls		0.7	0.5	0.3	0.7	0.5	0.3	0.7	0.5	0.3
Floor		0.2	0.2	0.2	0.2	0.2	0.2	0.2	0.2	0.2
k					Transfer factors to frieze plane				
0.60	TF(C,S)	0.5837	0.4643	0.3917	0.4955	0.3995	0.3398	0.3340	0.2761	0.2387
	TF(S,S)	1.5031	1.2866	1.1454	1.4705	1.2717	1.1390	1.4109	1.2433	1.1264
	TF(W,S)	0.2720	0.1354	0.0610	0.2540	0.1281	0.0582	0.2209	0.1143	0.0528
	TF(F,S)	0.0843	0.0563	0.0405	0.0780	0.0527	0.0381	0.0664	0.0459	0.0336
0.80	TF(C,S)	0.4575	0.3713	0.3164	0.3883	0.3192	0.2743	0.2616	0.2203	0.1923
	TF(S,S)	1.4019	1.2343	1.1203	1.3753	1.2216	1.1147	1.3265	1.1974	1.1039
	TF(W,S)	0.2396	0.1263	0.0592	0.2230	0.1190	0.0562	0.1925	0.1053	0.0505
	TF(F,S)	0.0797	0.0574	0.0438	0.0735	0.0536	0.0411	0.0622	0.0463	0.0361
1.00	TF(C,S)	0.3748	0.3097	0.2665	0.3181	0.2660	0.2308	0.2143	0.1834	0.1615
	TF(S,S)	1.3343	1.1987	1.1032	1.3120	1.1877	1.0983	1.2710	1.1668	1.0887
	TF(W,S)	0.2118	0.1163	0.0561	0.1967	0.1093	0.0531	0.1691	0.0961	0.0475
	TF(F,S)	0.0740	0.0561	0.0445	0.0681	0.0522	0.0417	0.0575	0.0450	0.0364

Table 11.11 Continued

k										
		Transfer factors to frieze plane								
1.25	TF(C,S)	0.3047	0.2567	0.2233	0.2587	0.2203	0.1932	0.1744	0.1517	0.1350
	TF(S,S)	1.2761	1.1675	1.0881	1.2576	1.1580	1.0837	1.2238	1.1401	1.0754
	TF(W,S)	0.1838	0.1047	0.0519	0.1705	0.0982	0.0491	0.1462	0.0859	0.0436
	TF(F,S)	0.0670	0.0531	0.0435	0.0617	0.0494	0.0408	0.0519	0.0424	0.0355
1.50	TF(C,S)	0.2561	0.2192	0.1927	0.2175	0.1881	0.1666	0.1467	0.1293	0.1162
	TF(S,S)	1.2352	1.1451	1.0771	1.2195	1.1368	1.0732	1.1908	1.1211	1.0658
	TF(W,S)	0.1620	0.0948	0.0480	0.1501	0.0888	0.0453	0.1285	0.0775	0.0400
	TF(F,S)	0.0608	0.0497	0.0418	0.0559	0.0462	0.0391	0.0470	0.0395	0.0339
2.00	TF(C,S)	0.1935	0.1699	0.1519	0.1645	0.1456	0.1311	0.1111	0.1000	0.0911
	TF(S,S)	1.1816	1.1149	1.0621	1.1697	1.1082	1.0589	1.1477	1.0958	1.0528
	TF(W,S)	0.1305	0.0794	0.0414	0.1210	0.0743	0.0390	0.1035	0.0645	0.0343
	TF(F,S)	0.0508	0.0433	0.0376	0.0467	0.0401	0.0351	0.0392	0.0342	0.0303
2.50	TF(C,S)	0.1552	0.1387	0.1257	0.1320	0.1189	0.1083	0.0893	0.0815	0.0752
	TF(S,S)	1.1482	1.0954	1.0523	1.1386	1.0899	1.0496	1.1209	1.0795	1.0443
	TF(W,S)	0.1094	0.0683	0.0363	0.1014	0.0638	0.0341	0.0867	0.0553	0.0299
	TF(F,S)	0.0434	0.0380	0.0337	0.0399	0.0352	0.0314	0.0335	0.0299	0.0270
3.00	TF(C,S)	0.1294	0.1173	0.1073	0.1101	0.1004	0.0924	0.0746	0.0689	0.0640
	TF(S,S)	1.1254	1.0818	1.0453	1.1174	1.0771	1.0429	1.1026	1.0682	1.0383
	TF(W,S)	0.0942	0.0599	0.0323	0.0874	0.0559	0.0303	0.0748	0.0484	0.0266
	TF(F,S)	0.0378	0.0337	0.0304	0.0347	0.0312	0.0283	0.0291	0.0265	0.0243
4.00	TF(C,S)	0.0970	0.0896	0.0833	0.0826	0.0767	0.0716	0.0560	0.0525	0.0495
	TF(S,S)	1.0962	1.0639	1.0359	1.0902	1.0602	1.0340	1.0791	1.0533	1.0303
	TF(W,S)	0.0739	0.0482	0.0266	0.0686	0.0450	0.0249	0.0589	0.0389	0.0217
	TF(F,S)	0.0299	0.0274	0.0252	0.0275	0.0253	0.0234	0.0231	0.0214	0.0200
5.00	TF(C,S)	0.0775	0.0725	0.0681	0.0661	0.0621	0.0585	0.0448	0.0425	0.0404
	TF(S,S)	1.0783	1.0526	1.0298	1.0735	1.0496	1.0282	1.0646	1.0439	1.0252
	TF(W,S)	0.0610	0.0404	0.0226	0.0567	0.0377	0.0211	0.0487	0.0327	0.0185
	TF(F,S)	0.0248	0.0230	0.0215	0.0228	0.0213	0.0199	0.0191	0.0180	0.0170

Table 11.12 Transfer factors to the ceiling for the four surface case

		Reflectances								
Ceiling		0.8	0.8	0.8	0.7	0.7	0.7	0.5	0.5	0.5
Walls		0.7	0.5	0.3	0.7	0.5	0.3	0.7	0.5	0.3
Floor		0.2	0.2	0.2	0.2	0.2	0.2	0.2	0.2	0.2
k		Transfer factors to ceiling								
0.60	TF(C,C)	1.2458	1.1354	1.0685	1.2086	1.1165	1.0594	1.1406	1.0806	1.0417
	TF(S,C)	0.4597	0.2611	0.1322	0.4460	0.2568	0.1311	0.4209	0.2485	0.1289
	TF(W,C)	0.2547	0.1272	0.0575	0.2471	0.1251	0.0570	0.2332	0.1211	0.0561
	TF(F,C)	0.0890	0.0630	0.0482	0.0863	0.0619	0.0478	0.0815	0.0599	0.0470

Table 11.12 Continued

k										
						Transfer factors to ceiling				
0.80	TF(C,C)	1.2482	1.1426	1.0756	1.2106	1.1226	1.0655	1.1419	1.0846	1.0459
	TF(S,C)	0.4804	0.2785	0.1424	0.4659	0.2736	0.1411	0.4395	0.2643	0.1385
	TF(W,C)	0.3002	0.1591	0.0749	0.2912	0.1563	0.0742	0.2747	0.1510	0.0729
	TF(F,C)	0.1112	0.0837	0.0668	0.1078	0.0822	0.0661	0.1017	0.0794	0.0649
1.00	TF(C,C)	1.2476	1.1484	1.0826	1.2101	1.1275	1.0716	1.1416	1.0878	1.0501
	TF(S,C)	0.4919	0.2903	0.1499	0.4771	0.2851	0.1483	0.4501	0.2750	0.1454
	TF(W,C)	0.3321	0.1833	0.0890	0.3221	0.1800	0.0880	0.3039	0.1737	0.0863
	TF(F,C)	0.1282	0.1007	0.0828	0.1244	0.0988	0.0819	0.1173	0.0954	0.0803
1.25	TF(C,C)	1.2450	1.1540	1.0909	1.2080	1.1322	1.0786	1.1402	1.0910	1.0549
	TF(S,C)	0.4998	0.3008	0.1570	0.4850	0.2951	0.1553	0.4578	0.2844	0.1518
	TF(W,C)	0.3601	0.2064	0.1029	0.3494	0.2025	0.1017	0.3298	0.1951	0.0995
	TF(F,C)	0.1444	0.1177	0.0994	0.1401	0.1155	0.0983	0.1322	0.1113	0.0962
1.50	TF(C,C)	1.2417	1.1583	1.0982	1.2053	1.1358	1.0849	1.1385	1.0934	1.0592
	TF(S,C)	0.5041	0.3083	0.1626	0.4894	0.3023	0.1606	0.4622	0.2910	0.1568
	TF(W,C)	0.3800	0.2239	0.1139	0.3689	0.2195	0.1125	0.3485	0.2113	0.1099
	TF(F,C)	0.1566	0.1312	0.1131	0.1520	0.1287	0.1117	0.1436	0.1239	0.1091
2.00	TF(C,C)	1.2353	1.1644	1.1106	1.2000	1.1410	1.0954	1.1351	1.0968	1.0664
	TF(S,C)	0.5080	0.3185	0.1708	0.4935	0.3121	0.1685	0.4668	0.3000	0.1640
	TF(W,C)	0.4063	0.2486	0.1302	0.3947	0.2436	0.1284	0.3734	0.2341	0.1250
	TF(F,C)	0.1736	0.1510	0.1339	0.1687	0.1480	0.1321	0.1595	0.1423	0.1286
2.50	TF(C,C)	1.2298	1.1687	1.1202	1.1955	1.1445	1.1037	1.1323	1.0991	1.0719
	TF(S,C)	0.5092	0.3251	0.1767	0.4950	0.3184	0.1741	0.4688	0.3058	0.1691
	TF(W,C)	0.4228	0.2652	0.1417	0.4110	0.2597	0.1396	0.3892	0.2494	0.1356
	TF(F,C)	0.1848	0.1647	0.1489	0.1797	0.1613	0.1467	0.1702	0.1549	0.1424
3.00	TF(C,C)	1.2255	1.1717	1.1279	1.1919	1.1471	1.1102	1.1299	1.1008	1.0763
	TF(S,C)	0.5096	0.3298	0.1811	0.4956	0.3229	0.1783	0.4698	0.3098	0.1728
	TF(W,C)	0.4341	0.2772	0.1503	0.4222	0.2714	0.1479	0.4003	0.2604	0.1434
	TF(F,C)	0.1928	0.1747	0.1601	0.1875	0.1711	0.1576	0.1777	0.1642	0.1528
4.00	TF(C,C)	1.2190	1.1758	1.1393	1.1865	1.1506	1.1198	1.1265	1.1031	1.0827
	TF(S,C)	0.5093	0.3359	0.1874	0.4957	0.3287	0.1842	0.4706	0.3152	0.1781
	TF(W,C)	0.4486	0.2934	0.1622	0.4367	0.2871	0.1595	0.4146	0.2752	0.1542
	TF(F,C)	0.2032	0.1883	0.1758	0.1978	0.1843	0.1727	0.1878	0.1767	0.1670
5.00	TF(C,C)	1.2145	1.1785	1.1472	1.1828	1.1528	1.1265	1.1241	1.1046	1.0872
	TF(S,C)	0.5087	0.3398	0.1916	0.4954	0.3324	0.1882	0.4708	0.3185	0.1816
	TF(W,C)	0.4576	0.3038	0.1702	0.4457	0.2972	0.1672	0.4236	0.2848	0.1613
	TF(F,C)	0.2097	0.1971	0.1862	0.2042	0.1928	0.1828	0.1941	0.1848	0.1764

Chapter 12: Photometric datasheets

Whilst most lighting design is based on computer simulation, photometric datasheets are still provided for luminaires as they can provide a quick and efficient way of picking a luminaire that is suitable for a given application. This chapter is split into three sections. The first section gives information on the measurement of luminaires and the production of intensity tables. The second lists a number of elements that may be found in a datasheet and gives an explanation where necessary on how to use this information. The third section gives information on the calculation of certain elements.

12.1 Photometric measurement

To ensure common standards of accuracy of measurement, the procedures used to measure the photometric performance of a luminaire are covered in BS EN 13032-1 (BSI, 2004a). The standard covers a number of areas that are necessary to ensure accuracy of the testing process but the definitions of angular systems to be used are also covered and they impact on the meaning of the data produced.

There are two systems of photometric angles defined in BS EN 13032-1: the C-γ and the B-β system. The vast majority of luminaire data produced is based on the C-γ system and the B-β system is only occasionally used for some types of floodlights. Note that, throughout the world, the C-γ and B-β systems are the most common angular systems used, however, in North America, they use other systems that are slight variants of them.

12.1.1 The C-γ system

Figure 12.1 shows the luminaire orientation for C-γ photometry.

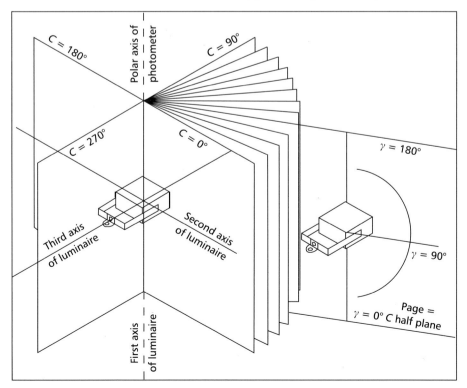

Figure 12.1 The C-γ system

The totality of the C-planes is the group of planes for which the line of intersection (polar axis) is the vertical line through the photometric centre of the luminaire. C-planes are marked by angles C_X of $0° \leq C_X \leq 360°$. Within a plane, the directions are given by the angle γ with $0° \leq \gamma \leq 180°$. Any two of these C-planes with an angular difference of $180°$ combine to a single plane in the mathematical sense.

The system of C-planes is orientated rigidly in space and does not follow a tilt in the luminaire. The polar axis does not necessarily coincide with the first axis of the luminaire. The first axis is mostly the axis going through the photometric centre and perpendicular to the light emitting area.

If the luminaire is tilted during measurement (the polar axis is not coincident with the first axis of the luminaire), the angle of tilt should be declared.

Notes:
1. For street luminaires, the direction or the road is located in the C_0/C_{180}-plane and the pole in the C_{270}-plane.
2. For indoor luminaires, the C_0/C_{180}-plane is the symmetry plane of the luminous intensity distribution with the highest degree of symmetry. For indoor luminaires with different luminous intensity distributions in the C_0 and C_{180} half-planes, the main direction of light output is orientated in the C_0-plane.

12.1.2 The B-β system

Figure 12.2 shows the luminaire orientation for B-β photometry.

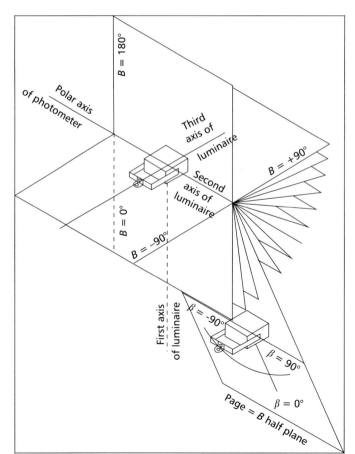

Figure 12.2 The B-β system

The totality of the B-planes is the group of planes for which the line of intersection (polar axis) goes through the photometric centre and is parallel to the second axis of the luminaire. B-planes are marked by angles B_X of $-180° \leq B_X \leq 180°$. Within a plane, directions are given by the angle β_X with $-90° \leq \beta_X \leq 90°$.

Note: The system of B-planes is coupled rigidly to the light source and follows its tilt if the luminaire is tilted. The first axis is mostly the axis going through the photometric centre and perpendicular to the emitting area. The second axis is coincident with the transversal axis (if any) of the emitting area or with the spigot axis of the luminaire.

12.1.3   Relationships between the two angular co-ordinate systems

The angular values of one plane system can be converted into the corresponding angular values of another plane system if the relations given in Table 12.1 are used. The relations are only valid if the tilt angle of the luminaire in the C-plane system is zero.

Table 12.1 Conversion between B-β and C-γ

Direction		Conversion formulae	
Given	Wanted	For planes	For angles
B-β	C-γ	$\tan C = \dfrac{\sin B}{\tan \beta}$	$\cos \gamma = \cos B \times \cos \beta$
C-γ	B-β	$\tan B = \sin C \times \tan \gamma$	$\sin \beta = \cos C \times \sin \gamma$

Care needs to be taken when applying the above formulae as the use of tangents means that values can go to infinity and so the calculations may cause problems if implemented in a software routine that cannot handle this sort of problem.

12.1.4   Photometric centre

Both the C-γ and the B-β systems of photometric angles are based around a central point known as the photometric centre. The position of the photometric centre of a luminaire is determined with the following rules and examples as shown in Figure 12.3.

● **Luminaires with substantially opaque sides:** At the centre of the main luminaire opening (or diffusing/prismatic member across the opening) if the lamp compartment is substantially white or luminous but at the lamp photometric centre if it is outside the plane of the opening, or if the lamp compartment is substantially black or non-luminous.

● **Luminaires with diffusing/prismatic sides:** At the centre of the solid figure bounded in outline by the luminous surfaces but at the lamp photometric centre, if it is outside this solid figure.

● **Luminaires with transparent sides or without side members:** At the lamp photometric centre.

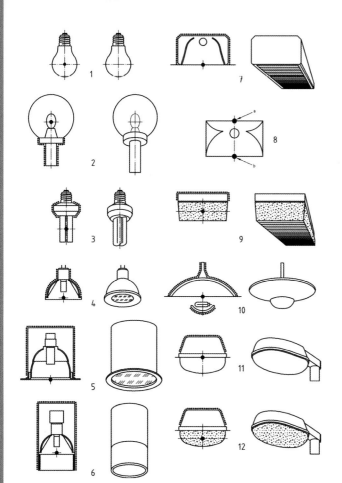

Figure 12.3 Photometric centres of luminaires

In Figure 12.3, the following drawing elements mean:

 Photometric centre
opaque, substantially black
/// /// /// opaque, diffuse or specular reflectant
translucent, clear
•••••••••••••• compartment

1) Incandescent lamp
2) With a clear cover
3) Compact fluorescent lamp
4) Reflector lamp
5) Luminaire with reflecting mirror
6) Luminaire with shield, substantially black
7) Luminaire with opaque sides
8) Direct-indirect luminaire
   a) Luminant area 1 with photometric centre 1
   b) Luminant area 2 with photometric centre 2
9) Luminaire with diffusing/prismatic sides
10) Indirect luminaire with secondary reflector
11) Outdoor luminaire with clear cover
12) Outdoor luminaire with diffusing/prismatic cover

12.2   Elements of a datasheet

There are many elements to any datasheet for a luminaire. These may cover mechanical and electrical aspects, and appearance of the luminaire as well as its photometric performance. The data that need to be produced for a normal luminaire for use in workplaces is covered in British Standard BS EN 13032-2 (BSI, 2004b). There is also a separate standard provided for emergency luminaires: BS EN 13032-3 (BSI, 2007d). A significant amount of the data to be presented in a datasheet is based on gonio-photometric measurement of the luminaire. These measurement results produce a normalised intensity table. The basis of these measurements is covered in section 12.1.

The following sections cover some of the key elements covering lighting performance.

12.2.1   Normalised intensity table

The normalised intensity table gives a list of the intensities in different directions for a luminaire where the total lamp flux is assumed to be 1000 lumens. Table 12.2 is an example of a typical table for a luminaire with two planes of symmetry; data values for C planes greater than 90° are not shown as they may be derived by symmetry. For example, the values for 60° are the same as the values for 120°, 240° and 300°.

Table 12.2 A normalised intensity table

γ angles	C planes						
	0	15	30	45	60	75	90
0	281.8	281.8	281.8	281.8	281.8	281.8	281.8
5	281.0	280.5	280.1	280.3	280.6	281.0	280.8
10	279.7	279.3	278.3	277.5	277.0	277.2	276.9
15	276.9	276.2	274.5	273.0	271.6	270.9	270.5
20	271.8	271.1	268.8	266.2	263.7	261.9	261.1
25	263.5	262.9	260.2	256.8	252.8	250.2	248.6
30	249.2	249.4	247.5	243.5	239.0	234.6	232.1
35	225.2	226.0	226.3	225.0	218.3	212.8	210.1
40	199.7	200.2	199.3	197.0	192.8	186.6	184.4
45	169.2	164.7	160.9	158.3	156.2	155.0	155.5
50	131.5	132.0	131.5	127.6	126.2	122.8	120.7
55	98.0	98.8	98.1	94.3	94.7	95.7	94.0
60	73.1	69.9	64.9	62.3	63.9	68.8	71.6
65	55.0	48.5	39.6	36.4	38.7	48.3	54.6
70	41.6	36.8	29.1	24.8	28.5	37.3	41.5
75	30.9	27.9	23.9	22.8	23.6	27.6	30.6
80	19.5	20.9	19.3	18.0	19.3	20.9	18.9
85	11.2	11.4	11.0	12.1	11.2	11.4	10.9
90	1.1	1.1	0.9	0.7	0.6	0.4	0.4

To use the data in a lighting calculation, it is first necessary to calculate the C and γ angles at the luminaire of the line joining the luminaire to the point where the illuminance is being calculated; usually the calculated angles will not line up exactly with a point in the table and some interpolation will be needed. The value obtained needs to be multiplied by the total lamp flux in kilolumens to give the absolute intensity in the direction of interest.

12.2.2 Intensity diagram

There are a number of graphical representations of the intensity data. The most common of these is the polar curve. Figure 12.4 shows a polar curve plotted for the data in Table 12.2.

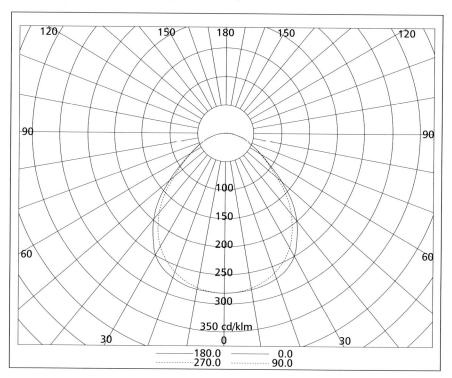

Figure 12.4 A polar curve

The polar curve contains a number of plotted lines, each one representing a different C plane. The γ values are plotted around the centre of the diagram and the distance of the line from the centre at any point represents the intensity at the given γ angle in terms of candelas per kilolumen.

12.2.3 Light output ratios

A light output ratio (LOR) of a luminaire gives the ratio of the total flux leaving the luminaire under standard conditions compared with the total flux of lamps used in the luminaire operated under standard conditions. As one might expect, the value of LOR is usually less than 1. However, with certain lamps, their maximum light output is not reached under standard conditions (for which their lumen value is declared) so inside a luminaire they may give out more light than under standard conditions. Where this happens, the LOR is higher than would be expected for the optical efficiency of the luminaire and may be greater than 1.

There are also a number of light output ratios that look at only part of the luminaire distribution.

- **Downward Light Output Ratio (DLOR):** this is the ratio of the light emitted below the horizontal to the total lamp flux.

- **Upward Light Output Ratio (ULOR):** this is the ratio of the light emitted above the horizontal to the total lamp flux.

- **Kerbside Light Output Ratio:** this is the ratio between the light emitted on the kerb side of a road lantern to the total light emitted from the lantern.

- **Roadside Light Output Ratio:** this is the ratio between the light emitted on the road side of a road lantern to the total light emitted from the lantern.

LOR may be used as a form of quality metric for luminaires; for two luminaires of similar distribution, then the one with higher LOR is more efficient. However, LOR should not be used to compare luminaires of dissimilar distributions as, in general, the more controlled the distribution of light from a luminaire, then the lower the LOR will be.

It is sometimes possible to use the ratio of the DLOR to ULOR to describe the general properties of a luminaire. For example, if the ULOR is much greater than the DLOR then it makes sense to describe the luminaire as an uplighter.

12.2.4 Spacing to height ratio (SHR)

SHR is the ratio of spacing between luminaires to the height of the geometric centres of the luminaires above the reference plane. The value given in most datasheets is the maximum permitted value of SHR that enables the lighting scheme to achieve an illuminance uniformity (minimum divided by the average) of 0.7.

The SHR is useful to have if you need to ensure that an area that is being lit will have good uniformity. Provided the spacing between the luminaires does not cause the SHR of the installation to be greater than the maximum permitted SHR given on the datasheet, then, in most cases, uniformity will be assured.

12.2.5 Utilisation factor (UF) tables

Utilisation factors are one of the key elements of what is known as the average lumen method of lighting design. The method is based on the fact that in any regular array lighting installation, it is possible to think about the amount of light incident on the reference plane as a certain fraction of the flux of the lamps in the installation. This fraction is known as the utilisation factor (UF). The reference plane is the plane in the room where you want to know what the illuminance is, for example, in an office, this is the height of the desktops, usually taken to be 0.8 m above the floor. The UF of an installation depends on three factors: the luminaire used, the shape of the room and the surface finishes of the room. The room shape is characterised by a metric called the room index (K). Equation 12.1 gives the formula for calculating the room index from the length (L) and width (W) of the room together with the height that the luminaires are mounted above the reference plane (h_m)

$$K = \frac{L \times W}{(L+W)h_m} \tag{12.1}$$

The UF table for a given luminaire provides a series of utilisation factors for the luminaire when used in a range of room shapes and with different surface finishes.

Table 12.3 is a typical *UF* table.

Table 12.3 Typical *UF* table

Reflectance values

	Ceiling	0.8	0.8	0.8	0.7	0.7	0.7	0.5	0.5	0.5
	Walls	0.7	0.5	0.3	0.7	0.5	0.3	0.7	0.5	0.3
	Floor cavity	0.2	0.2	0.2	0.2	0.2	0.2	0.2	0.2	0.2
Room index	0.60	0.64	0.55	0.50	0.63	0.55	0.50	0.61	0.54	0.49
	0.80	0.74	0.67	0.62	0.73	0.66	0.61	0.71	0.65	0.61
	1.00	0.80	0.74	0.69	0.79	0.73	0.69	0.77	0.72	0.68
	1.25	0.85	0.80	0.76	0.84	0.79	0.75	0.81	0.77	0.74
	1.50	0.89	0.84	0.80	0.87	0.83	0.79	0.84	0.81	0.78
	2.00	0.91	0.87	0.84	0.90	0.86	0.83	0.87	0.84	0.81
	2.50	0.93	0.90	0.87	0.91	0.88	0.86	0.88	0.86	0.83
	3.00	0.94	0.91	0.88	0.92	0.90	0.87	0.89	0.87	0.85
	4.00	0.96	0.93	0.91	0.94	0.91	0.89	0.90	0.88	0.87
	5.00	0.97	0.94	0.92	0.95	0.93	0.91	0.91	0.89	0.88

Note: The reflectance of the floor cavity is the effective reflectance of the room volume below the working plane.

As the concept of *UF* is simple, it is very easy to calculate the number of luminaires needed to provide a given average illuminance in a given room. The number of luminaires (*N*) is a function of the required illuminance (*E*), the flux in each lamp (*F*) and the number of lamps per luminaire (*n*). Equation 12.2 gives the calculation

$$N = \frac{E \times L \times W}{F \times n \times UF \times MF} \tag{12.2}$$

In equation 12.2, the term *MF* stands for maintenance factor. It is a factor that allows for the fact that, as lighting systems age, the amount of light they deliver decreases. The use of *MF* ensures that the system is slightly over-specified when new but is still delivering the correct illuminance when the system is maintained. See Chapter 18: Predicting maintenance factor.

12.2.6 Shielding angle

It is important that bright light sources do not appear in the centre of the field of view. To block the view of lamps or bright parts of luminaires, it is normal to recess them into the luminaire. The shielding angle is a guide to how far the bright objects have been taken out of the line of sight. Figure 12.5 shows the shielding angle.

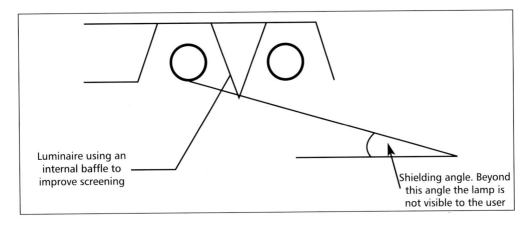

Figure 12.5 Shielding angle

For indoor workplaces, there are recommendations for the minimum shielding angles to be used for various levels of high luminance sources. See section 2.1.5.3: Shielding against glare.

12.2.7 Normalised luminance table

With certain types of display screen, it is possible that the reflection of a high luminance luminaire may make it hard to read the information being displayed. To ensure that this problem is minimised, most screens have a maximum rated luminance that they are able to accept from a luminaire. The use of average luminance tables enables the designer to check if there may be problems with the combination of luminaires and displays used in a given area. It is only necessary to consider this problem if there is not the flexibility in the layout of screens in the room that will permit their positioning so that reflections can be avoided. For most potential problem areas, it is only light above a γ angle of 65° that causes a problem, however, in certain instances of high levels of screen use, it may be necessary to consider the luminance at 55°.

Table 12.4 Normalised luminance

		C planes					
		0	15	30	45	60	75	90
γ angle	55	714	835	880	1412	1913	1662	1473
	65	41	62	124	186	83	31	21
	75	0	17	0	0	0	0	0
	85	0	0	0	0	0	0	0

Table 12.4 is an example of a normalised luminance table. Note that, to use this table, it is necessary to multiply the values by the total lamp flux per luminaire in kilolumens.

12.2.8   Unified Glare Rating (*UGR*) table

UGR is the metric used to control discomfort glare in interior workplaces. The equation used to evaluate *UGR* is given in CIE 117 (CIE, 1995a), however, the equation is complex to evaluate and its results have only been validated in a limited range of circumstances so it is better to use the *UGR* table as a basis for the calculation. The *UGR* table may be used to assess the glare experienced by a person looking into the room from the midpoint of a wall with eye height 1.2 m.

Table 12.5 shows a typical *UGR* table. It shows *UGR* for a given luminaire as a function of room size expressed in multiples of the luminaire height above eye height and the room surface reflectances. The *UGR* values in the table are based on a total lamp flux of the luminaire of 1000 lumens and it is necessary to correct the values for the actual lamp flux used.

To use the table, it is first necessary to calculate the size of the room in terms of the mounting height of the luminaires above eye level (1.2 m). For example, if a room is 12 m long and the luminaires are mounted 2.7 m above the floor, then the luminaire would be 1.5 m above eye height and the length could be expressed as 8H. Figure 12.6 gives the location of the *X* and *Y* dimensions of the room for the views of observers on the long and short walls of a room for both crosswise and endwise views. Selection of endwise or crosswise views depends on the orientation of the luminaires within the room.

Table 12.5 *UGR* table

Uncorrected *UGR* values calculated at $SHR = 1.0$

		Crosswise view									Endwise view								
Ceiling		0.8	0.8	0.8	0.7	0.7	0.7	0.5	0.5	0.5	0.8	0.8	0.8	0.7	0.7	0.7	0.5	0.5	0.5
Wall		0.7	0.5	0.3	0.7	0.5	0.3	0.7	0.5	0.3	0.7	0.5	0.3	0.7	0.5	0.3	0.7	0.5	0.3
Floor		0.2	0.2	0.2	0.2	0.2	0.2	0.2	0.2	0.2	0.2	0.2	0.2	0.2	0.2	0.2	0.2	0.2	0.2
X	Y																		
2	2	8.4	9.8	11.2	8.6	10.0	11.4	9.1	10.5	11.9	5.1	6.5	8.0	5.4	6.8	8.2	5.9	7.2	8.7
2	3	10.8	12.0	13.3	11.0	12.3	13.6	11.6	12.8	14.1	6.3	7.6	8.9	6.6	7.8	9.1	7.1	8.3	9.6
2	4	11.7	12.9	14.2	12.0	13.2	14.4	12.5	13.7	14.9	6.8	8.0	9.2	7.1	8.2	9.5	7.6	8.8	10.0
2	6	12.5	13.6	14.8	12.8	13.9	15.1	13.4	14.4	15.6	7.2	8.3	9.5	7.5	8.6	9.7	8.1	9.1	10.3
2	8	12.9	14.0	15.1	13.2	14.3	15.4	13.8	14.8	15.9	7.4	8.5	9.6	7.7	8.7	9.8	8.2	9.3	10.4
2	12	13.3	14.3	15.4	13.6	14.6	15.7	14.2	15.1	16.2	7.5	8.6	9.6	7.8	8.8	9.9	8.4	9.4	10.4
4	2	8.8	10.0	11.2	9.0	10.2	11.5	9.6	10.7	12.0	6.1	7.3	8.6	6.4	7.6	8.8	7.0	8.1	9.3
4	3	11.4	12.5	13.5	11.7	12.7	13.8	12.3	13.3	14.3	7.7	8.7	9.8	8.0	9.0	10.0	8.5	9.5	10.6
4	4	12.6	13.5	14.5	12.9	13.8	14.8	13.5	14.4	15.3	8.3	9.3	10.2	8.6	9.5	10.5	9.2	10.1	11.0
4	6	13.7	14.5	15.4	14.0	14.8	15.6	14.6	15.4	16.2	8.9	9.7	10.6	9.2	10.0	10.9	9.8	10.6	11.4
4	8	14.2	15.0	15.8	14.5	15.3	16.1	15.1	15.8	16.6	9.1	9.9	10.7	9.4	10.2	11.0	10.0	10.8	11.6
4	12	14.8	15.5	16.2	15.0	15.7	16.5	15.7	16.3	17.1	9.3	10.0	10.8	9.6	10.3	11.1	10.2	10.9	11.7
8	4	12.9	13.7	14.5	13.2	13.9	14.7	13.8	14.5	15.3	9.2	10.0	10.8	9.5	10.2	11.0	10.1	10.8	11.6
8	6	14.2	14.8	15.5	14.5	15.1	15.8	15.1	15.7	16.4	10.0	10.6	11.3	10.3	10.9	11.6	10.9	11.5	12.2
8	8	14.9	15.5	16.1	15.2	15.7	16.3	15.8	16.4	17.0	10.4	10.9	11.5	10.7	11.2	11.8	11.3	11.9	12.4
8	12	15.7	16.2	16.7	16.0	16.5	17.0	16.6	17.1	17.6	10.7	11.2	11.7	11.0	11.5	12.0	11.6	12.1	12.6
12	4	12.9	13.7	14.4	13.2	13.9	14.7	13.8	14.5	15.3	9.4	10.1	10.9	9.7	10.4	11.1	10.3	11.0	11.7
12	6	14.3	14.9	15.5	14.6	15.1	15.7	15.2	15.8	16.4	10.4	10.9	11.5	10.6	11.2	11.8	11.3	11.8	12.4
12	8	15.0	15.6	16.1	15.3	15.8	16.4	16.0	16.5	17.0	10.8	11.3	11.9	11.1	11.6	12.2	11.8	12.3	12.8

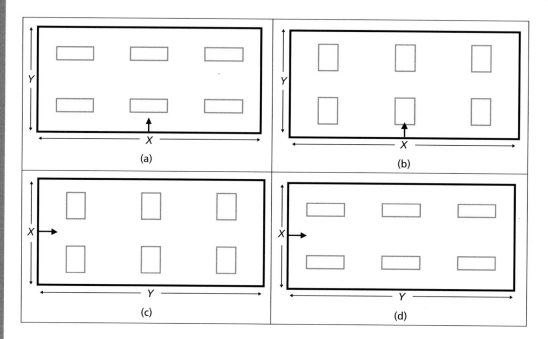

Figure 12.6 Orientation of luminaires; (a) crosswise view from long wall; (b) endwise view from long wall; (c) crosswise view from short wall; (d) endwise view from short wall

Once the X and Y values have been determined for the long and short wall views of the room, it is possible to look up the uncorrected *UGR* value in the table. These values must then be corrected for true lamp flux by adding a correction term to the values.

The correction term may be calculated using equation 12.3

$$\text{Correction term} = 8 \log_{10}\left(\frac{\text{Total flux of lamps in luminaire}}{1000}\right) \tag{12.3}$$

Table 12.6 gives tabulated values of the correction factor.

Table 12.6 Tabulated values of correction factor

Lamp flux	Correction factor	Lamp flux	Correction factor	Lamp flux	Correction factor	Lamp flux	Correction factor	Lamp flux	Correction factor
		1000	0.0	2000	2.4	3000	3.8	4000	4.8
100	−8.0	1100	0.3	2100	2.6	3100	3.9	4100	4.9
200	−5.6	1200	0.6	2200	2.7	3200	4.0	4200	5.0
300	−4.2	1300	0.9	2300	2.9	3300	4.1	4300	5.1
400	−3.2	1400	1.2	2400	3.0	3400	4.3	4400	5.1
500	−2.4	1500	1.4	2500	3.2	3500	4.4	4500	5.2
600	−1.8	1600	1.6	2600	3.3	3600	4.5	4600	5.3
700	−1.2	1700	1.8	2700	3.5	3700	4.5	4700	5.4
800	−0.8	1800	2.0	2800	3.6	3800	4.6	4800	5.4
900	−0.4	1900	2.2	2900	3.7	3900	4.7	4900	5.5

12.2.9 Luminaire maintenance factor (*LMF*)

As luminaires are used, dirt accumulates on the optical surfaces and over time, this reduces the efficiency of the luminaire. It is important for the designer to know how much light output will drop with time so that a maintenance schedule can be worked out. Also *LMF* is one element of overall maintenance factor as used in section 12.2.5. Luminaire manufacturers may choose to test their luminaires over a period of a few years whilst they are operating under different conditions and present the results on the datasheet or they may just state the maintenance class of the luminaire. See Chapter 18: Predicting maintenance factor.

12.2.10 Spacing tables (Emergency Lighting)

Spacing tables are used when designing emergency lighting for corridors and defined escape routes. They provide a convenient way of calculating the layout of luminaires required. The layout is calculated for strips between luminaires to ensure that the illuminance on the centre line of the escape route does not fall below the required illuminance and that the edges of the strip have at least half that illuminance. The luminaire spacing is calculated for five conditions:

- from a luminaire that is mounted transverse to the escape route and an end wall (S_{TW})

- between luminaires that are mounted transverse to the escape route (S_{TT})

- between a luminaire that is mounted transverse to the escape route and a luminaire mounted axial to the escape route (S_{TA})

- between luminaires that are mounted axial to the escape route (S_{AA})

- from a luminaire that is mounted axial to the escape route and an end wall (S_{AW}).

Figure 12.7 illustrates these options.

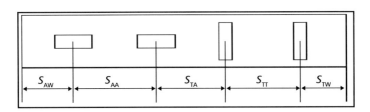

Figure 12.7 Types of spacing

Table 12.7 is an example of a spacing table.

Table 12.7 Spacing table

Mounting height / m	S_{TW} transverse to wall / m	S_{TT} transverse to transverse / m	S_{TA} transverse to axial / m	S_{AA} axial to axial / m	S_{AW} axial to wall / m
2	2.3	5.9	7	8.2	3
2.2	2.3	6.1	7.2	8.5	3.1
2.4	2.3	6.3	7.3	8.5	3.1
2.6	2.3	6.4	7.4	8.6	3
2.8	2.3	6.5	7.5	8.7	3
3	2.3	6.6	7.6	8.8	3
3.2	2.3	6.7	7.7	8.8	3
3.2	2.3	6.7	7.7	8.8	2.9
3.4	2.2	6.7	7.7	8.7	2.8
3.6	2.1	6.7	7.7	8.7	2.7
3.8	2	6.7	7.6	8.7	2.5
4	1.9	6.7	7.6	8.6	2.3

The table gives the maximum permitted spacing when using the given luminaire. SLL Lighting Guide 12 (SLL, 2004) gives more information on the design of emergency lighting and calculation of values for datasheets.

12.3   Calculations for datasheets

Most of the calculations in this section are covered in BS EN 13032-2 (BSI, 2004b), however, the first section on the method of calculating flux from intensity is not.

12.3.1   Flux calculations

From the basis of photometry, intensity is flux per unit solid angle so, by multiplying intensity values by the solid angle over which they are valid, flux may be calculated. With the C-γ coordinate system, the size of zone (Ω) for a complete band of a range of γ angles (θ' to θ'') may be calculated with equation 12.4 and the calculation is illustrated by Figure 12.8.

$$\Omega = 2\pi\left(\cos\theta' - \cos\theta''\right)$$

(12.4)

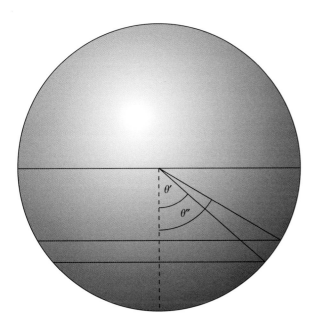

Figure 12.8 Calculation of zone size

The fraction of the zone associated with any given intensity value is the width of the zone in degrees divided by 360.

For example, consider the intensity for $\gamma = 35°$ and $C = 15°$ in Table 12.2. The value of the intensity is 226.0 cd·klm^{-1}; this value is the best estimate we have for intensity in the range γ 32.5–37.5°; these zone limits are half way to the next points that have intensity values. Similarly, if in the C direction, the value is the best estimate for intensity in the range C 7.5–22.5°.

The full zone in the range γ 32.5–37.5° is 0.3144 sr. The zone required runs from $C = 7.5°$ to $C = 22.5°$ and is thus 15° in width. It is thus 0.04167 of the full zone. Thus the zone for the intensity value in question is 0.0131 sr so the flux in the zone is 2.96 lumens per kilolumen of total lamp flux.

This calculation forms the basis of light output ratio calculations where all or a subset of the intensity values in the table have their flux calculated for them and the results are summed and the result divided by 1000.

12.3.2 Calculation of spacing to height ratio

To calculate the spacing to height ratio for a rotationally symmetric or dissymmetric luminaire, it is necessary to calculate the direct illuminance on each point of a grid illuminated by a 4 × 4 square array of luminaires. The general layout for the luminaires and calculation points is shown in Figure 12.9.

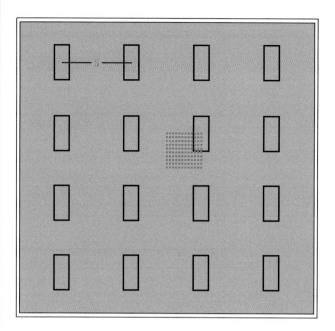

Figure 12.9 General layout of luminaires and calculation points

The luminaires are taken to be 1 m above the grid of points and the illuminance calculation treats them as point sources. This grid of points is placed so that it has one corner in the centre of the array of luminaires and the opposite corner directly under the centre of one of the luminaires. The spacing between each of the grid points is one twentieth of the spacing (S) between the luminaires; the locations of the grid points are shown in Figure 12.10.

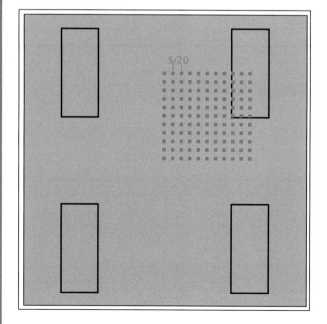

Figure 12.10 Grid of calculation points

Once the illuminances at each of the grid points have been calculated, it is then necessary to find the minimum value (E_{min}) and the average value (E_{av}). The uniformity (U) may then be calculated using the equation:

$$U = \frac{E_{min}}{E_{av}} \tag{12.5}$$

The spacing (S) is adjusted until a uniformity value of 0.7 is obtained and this spacing is then determined as being SHR_{max}.

12.3.3 Calculation of utilisation factors

The process of utilisation factor calculation involves the calculation of the amount of light received by each of the room surfaces directly from the luminaires and then a calculation of the inter-reflected light in the space.

The calculation requires the $ULOR$ and the $DLOR$ for the luminaire together with the total flux for the following zones: 0° to 41.4° ($FCL1$), 0° to 60° ($FCL2$), 0° to 75.5° ($FCL3$) and 0° to 90° ($FCL4$). In addition, if the luminaire has a significant amount of upward light and is designed to be suspended below the ceiling rather than mounted on it, then four further total fluxes are required; they are: 138.6° to 180° ($FCU1$), 120° to 180° ($FCU2$), 104.5° to 180° ($FCU3$) and 90° to 180° ($FCU4$). All of these values may be calculated using the method set out in section 12.3.1.

The direct flux on to the floor (DFL) may be calculated using equation 12.6:

$$DFL = GML1 \times FCL1 + GML2 \times FCL2 + GML3 \times FCL3 + GML4 \times FCL4 \qquad (12.6)$$

where $GML1$, $GML2$, $GML3$ and $GML4$ are geometric multipliers that are a function of spacing to height ratio and room index. Table 12.8 lists their values.

In a similar way, the flux received directly by the ceiling (DFU) may be calculated using equation 12.7:

$$DFU = GMU1 \times FCU1 + GMU2 \times FCU2 + GMU3 \times FCL3 + GMU4 \times FCU4 \qquad (12.7)$$

where $GMU1$, $GMU2$, $GMU3$ and $GMU4$ are also geometric multipliers that are a function of spacing to height ratio and room index. Table 12.9 lists their values.

Table 12.8 Geometric multipliers for the lower hemisphere

Room index / K	0.6	0.8	1	1.25	1.5	2	2.5	3	4	5
					$SHRNOM = 1.00$					
GML1	0.943	0.752	0.636	0.510	0.429	0.354	0.286	0.258	0.236	0.231
GML2	−0.317	−0.033	0.121	0.238	0.275	0.248	0.190	0.118	−0.006	−0.099
GML3	0.145	0.081	0.088	0.131	0.202	0.350	0.470	0.563	0.684	0.748
GML4	−0.027	−0.016	−0.015	−0.016	−0.018	−0.015	−0.003	0.016	0.060	0.107
					$SHRNOM = 1.25$					
GML1	1.013	0.893	0.692	0.569	0.498	0.355	0.317	0.268	0.242	0.234
GML2	−0.338	−0.112	0.151	0.256	0.274	0.284	0.184	0.132	0.005	−0.091
GML3	0.144	0.102	0.065	0.119	0.197	0.337	0.471	0.563	0.685	0.751
GML4	−0.026	−0.019	−0.011	−0.014	−0.017	−0.013	−0.002	0.016	0.061	0.108

Table 12.8 Continued

Room index / K	0.6	0.8	1	1.25	1.5	2	2.5	3	4	5
					SHRNOM = 1.50					
GML1	1.070	0.934	0.774	0.591	0.435	0.353	0.293	0.279	0.236	0.231
GML2	−0.340	−0.045	0.168	0.342	0.420	0.334	0.254	0.161	0.026	−0.076
GML3	0.133	0.058	0.038	0.062	0.141	0.316	0.447	0.541	0.679	0.746
GML4	−0.024	−0.011	−0.006	−0.004	−0.009	−0.009	0.002	0.020	0.062	0.109
					SHRNOM = 1.75					
GML1	NA	NA	0.717	0.529	0.413	0.331	0.249	0.238	0.229	0.219
GML2	NA	NA	0.316	0.480	0.494	0.386	0.313	0.209	0.054	−0.051
GML3	NA	NA	−0.029	0.003	0.101	0.292	0.436	0.538	0.664	0.737
GML4	NA	NA	0.005	0.005	−0.002	−0.006	0.003	0.021	0.065	0.112
					SHRNOM = 2.00					
GML1	NA	NA	0.644	0.459	0.319	0.217	0.206	0.176	0.192	0.204
GML2	NA	NA	0.441	0.588	0.657	0.529	0.407	0.319	0.100	−0.015
GML3	NA	NA	−0.080	−0.032	0.037	0.266	0.390	0.491	0.658	0.718
GML4	NA	NA	0.012	0.009	0.006	−0.004	0.010	0.028	0.067	0.114

Table 12.9 Geometric multipliers for the upper hemisphere

Room index / K	0.6	0.8	1	1.25	1.5	2	2.5	3	4	5
					SHRNOM = 1.00					
GMU1	0.124	0.176	0.136	0.148	0.134	0.102	0.138	0.148	0.156	0.152
GMU2	0.753	0.463	0.373	0.199	0.125	0.009	−0.097	−0.157	−0.222	−0.247
GMU3	0.130	0.363	0.475	0.610	0.666	0.749	0.765	0.764	0.738	0.702
GMU4	0.003	0.007	0.030	0.057	0.093	0.158	0.219	0.271	0.355	0.420

Table 12.9 Continued

Room index / K	0.6	0.8	1	1.25	1.5	2	2.5	3	4	5
					SHRNOM = 1.25					
GMU1	0.028	−0.008	0.028	0.023	0.028	0.077	0.099	0.118	0.133	0.138
GMU2	0.826	0.645	0.416	0.276	0.170	−0.012	−0.087	−0.166	−0.230	−0.256
GMU3	0.152	0.360	0.541	0.656	0.726	0.797	0.792	0.801	0.767	0.723
GMU4	−0.004	0.000	0.050	0.050	0.083	0.153	0.215	0.269	0.355	0.421
					SHRNOM = 1.50					
GMU1	−0.036	−0.094	−0.071	−0.026	0.023	0.061	0.091	0.103	0.128	0.133
GMU2	0.840	0.633	0.422	0.217	0.066	−0.059	−0.144	−0.187	−0.253	−0.273
GMU3	0.205	0.464	0.639	0.774	0.841	0.863	0.860	0.842	0.796	0.745
GMU4	−0.016	−0.017	−0.001	0.032	0.076	0.147	0.211	0.262	0.354	0.420
					SHRNOM = 1.75					
GMU1	NA	NA	−0.073	−0.019	0.023	0.063	0.100	0.113	0.127	0.134
GMU2	NA	NA	0.322	0.119	0.015	−0.095	−0.195	−0.236	−0.273	−0.295
GMU3	NA	NA	0.750	0.873	0.898	0.900	0.901	0.877	0.819	0.766
GMU4	NA	NA	−0.010	0.025	0.069	0.144	0.213	0.266	0.351	0.420
					SHRNOM = 2.00					
GMU1	NA	NA	−0.056	−0.003	0.043	0.092	0.108	0.127	0.137	0.137
GMU2	NA	NA	0.245	0.046	−0.091	−0.208	−0.254	−0.315	−0.315	−0.315
GMU3	NA	NA	0.815	0.932	0.990	0.981	0.956	0.933	0.849	0.786
GMU4	NA	NA	−0.013	0.025	0.066	0.151	0.209	0.264	0.355	0.417

To calculate the distribution factors for the floor [DF(F)], walls [DF(W)] and ceiling [DF(C)] for surface mounted and recessed luminaires, use equations 12.8 to 12.10:

$$DF(F) = \frac{DFL}{1000} \tag{12.8}$$

$$DF(W) = DLOR - DF(F) \tag{12.9}$$

$$DF(C) = ULOR \tag{12.10}$$

When using suspended luminaires, the distribution factors for the walls and floor remain the same but equation 12.11 is used to evaluate the distribution factor for the ceiling, and the distribution factor for the wall above the luminaires, known as the frieze [DF(S)], is calculated with equation 12.12:

$$DF(C) = \frac{DFU}{1000} \tag{12.11}$$

$$DF(S) = ULOR - DF(C) \tag{12.12}$$

With ceiling mounted and recessed luminaires, it is possible to calculate the utilisation factor on the floor [UF(F)] using equation 12.13:

$$UF(F) = DF(F) \cdot TF(F,F) + DF(W) \cdot TF(W,F) + DF(C) \cdot TF(C,F) \tag{12.13}$$

For suspended luminaires, the UF is calculated with equation 12.14:

$$UF(F) = DF(F) \cdot TF(F,F) + DF(W) \cdot TF(W,F) \\ + DF(S) \cdot TF(S,F) + DF(C) \cdot TF(C,F) \tag{12.14}$$

In the above equation, TF(F,F) is the transfer factor from floor to floor, TF(W,F) is the transfer factor from walls to floor, TF(S,F) is the transfer factor from frieze to floor and TF(C,F) is the transfer factor from ceiling to floor. The tables needed are given in section 11.3.5, Tables 11.6 to 11.12.

12.3.4 Calculation of normalised luminance tables

The normalised luminance of a luminaire is a function of the normalised intensity in a particular direction and the projected luminous area of the luminaire in that direction. Thus, to calculate the normalised luminance, equation 12.15 can be used:

$$L(C,\gamma) = \frac{I(C,\gamma)}{A_b \cos\gamma + A_S \sin\gamma\cos C + A_e \sin\gamma\sin C} \tag{12.15}$$

where

- $I(C, \gamma)$ is the normalised intensity at elevation angle γ and azimuth plane C

- A_b is the luminous area of the base of the luminaire (the area when viewed from $\gamma = 0$)

- A_s is the luminous area of the side of the luminaire (the area when viewed from $\gamma = 90, C=0$)

- A_e is the luminous area of the end of the luminaire (the area when viewed from $\gamma = 90, C=90$)

12.3.5 Calculation of UGR tables

The Discomfort Glare Rating of a lighting installation is determined by the CIE Unified Glare Rating (UGR) tabular method (CIE, 1995a) based on the basic equation 12.16

$$UGR = 8\log_{10}\left[\sum \frac{0.25}{L_b} \frac{L^2\omega}{p^2}\right] \tag{12.16}$$

where

- L_b is the background luminance (cd/m²), calculated as $\frac{E_{ind}}{\pi}$, in which E_{ind} is the vertical indirect illuminance at the observer eye and thus, the average luminance that created this illuminance is $\frac{E_{ind}}{\pi}$.

- L is the luminance of the luminous parts of each luminaire in the direction of the observer's eye (cd/m^2).

- ω is the solid angle of the luminous parts of each luminaire at the observer's eye (steradian).

- p is the Guth Position Index for each individual luminaire, which relates to its displacement from the line of sight.

Note that there is a lower limit of ω for which the equation works, but the eye has limited ability to focus on small objects. However, no allowance is made for this in these calculations which are based on the procedure used in CIE 190 (CIE, 2010).

Standard conditions are defined for the calculation of UGR tables; the conditions are as follows:

- The position of the complete array of luminaires is shown in Figure 12.11.

- The observer is located at the mid-point marked O of a wall and has a horizontal line of sight towards the centre of the opposite wall.

- The height of the luminaires' centre above the observer's eye level is $H = 2$ m.

- The spacing of the luminaires is 2 m in both x_T and y_R directions where x_T is the horizontal distance between vertical planes through the luminaire centre and through the observer's eye position. Both are parallel to the direction of view, and y_R is the horizontal distance parallel to the viewing direction from the observer's eye position to the vertical plane. This is perpendicular to the viewing direction through the luminaire centre. See Figures 12.12 and 12.13. The spacing to height ratio (SHR) is 1:1. The height of the wall is 2 m.

- The horizontal reference plane is at the observer eye level at 1.2 m above the floor.

- The room dimensions X and Y are expressed in terms of H (the mounting height) and where the X dimension is perpendicular to the line of sight and the Y dimension is parallel to the line of sight.

- The luminous intensity distribution of the luminaire (I table) is provided in the normalised form of cd/1000 lm.

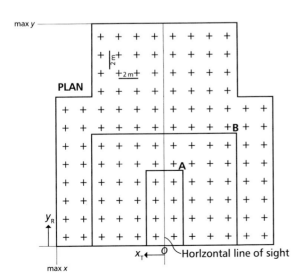

Figure 12.11 The position of luminaires in the standard array area, with examples of area A, size $2H \times 4H$ and area B, size $8H \times 6H$

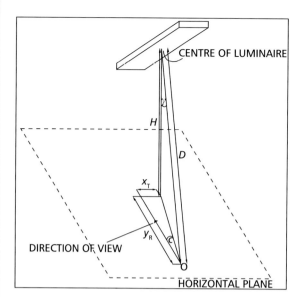

Figure 12.12 The position of luminaire centre relative to observer

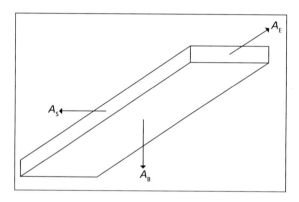

Figure 12.13 The projected areas for linear luminaire

The table is generated with the basic equation (12.16) but rearranged and simplified using preset conditions and values. This may be expressed in terms of the apparent luminaire surface area, the distance to the luminaire, the intensity towards the observer, the position index and the indirect illuminance on the wall produced by the luminaire. The revised equation is stated as 12.17

$$UGR = 8\log_{10}\left[\sum \frac{0.25\pi}{E_{\mathrm{WID}}} \frac{I_{C\gamma}^2}{A^2} \frac{A}{D^2} \frac{1}{p^2}\right]$$ (12.17)

This equation can be further simplified for the standard luminaire arrangements to 12.18

$$UGR = 8\log_{10}\left[\sum \frac{K}{E_{\mathrm{WID}}} \frac{I_{C\gamma}^2}{A}\right]$$

$$UGR = 8\log_{10}\left[\sum \frac{KI_{C\gamma}^2}{A}\right] - 8\log_{10} E_{\mathrm{WID}}$$ (12.18)

where:

- $K = \dfrac{\pi}{4p^2D^2}$

- $I_{C\gamma}$ is the luminous intensity of the source at the angles to the downward vertical γ and of azimuth C, appropriate to the eye position O of the observer and the viewing direction – see Figure 12.11

- $C = \tan^{-1} \dfrac{x_T}{y_R}$

- $\gamma = \cos^{-1} \dfrac{H}{D}$

- A is the projected luminous area of the source (m²) from the observer position O (see Figure 12.11).

Note: A is dependent on the viewing direction and may be calculated using equations 12.19 to 12.21. For crosswise viewing, equation 12.19 should be used and for endwise view, equation 12.20.

For crosswise view: $\qquad A = A_B \dfrac{H}{D} + A_S \dfrac{y_R}{D} + A_E \dfrac{x_T}{D}$ $\qquad\qquad$ (12.19)

For endwise view: $\qquad\quad A = A_B \dfrac{H}{D} + A_S \dfrac{x_T}{D} + A_E \dfrac{y_R}{D}$ $\qquad\qquad$ (12.20)

$\qquad D = \sqrt{H^2 + x_T^2 + y_R^2}$ $\qquad\qquad\qquad\qquad\qquad\qquad\qquad\qquad$ (12.21)

- E_{WID} is the indirect component of the illuminance on the walls

Note: The pre-calculated parameters of K, azimuth (C), elevation (γ), $\dfrac{H}{D}$, $\dfrac{x_T}{D}$, $\dfrac{y_R}{D}$, are given in Table 12.10.

Table 12.10 Pre-calculated parameters for luminaires in the standard array

$\dfrac{x_T}{H} = 0.5$

$\dfrac{y_R}{H}$	C (°)	γ (°)	K	$\dfrac{H}{D}$	$\dfrac{y_R}{D}$	$\dfrac{x_T}{D}$
0.5	45.00	35.26	n/a	0.8165	0.4082	0.4082
1.5	18.43	57.69	0.00412	0.5345	0.8018	0.2673
2.5	11.31	68.58	0.00541	0.3651	0.9129	0.1826
3.5	8.13	74.21	0.00473	0.2722	0.9526	0.1361
4.5	6.34	77.55	0.00386	0.2157	0.9705	0.1078
5.5	5.19	79.74	0.00308	0.1782	0.9800	0.0891
6.5	4.40	81.28	0.00243	0.1516	0.9855	0.0758
7.5	3.81	82.42	0.00197	0.1319	0.9891	0.0659
8.5	3.37	83.30	0.00163	0.1166	0.9915	0.0583
9.5	3.01	84.00	0.00137	0.1045	0.9931	0.0523
10.5	2.73	84.57	0.00116	0.0947	0.9944	0.0474
11.5	2.49	85.03	0.00100	0.0865	0.9953	0.0433

$\dfrac{x_T}{H} = 1.5$

$\dfrac{y_R}{H}$	C (°)	γ (°)	K	$\dfrac{H}{D}$	$\dfrac{y_R}{D}$	$\dfrac{x_T}{D}$
0.5	71.57	57.69	n/a	0.5345	0.2673	0.8018
1.5	45.00	64.76	0.00155	0.4264	0.6396	0.6396
2.5	30.96	71.07	0.00294	0.3244	0.8111	0.4867
3.5	23.20	75.29	0.00329	0.2540	0.8890	0.3810
4.5	18.43	78.10	0.00292	0.2063	0.9283	0.3094
5.5	15.26	80.05	0.00249	0.1728	0.9503	0.2592
6.5	12.99	81.47	0.00209	0.1482	0.9636	0.2224
7.5	11.31	82.55	0.00177	0.1296	0.9723	0.1945
8.5	10.01	83.39	0.00150	0.1151	0.9782	0.1726
9.5	8.97	84.06	0.00129	0.1034	0.9825	0.1551
10.5	8.13	84.61	0.00111	0.0939	0.9856	0.1408
11.5	7.43	85.07	0.00097	0.0859	0.9879	0.1289

$\dfrac{x_T}{H} = 2.5$

$\dfrac{y_R}{H}$	C (°)	γ (°)	K	$\dfrac{H}{D}$	$\dfrac{y_R}{D}$	$\dfrac{x_T}{D}$
0.5	78.69	68.58	n/a	0.3651	0.1826	0.9129
1.5	59.04	71.07	0.00053	0.3244	0.4867	0.8111
2.5	45.00	74.21	0.00119	0.2722	0.6804	0.6804
3.5	35.54	76.91	0.00166	0.2265	0.7926	0.5661

$\dfrac{x_T}{H} = 3.5$

$\dfrac{y_R}{H}$	C (°)	γ (°)	K	$\dfrac{H}{D}$	$\dfrac{y_R}{D}$	$\dfrac{x_T}{D}$
0.5	81.87	74.21	n/a	0.2722	0.1361	0.9526
1.5	66.80	75.29	0.00024	0.2540	0.3810	0.8890
2.5	54.46	76.91	0.00053	0.2265	0.5661	0.7926
3.5	45.00	78.58	0.00083	0.1980	0.6931	0.6931

Table 12.10 Continued

2.5

$\frac{y_R}{H}$	$C\,(°)$	$\gamma\,(°)$	K	$\frac{H}{D}$	$\frac{y_R}{D}$	$\frac{x_T}{D}$
4.5	29.05	79.01	0.00183	0.1907	0.8581	0.4767
5.5	24.44	80.60	0.00176	0.1633	0.8981	0.4082
6.5	21.04	81.83	0.00159	0.1421	0.9239	0.3553
7.5	18.43	82.79	0.00140	0.1255	0.9412	0.3137
8.5	16.39	83.56	0.00124	0.1122	0.9533	0.2804
9.5	14.74	84.19	0.00109	0.1013	0.9621	0.2532
10.5	13.39	84.71	0.00096	0.0923	0.9687	0.2306
11.5	12.26	85.14	0.00084	0.0847	0.9737	0.2117

3.5

$\frac{y_R}{H}$	$C\,(°)$	$\gamma\,(°)$	K	$\frac{H}{D}$	$\frac{y_R}{D}$	$\frac{x_T}{D}$
4.5	37.87	80.05	0.00105	0.1728	0.7775	0.6047
5.5	32.47	81.28	0.00115	0.1516	0.8339	0.5307
6.5	28.30	82.29	0.00113	0.1342	0.8725	0.4698
7.5	25.02	83.11	0.00106	0.1200	0.8996	0.4198
8.5	22.38	83.79	0.00099	0.1081	0.9193	0.3785
9.5	20.22	84.36	0.00090	0.0983	0.9338	0.3440
10.5	18.43	84.84	0.00081	0.0900	0.9448	0.3149
11.5	16.93	85.24	0.00073	0.0829	0.9534	0.2902

4.5

$\frac{y_R}{H}$	$C\,(°)$	$\gamma\,(°)$	K	$\frac{H}{D}$	$\frac{y_R}{D}$	$\frac{x_T}{D}$
0.5	83.66	77.55	n/a	0.2157	0.1078	0.9705
1.5	71.57	78.10	0.00015	0.2063	0.3094	0.9283
2.5	60.95	79.01	0.00027	0.1907	0.4767	0.8581
3.5	52.13	80.05	0.00045	0.1728	0.6047	0.7775
4.5	45.00	81.07	0.00059	0.1552	0.6985	0.6985
5.5	39.29	81.99	0.00072	0.1393	0.7664	0.6271
6.5	34.70	82.79	0.00077	0.1255	0.8157	0.5647
7.5	30.96	83.48	0.00078	0.1136	0.8519	0.5112

5.5

$\frac{y_R}{H}$	$C\,(°)$	$\gamma\,(°)$	K	$\frac{H}{D}$	$\frac{y_R}{D}$	$\frac{x_T}{D}$
0.5	84.81	79.74	n/a	0.1782	0.0891	0.9800
1.5	74.74	80.05	n/a	0.1728	0.2592	0.9503
2.5	65.56	80.60	0.00017	0.1633	0.4082	0.8981
3.5	57.53	81.28	0.00026	0.1516	0.5307	0.8339
4.5	50.71	81.99	0.00036	0.1393	0.6271	0.7664
5.5	45.00	82.67	0.00044	0.1275	0.7013	0.7013
6.5	40.24	83.30	0.00052	0.1166	0.7582	0.6415
7.5	36.25	83.86	0.00056	0.1069	0.8018	0.5880

The indirect component of the illuminance on the walls can be calculated by the method given below using equation 12.22

$$E_{\text{WID}} = \frac{UF_{\text{WID}} \; N \; \Phi_0}{A_{\text{W}}}$$

(12.22)

where:

- UF_{WID} = indirect utilisation factor for walls

- N = number of luminaires

- A_{W} = total area of walls (m^2) between reference plane and luminaire plane

- Φ_0 = 1000 lm

This may be simplified to equation 12.23

$$E_{\text{WID}} = UF_{\text{WID}} \; B$$

(12.23)

where:

$$B = \frac{1000 \; N}{A_{\text{W}}}$$

(12.24)

$$UF_{\text{WID}} = DF(F)TF(F,W) + DF(W)(TF(W,W) - 1) + DF(C)TF(C,W)$$

(12.25)

The distribution factors $DF(F)$, $DF(W)$ and $DF(C)$ may be calculated using a similar method to that set out in section 12.3.3, however, the geometric multipliers $GML1$, $GML2$, $GML3$ and $GML4$ should be taken from Table 12.11, which also gives the values of B. Table 12.12 gives the transfer factors needed.

Table 12.11 Values of B and geometric multipliers

X dimension	Y dimension	B	GM1	GM2	GM3	GM4
2H	2H	125.00	0.690	0.109	0.085	–0.016
	3H	150.00	0.578	0.200	0.127	–0.018
	4H	166.67	0.528	0.218	0.170	–0.017
	6H	187.50	0.485	0.215	0.222	–0.012
	8H	200.00	0.466	0.207	0.249	–0.006
	12H	214.29	0.448	0.198	0.272	0.005
4H	2H	166.67	0.528	0.218	0.170	–0.017
	3H	214.29	0.394	0.275	0.268	–0.020
	4H	250.00	0.338	0.257	0.351	–0.018
	6H	300.00	0.296	0.203	0.449	–0.006
	8H	333.33	0.280	0.165	0.499	0.006
	12H	375.00	0.264	0.125	0.541	0.027
8H	4H	333.33	0.280	0.165	0.499	0.006
	6H	428.57	0.248	0.058	0.628	0.032
	8H	500.00	0.239	–0.012	0.690	0.058
	12H	600.00	0.232	–0.084	0.740	0.098
12H	4H	375.00	0.264	0.125	0.541	0.027
	6H	500.00	0.238	–0.003	0.677	0.063
	8H	600.00	0.232	–0.084	0.740	0.098

Table 12.12 Transfer factors

			Reflectances								
		c	0.5	0.5	0.5	0.7	0.7	0.7	0.8	0.8	0.8
		w	0.3	0.5	0.7	0.3	0.5	0.7	0.7	0.5	0.7
XH	YH	*f*	0.2	0.2	0.2	0.2	0.2	0.2	0.2	0.2	0.2
2	2	TF(F,W)	0.1722	0.1984	0.2341	0.1882	0.2199	0.2645	0.1964	0.2313	0.2811
		TF(W,W)−1	0.1982	0.3807	0.6288	0.2165	0.4218	0.7103	0.2260	0.4435	0.7547
		TF(C,W)	0.3861	0.4449	0.5248	0.5526	0.6459	0.7770	0.6388	0.7521	0.9142
	3	TF(F,W)	0.1565	0.1777	0.2054	0.1727	0.1987	0.2339	0.1811	0.2098	0.2494
		TF(W,W)−1	0.1785	0.3377	0.5467	0.1962	0.3762	0.6200	0.2054	0.3966	0.6599
		TF(C,W)	0.3462	0.3930	0.4544	0.4966	0.5714	0.6726	0.5746	0.6658	0.7913
	4	TF(F,W)	0.1476	0.1661	0.1900	0.1637	0.1867	0.2171	0.1721	0.1975	0.2319
		TF(W,W)−1	0.1672	0.3137	0.5023	0.1845	0.3506	0.5709	0.1934	0.3702	0.6083
		TF(C,W)	0.3241	0.3648	0.4172	0.4654	0.5307	0.6173	0.5389	0.6186	0.7262
	6	TF(F,W)	0.1380	0.1538	0.1738	0.1538	0.1736	0.1993	0.1621	0.1842	0.2132
		TF(W,W)−1	0.1545	0.2872	0.4542	0.1713	0.3222	0.5178	0.1800	0.3408	0.5524
		TF(C,W)	0.3007	0.3352	0.3787	0.4323	0.4880	0.5602	0.5008	0.5691	0.6589
	8	TF(F,W)	0.1329	0.1474	0.1654	0.1485	0.1668	0.1901	0.1567	0.1771	0.2035
		TF(W,W)−1	0.1474	0.2725	0.4282	0.1639	0.3066	0.4892	0.1724	0.3247	0.5224
		TF(C,W)	0.2884	0.3199	0.3590	0.4150	0.4659	0.5311	0.4810	0.5435	0.6246
	12	TF(F,W)	0.1276	0.1407	0.1568	0.1431	0.1596	0.1806	0.1511	0.1697	0.1935
		TF(W,W)−1	0.1397	0.2567	0.4005	0.1558	0.2898	0.4589	0.1642	0.3073	0.4905
		TF(C,W)	0.2760	0.3043	0.3391	0.3973	0.4434	0.5015	0.4607	0.5173	0.5898
4	2	TF(F,W)	0.1476	0.1661	0.1900	0.1637	0.1867	0.2171	0.1721	0.1975	0.2319
		TF(W,W)−1	0.1672	0.3137	0.5023	0.1845	0.3506	0.5709	0.1934	0.3702	0.6083
		TF(C,W)	0.3241	0.3648	0.4172	0.4654	0.5307	0.6173	0.5389	0.6186	0.7262
	3	TF(F,W)	0.1263	0.1396	0.1560	0.1417	0.1585	0.1797	0.1497	0.1685	0.1925

Table 12.12 Continued

XH	YH	f	Reflectances								
		c	0.8	0.8	0.8	0.7	0.7	0.7	0.5	0.5	0.5
		w	0.7	0.5	0.3	0.7	0.5	0.3	0.7	0.5	0.3
		f	0.2	0.2	0.2	0.2	0.2	0.2	0.2	0.2	0.2
4	4	TF(W,W)–1	0.5003	0.3126	0.1667	0.4690	0.2954	0.1585	0.4115	0.2630	0.1428
		TF(C,W)	0.5855	0.5122	0.4553	0.4980	0.4391	0.3927	0.3369	0.3014	0.2727
		TF(F,W)	0.1707	0.1516	0.1363	0.1590	0.1422	0.1286	0.1375	0.1246	0.1139
	6	TF(W,W)–1	0.4421	0.2804	0.1513	0.4144	0.2646	0.1436	0.3632	0.2350	0.1289
		TF(C,W)	0.5108	0.4535	0.4078	0.4347	0.3887	0.3514	0.2943	0.2666	0.2437
		TF(F,W)	0.1474	0.1329	0.1211	0.1370	0.1244	0.1139	0.1180	0.1084	0.1002
	8	TF(W,W)–1	0.3794	0.2445	0.1336	0.3553	0.2304	0.1266	0.3110	0.2040	0.1132
		TF(C,W)	0.4337	0.3913	0.3565	0.3692	0.3352	0.3069	0.2502	0.2297	0.2124
		TF(F,W)	0.1351	0.1230	0.1129	0.1256	0.1149	0.1060	0.1079	0.0999	0.0929
	12	TF(W,W)–1	0.3452	0.2244	0.1236	0.3231	0.2113	0.1169	0.2823	0.1866	0.1042
		TF(C,W)	0.3944	0.3590	0.3294	0.3358	0.3074	0.2835	0.2276	0.2106	0.1960
		TF(F,W)	0.1225	0.1126	0.1042	0.1137	0.1051	0.0977	0.0976	0.0910	0.0853
8	4	TF(W,W)–1	0.3080	0.2022	0.1123	0.2880	0.1900	0.1060	0.2509	0.1672	0.0940
		TF(C,W)	0.3546	0.3259	0.3015	0.3020	0.2790	0.2593	0.2048	0.1911	0.1791
		TF(F,W)	0.1351	0.1230	0.1129	0.1256	0.1149	0.1060	0.1079	0.0999	0.0929
	6	TF(W,W)–1	0.3452	0.2244	0.1236	0.3231	0.2113	0.1169	0.2823	0.1866	0.1042
		TF(C,W)	0.3944	0.3590	0.3294	0.3358	0.3074	0.2835	0.2276	0.2106	0.1960
		TF(F,W)	0.1088	0.1007	0.0937	0.1009	0.0939	0.0877	0.0864	0.0810	0.0763
	8	TF(W,W)–1	0.2809	0.1858	0.1037	0.2632	0.1749	0.0981	0.2304	0.1544	0.0872
		TF(C,W)	0.3117	0.2886	0.2686	0.2656	0.2470	0.2309	0.1802	0.1691	0.1593
		TF(F,W)	0.0949	0.0887	0.0832	0.0880	0.0826	0.0777	0.0752	0.0710	0.0674
		TF(W,W)–1	0.2466	0.1646	0.0926	0.2311	0.1549	0.0875	0.2025	0.1367	0.0778

Table 12.12 Continued

XH	YH		Reflectances								
		c	0.8	0.8	0.8	0.7	0.7	0.7	0.5	0.5	0.5
		w	0.7	0.5	0.3	0.7	0.5	0.3	0.7	0.5	0.5
		f	0.2	0.2	0.2	0.2	0.2	0.2	0.2	0.2	0.3
8	12	TF(C,W)	0.2695	0.2517	0.2362	0.2297	0.2155	0.2029	0.1560	0.1474	0.1398
		TF(F,W)	0.0807	0.0761	0.0720	0.0747	0.0707	0.0672	0.0637	0.0607	0.0580
		TF(W,W)−1	0.2093	0.1411	0.0801	0.1962	0.1327	0.0756	0.1718	0.1170	0.0671
	6	TF(C,W)	0.2267	0.2139	0.2025	0.1933	0.1831	0.1738	0.1314	0.1252	0.1196
		TF(F,W)	0.1225	0.1126	0.1042	0.1137	0.1051	0.0977	0.0976	0.0910	0.0853
		TF(W,W)−1	0.3080	0.2022	0.1123	0.2880	0.1900	0.1060	0.2509	0.1672	0.0940
12	6	TF(C,W)	0.3546	0.3259	0.3015	0.3020	0.2790	0.2593	0.2048	0.1911	0.1791
		TF(F,W)	0.0951	0.0889	0.0835	0.0881	0.0828	0.0780	0.0753	0.0712	0.0676
		TF(W,W)−1	0.2432	0.1625	0.0915	0.2277	0.1527	0.0864	0.1989	0.1345	0.0766
	8	TF(C,W)	0.2700	0.2525	0.2370	0.2301	0.2161	0.2036	0.1563	0.1479	0.1403
		TF(F,W)	0.0807	0.0761	0.0720	0.0747	0.0707	0.0672	0.0637	0.0607	0.0580
		TF(W,W)−1	0.2093	0.1411	0.0801	0.1962	0.1327	0.0756	0.1718	0.1170	0.0671
		TF(C,W)	0.2267	0.2139	0.2025	0.1933	0.1831	0.1738	0.1314	0.1252	0.1196

Chapter 13: Indoor lighting calculations

The main calculation required in indoor lighting is that of calculating the illuminance on any given plane, whether this is on a real solid object or a plane in space. This sort of calculation is quite straightforward and methods are given in Chapters 10 and 11, the only complexity being dealing with the geometry of any given situation.

However, the approach of only considering planes on which light is falling is very limiting if one is trying to think about objects that might appear in the space. To obtain an understanding of this, it is necessary to consider the light arriving at the object from all possible directions. Hence, the concept of cubic illuminance (Cuttle, 1997) has been developed.

13.1 Introduction

In cubic illumination, the illuminance on the six faces of a small virtual cube is considered. From these illuminances, it is possible to consider the flow of light at the centre of the cube as a vector and the x, y and z components of the vector give both its magnitude and direction. To understand how the various relationships work, it is first necessary to know a little of the mathematics of vectors.

To start the consideration of vectors, it is first necessary to look at how space is defined in terms of x, y and z. Figure 13.1 shows a set of orthogonal axes.

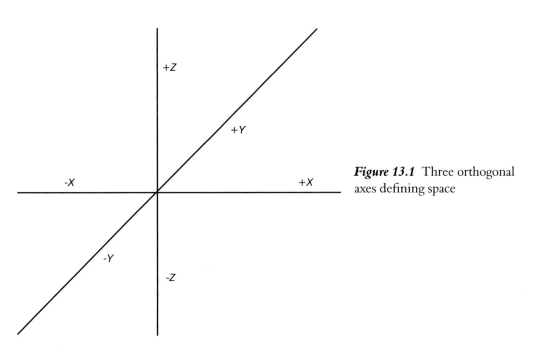

Figure 13.1 Three orthogonal axes defining space

Given that space is defined by the set of axes, it is now possible to define any point in space relative to the origin of the axes by x, y and z values of the point. It is also possible to define the point by its distance from the origin and the x, y and z components of that distance.

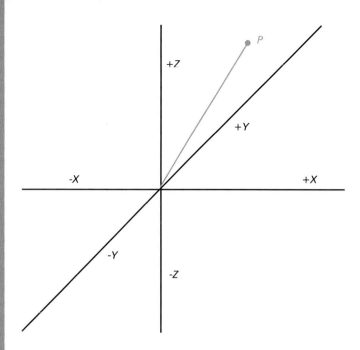

Figure 13.2 A point in space

Consider point P in Figure 13.2. We can define the position vector of point P (P) as:

$$P = \left(P_{(x)}, P_{(y)}, P_{(z)}\right) \tag{13.1}$$

where $P_{(x)}$, $P_{(y)}$ and $P_{(z)}$ are the x, y and z coordinates of point P relative to the origin.

The distance of point P from the origin or magnitude of vector P is given by equation 13.2:

$$|P| = \sqrt{P_{(x)}^2 + P_{(y)}^2 + P_{(z)}^2} \tag{13.2}$$

The unit vector of P that defines the direction but not the distance is defined by equation 13.3:

$$p = \left(p_{(x)}, p_{(y)}, p_{(z)}\right) = \left(\frac{P_{(x)}}{|P|}, \frac{P_{(y)}}{|P|}, \frac{P_{(z)}}{|P|}\right) \tag{13.3}$$

One other useful tool of vector mathematics is the dot product; it is useful for calculating the component of a given vector on another. In Figure 13.3, a unit vector p is show next to the unit vector n that represents the normal to the surface.

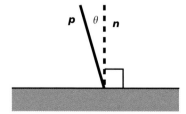

Figure 13.3 The dot product of unit vectors

The dot product ($p.n$) of the two vectors is calculated using equation 13.4:

$$p.n = p_{(x)}.n_{(x)} + p_{(y)}.n_{(y)} + p_{(z)}.n_{(z)} \qquad (13.4)$$

When considering the case shown in Figure 13.3, then the dot product $p.n$ is cos θ.

13.2 The illumination vector

The illuminance received at a point may be considered as a vector. Figure 13.4 shows the illuminance due to a source S. The vector E is the illumination vector and the distance from the origin of the axes and the circle indicates the relative illuminance falling on a plane normal to the line joining the origin to the circle.

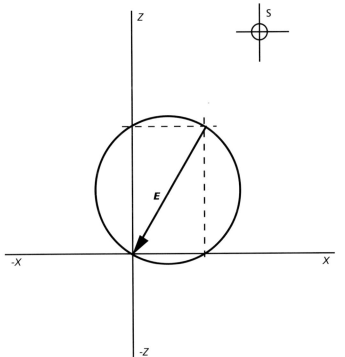

Figure 13.4 The illumination vector

The situation where there is only one light source illuminating a point is quite unusual and in general, there is light falling from a number of different sources on to the point. Thus the illumination vector would become the sum of two or more vectors. As a vector may be a sum of more than one component, it is also possible to analyse vectors into a series of components.

13.3 Cubic illuminance

To characterise the pattern of illuminance at any given point, it is necessary to measure or calculate the illuminance falling on the six faces of a small cube, see Figure 13.5.

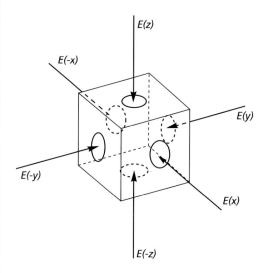

E(z)

E(-x)

E(y)

E(-y)

E(x)

E(-z)

Figure 13.5 The illuminance on six faces of a small cube

From such an assessment of illuminance, the six values obtained are $E_{(+x)}$, $E_{(-x)}$, $E_{(+y)}$, $E_{(-y)}$, $E_{(+z)}$ and $E_{(-z)}$. From these values, it is common to calculate the vector ($'E$) and symmetric ($\sim E$) components. The process is given in equation 13.5:

$$'E = \left('E_{(x)}, 'E_{(y)}, 'E_{(z)}\right) \qquad \sim E = \left(\sim E_{(x)}, \sim E_{(y)}, \sim E_{(z)}\right)$$

$$'E_{(X)} = E_{(+X)} - E_{(-X)} \qquad \sim E_{(X)} = \min\left(E_{(+X)}, E_{(-X)}\right)$$

$$'E_{(Y)} = E_{(+Y)} - E_{(-Y)} \qquad \sim E_{(Y)} = \min\left(E_{(+Y)}, E_{(-Y)}\right) \qquad (13.5)$$

$$'E_{(Z)} = E_{(+Z)} - E_{(-Z)} \qquad \sim E_{(x)} = \min\left(E_{(+Z)}, E_{(-Z)}\right)$$

The symmetric component of the vector in this analysis is the amount of light that is equally received on each side of the cube and the $'E$ vector components are the differences between each pair of opposite sides of the cube.

The application of equation 13.5 is illustrated in Table 13.1.

Table 13.1 Example of cubic illuminance data and some derived values

+x	480	−x	210	'x	270	~x	210
+y	270	−y	270	'y	0	~y	270
+z	660	−z	135	'z	525	~z	135

The magnitude of the illuminance vector $|E|$ is 590 and the unit vector p = (0.457, 0.000, 0.889).

From the data, it also possible to calculate the illuminance solids for the point in question. There are three lines in Figure 13.6, one for the vector component, one for the symmetric component and one for the total.

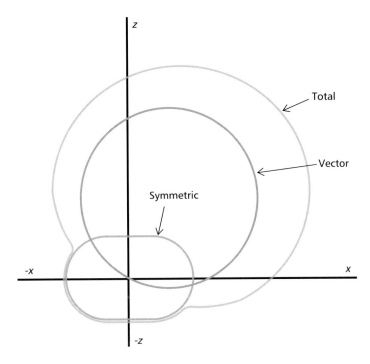

Figure 13.6 The illumination solids for the data in Table 13.1

13.4 Derived values

Given that cubic illuminances carry a lot of information about the illumination at a point, it is not surprising that it is possible to derive some other parameters that describe the illuminance from them.

Planar illuminance
The planar illuminance (E_{pr}) may be calculated using equation 13.6 where n is the unit vector of the normal to the plane:

$$E_{pr} = |E| e.n + \sim E_{(x)} n^2_{(x)} + \sim E_{(y)} n^2_{(y)} + \sim E_{(z)} n^2_{(z)} \qquad (13.6)$$

Scalar illuminance
Scalar illuminance E_{sr} is the average illuminance on the surface of a small sphere and is given by equation 13.7:

$$E_{sr} = \frac{|E|}{4} + \frac{\left(\sim E_{(x)} + \sim E_{(y)} + \sim E_{(z)} \right)}{3} \qquad (13.7)$$

Hemispherical illuminance
Hemispherical illuminance E_{hs}, is the average illuminance falling onto the curved side of a hemisphere. The normal to the centre of the flat side through the curved side has the unit vector n, and the value may be calculated with equation 13.8:

$$E_{hs} = \frac{|E|(1 + e.n)}{4} + \frac{\left(\sim E_{(x)} + \sim E_{(y)} + \sim E_{(z)} \right)}{3} \qquad (13.8)$$

Cylindrical illuminance

Cylindrical illuminance E_{cl} is the average illuminance on the curved surface of a cylinder, the axis of the cylinder is vertical. Equation 13.9 may be used to calculate cylindrical illuminance:

$$E_{cl} = \frac{|E|e.e_{(x,y)}}{\pi} + \frac{\left(\sim E_{(x)} + \sim E_{(y)}\right)}{2}$$

(13.9)

Semi-cylindrical illuminance

Semi-cylindrical illuminance E_{scl}, is the average illuminance on the curved surface of a semi-cylinder. It is assumed that the axis of the semi-cylinder through the middle of the flat side is vertical. The direction of the curved surface is defined by the unit vector n is the normal to the flat side and passes through the curved side. Semi-cylindrical illuminance may be calculated with equation 13.10:

$$E_{scl} = \frac{|E|\left(e.e_{(x,y)}\right)\left(1 + e_{(x,y)}.n_{(x,y)}\right)}{\pi} + \frac{\left(\sim E_{(x)} + \sim E_{(y)}\right)}{2}$$

(13.10)

Chapter 14: Outdoor lighting calculations

This chapter on outdoor lighting calculations takes many of the calculation techniques from the European Standard method of road lighting calculation (BSI, 2003c). There are many other standard methods of calculation available, most notably CIE (2000). The differences in the methods are minor and generally the difference in results between the methods small. In virtually all outdoor lighting calculations, inter-reflected light is neglected. The difficult part of most calculations is finding the intensity towards a given point from a luminaire; this is complex as the luminaires may be aimed in a variety of ways making it hard to calculate the photometric angles C and γ. The first section of this chapter covers the calculation of intensity towards a point.

The main things that need to be calculated are illuminance (sometimes including semi-cylindrical and semi-spherical illuminance), road surface luminance and glare. The calculation of road surface luminance requires knowledge of the reflective properties of the road surface. Section 14.2 discusses the way that road surfaces are dealt with as this is required before the luminance calculation process can be considered.

14.1 Calculation of intensity towards a point

To determine the luminous intensity $I(C, \gamma)$ from a luminaire to a point, it is necessary to find the vertical photometric angle (γ) and photometric azimuth (C) of the light path to the point. To do this, account has to be taken of the tilt in application in relation to the tilt during measurement, the orientation, and rotation of the luminaire. For this purpose, mathematical sign conventions for measuring distances on the road and for rotations about axes have to be established. The system used is a right-handed Cartesian co-ordinate system as shown in Figure 14.1.

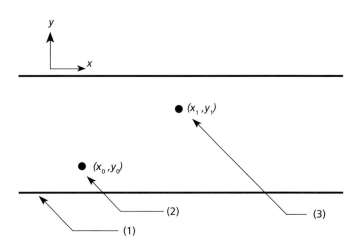

Figure 14.1 Geometry of calculation point (2), luminaire (3) and road (1)

Figure 14.1 is taken from EN 13201-3 (BSI, 2003c) and illustrates the calculations on a road; however, the same general principles apply to all outdoor calculations.

Equations 14.1 and 14.2 give the separation between the luminaire and the point in terms of X and Y:

$$X = X_p - X_1 \qquad (14.1)$$

$$Y = Y_p - Y_1 \qquad (14.2)$$

where X_p and Y_p are the coordinates of the point and X_1 and Y_1 are the coordinates of the luminaire. The luminaire may be rotated or tilted on each of three axes. Figure 14.2 show the various rotations.

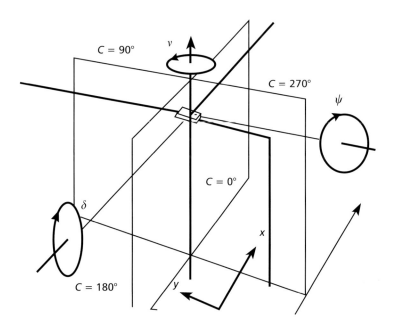

Figure 14.2 Rotation of the luminaire

14.1.1 Calculation of C and γ

There are four stages in the calculation of C and γ.

First substitute the values of X, Y, H, υ, γ and δ to obtain X', Y' and H' in equation 14.3.

$$
\begin{aligned}
X' &= X\left(\cos\upsilon\cos\psi - \sin\upsilon\sin\psi\right) + Y\left(\sin\upsilon\sin\psi - \cos\upsilon\sin\delta\sin\psi\right) \\
&\quad + H\cos\delta\sin\psi \\
Y' &= X\sin\upsilon\cos\delta + Y\cos\upsilon\cos\delta - H\sin\delta \\
H' &= H\cos\delta\cos\psi - X\left(\sin\upsilon\sin\delta\cos\psi - \cos\upsilon\sin\psi\right) \\
&\quad - Y\left(\sin\upsilon\sin\psi - \cos\upsilon\sin\delta\cos\psi\right)
\end{aligned} \qquad (14.3)
$$

where X and Y are the longitudinal and transverse distances between the calculation point and the luminaire, H is the height of the luminaire above the calculation point, X', Y' and H' are distances for the calculation of C and γ and may be regarded as intermediate variables, and v, δ, and ψ are tilts of the lantern as shown in Figure 14.2.

Note that, in a number of installations such as straight roads, car parks and some industrial yards, it is only angle δ that is fully variable, ψ is usually zero and v is either 0° or 180°. Under these constraints, equations 14.3 may be simplified to give equations 14.4:

$$X' = X \cos v$$
$$Y' = Y \cos v \cos \delta - H \sin \delta \qquad (14.4)$$
$$H' = H \cos \delta + Y \cos v \sin \delta$$

The second stage is the evaluation of the installation azimuth angle φ. It is given by equation 14.5:

$$\varphi = \arctan \left(\frac{Y'}{X'} \right) \qquad (14.5)$$

Care must be exercised when using 14.5 as when X' is zero, $\frac{Y'}{X'}$ becomes infinite so the φ becomes 90° if $Y' > 0$ and –90° if $Y' < 0$. If Y' is zero, then any value of φ may be used as the point must line up with the axis of the luminaire and the γ angle is zero.

A second problem with equation 14.5 is that the arctan function generally returns a value between –90° and +90° so it is necessary to check in which quadrant the point lies. It is also normal to work with the angle range 0 to 360°. The equations given in 14.6 should be used.

$$
\begin{array}{ll}
\text{For } X' > 0,\, Y' > 0 & \varphi = \arctan (Y'/X') \\
\text{For } X' < 0,\, Y' > 0 & \varphi = 180° + \arctan (Y'/X') \\
\text{For } X' < 0,\, Y' < 0 & \varphi = 180° + \arctan (Y'/X') \\
\text{For } X' > 0,\, Y' < 0 & \varphi = 360° + \arctan (Y'/X')
\end{array}
\qquad (14.6)
$$

Note: a number of calculation tools provide an arctan function that takes two arguments, for example ATAN2 in Excel; this type of function makes the calculation much simpler as it is only necessary to provide values of X' and Y'.

The third step is to calculate C with equation 14.7:

$$C = \varphi - v \qquad (14.7)$$

The fourth step is the calculation of γ using equation 14.8:

$$\gamma = \arctan \left(\frac{\sqrt{X'^2 + Y'^2}}{H'} \right) \qquad (14.8)$$

14.1.2 Finding the intensity value *I*

It is necessary to find the intensity value for the direction C and γ calculated for a given point. In general, this is not likely to coincide with an exact point in the measured intensity table for the luminaire so some form of interpolation is necessary.

If the values in the photometric table have been measured at close spacings ($\Delta\gamma < 2.5°$ and $\Delta C < 5°$), then linear interpolation may be used, however, at larger angular separations, quadratic interpolation should be used.

Linear interpolation

To estimate the luminous intensity $I(C, \gamma)$, it is necessary to interpolate between four values of luminous intensity lying closest to the direction. The situation is illustrated in Figure 14.3.

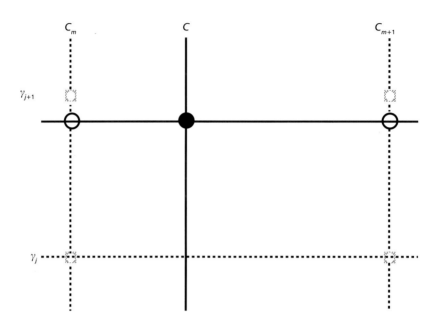

Figure 14.3 Angles required for linear interpolation

In Figure 14.3 and equations 14.9, the following symbols are used:
C is the azimuth, measured about the first photometric axis
γ is the vertical angle measured from the first photometric axis
$j, j+1, m, m+1$ are integers indicating the number of the column or row in the I-table.

The intensity $I(C, \gamma)$ may be found using equations 14.9

$$I(C,\gamma_j) = I(C_m,\gamma_j) + \frac{(C - C_m)(I(C_{m+1},\gamma_j) - I(C_m,\gamma_j))}{C_{m+1} - C_m}$$

$$I(C,\gamma_{j+1}) = I(C_m,\gamma_{j+1}) + \frac{(C - C_m)(I(C_{m+1},\gamma_{j+1}) - I(C_m,\gamma_{j+1}))}{C_{m+1} - C_m} \qquad (14.9)$$

$$I(C,\gamma) = I(C,\gamma_j) + \frac{(\gamma - \lambda_j)(I(C,\gamma_{j+1}) - I(C,\gamma_j))}{\gamma_{j+1} - \gamma_j}$$

Quadratic interpolation

Quadratic interpolation requires three values in the *I*-table for each interpolated value. Figure 14.4 shows the values needed. If a value of *I* is required at (C, γ), interpolation is first carried out down three adjacent columns of the *I*-table enclosing the point. This enables three values of *I* to be found at γ. Interpolation is then carried out across the table to find the required value at (C, γ). If preferred, this procedure may be reversed; that is, interpolation can be carried out across and then down the *I*-table without affecting the result.

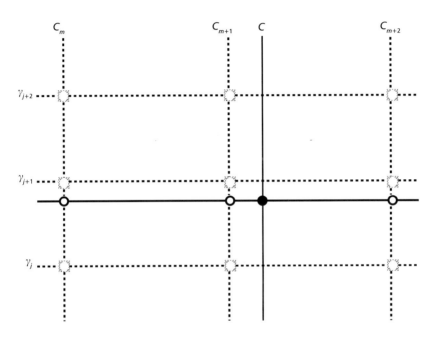

Figure 14.4 Angles required for quadratic interpolation

The following equations give the interpolation starting across the *C* planes; however, the procedure would work equally well working across the γ angles. Initially, calculate the constants K_1, K_2 and K_3 using equations 14.10.

$$K_1 = \frac{(C - C_{m+1})(C - C_{m+2})}{(C_m - C_{m+1})(C_{m+1} - C_{m+2})}$$

$$K_2 = \frac{(C - C_m)(C - C_{m+2})}{(C_{m+1} - C_m)(C_{m+1} - C_{m+2})} \qquad (14.10)$$

$$K_3 = \frac{(C - C_m)(C - C_{m+1})}{(C_{m+2} - C_m)(C_{m+2} - C_{m+1})}$$

It is now possible to calculate three intensity values at the desired value of C but differing at three different γ angles using equations 14.11:

$$I(C,\gamma_j) = K_1 I(C_m,\gamma_j) + K_2 I(C_{m+1},\gamma_j) + K_3 I(C_{m+2},\gamma_j)$$

$$I(C,\gamma_{j+1}) = K_1 I(C_m,\gamma_{j+1}) + K_2 I(C_{m+1},\gamma_{j+1}) + K_3 I(C_{m+2},\gamma_{j+1}) \tag{14.11}$$

$$I(C,\gamma_{j+2}) = K_1 I(C_m,\gamma_{j+2}) + K_2 I(C_{m+1},\gamma_{j+2}) + K_3 I(C_{m+2},\gamma_{j+2})$$

Then it is necessary to interpolate across the different γ angles, so three new constants (k_1, k_2, k_3) are needed.

$$k_1 = \frac{(\gamma - \gamma_{j+1})(\gamma - \gamma_{j+2})}{(\gamma_j - \gamma_{j+1})(\gamma_{j+1} - \gamma_{j+2})}$$

$$k_2 = \frac{(\gamma - C_j)(\gamma - \gamma_{j+2})}{(\gamma_{j+1} - \gamma_j)(\gamma_{j+1} - \gamma_{j+2})} \tag{14.12}$$

$$k_3 = \frac{(\gamma - \gamma_j)(\gamma - \gamma_{j+1})}{(\gamma_{j+2} - \gamma_j)(\gamma_{j+2} - \gamma_{j+1})}$$

Then the value of $I(C, \gamma)$ may be found using equation 14.13:

$$I(C,\lambda) = k_1 I(C,\gamma_j) + k_2 I(C,\gamma_{j+1}) + k_3 I(C,\gamma_{j+2}) \tag{14.13}$$

Note: When performing quadratic interpolation close to $\gamma = 0$, it may be necessary to take one or more of the values for interpolation from the opposite half plane. This means that photometric data files that are used must contain a full set of planes either actually in the data or implied by symmetry. The presence of orphaned half planes may make it impossible to use this sort of interpolation.

14.2   The reflective properties of road surfaces

The reflective properties of road surfaces are described by r-tables. The use of r-tables was developed when computers were less powerful than today and they were originally designed to make calculations easier. To calculate the luminance on a piece of road towards an observer, it would be normal to use equation 14.14:

$$L = q E \tag{14.14}$$

where L is the luminance of the surface, E is the illuminance on the surface and q the luminance coefficient. For a fixed viewing angle α, the value of the luminance coefficient varies with angle of incidence of the incoming light i and the angle of deflection β. Figure 14.5 shows these angles.

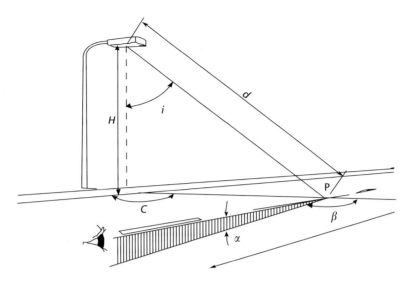

Figure 14.5 Angles in the definition of *r*-tables

The tables of reflectance properties are stored as a table of reduced luminance coefficients (*r*) as defined in equation 14.15:

$$r = q\cos^3 i \tag{14.15}$$

This way of working reduces the calculations needed as the calculation process simplifies as shown in equation 14.16:

$$L = q\,E \quad \text{and} \quad E = \frac{I\cos^3 i}{H^2}$$

$$\text{thus} \quad L = \frac{q\,I\cos^3 i}{H^2} \tag{14.16}$$

$$\therefore L = \frac{I\,r}{H^2}$$

Table 14.1 shows a typical *r*-table. The angles *β* and *i* are shown in Figure 14.5; note that in some *r*-tables, the symbol *γ* is used. However, as *γ* is also used as the photometric elevation angle, its use in this section has been avoided.

14.3 Calculation of illuminance and luminance

With the intensity calculated in section 14.1, it is now possible to calculate planar illuminance using equation 10.1, semi-cylindrical illuminance using equation 10.5 and hemispherical illuminance with equation 10.7. Most of the lighting standards set out rules that determine the number and position of points that must be calculated to ensure that the results fairly represent the lighting of the whole road. The same standards also set rules for the calculation of overall parameters of the lighting such as average values and uniformity.

Table 14.1 A typical r-table

β^0/tan i	0	2	5	10	15	20	25	30	35	40	45	60	75	90	105	120	135	150	165	180
0	329	329	329	329	329	329	329	329	329	329	329	329	329	329	329	329	329	329	329	329
0.25	362	358	371	364	371	369	362	357	351	349	348	340	328	312	299	294	298	288	292	281
0.5	379	368	375	373	367	359	350	340	328	317	306	280	266	249	237	237	231	231	227	235
0.75	380	375	378	365	351	334	315	295	275	256	239	218	198	178	175	176	176	169	175	176
1	372	375	372	354	315	277	243	221	205	192	181	152	134	130	125	124	125	129	128	128
1.25	375	373	352	318	265	221	189	166	150	136	125	107	91	93	91	91	88	94	97	97
1.5	354	352	336	271	213	170	140	121	109	97	87	76	67	65	66	66	67	68	71	71
1.75	333	327	302	222	166	129	104	90	75	68	63	53	51	49	49	47	52	51	53	54
2	318	310	266	180	121	90	75	62	54	50	48	40	40	38	38	38	41	41	43	45
2.5	268	262	205	119	72	50	41	36	33	29	26	25	23	24	25	24	26	27	29	28
3	227	217	147	74	42	29	25	23	21	19	18	16	16	17	18	17	19	21	21	23
3.5	194	168	106	47	30	22	17	14	13	12	12	11	10	11	12	13	15	14	15	14
4	168	136	76	34	19	14	13	11	10	10	10	8	8	9	10	9	11	12	11	13
4.5	141	111	54	21	14	11	9	8	8	8	8	8	8	8	8	8	8	10	10	11
5	126	90	43	17	10	8	8	7	6	6	7	7	7	6	6	7	8	8	8	9
5.5	107	79	32	12	8	7	7	7	6	5		6	7	6	6					
6	94	65	26	10	7	6	6	6	5											
6.5	86	56	21	8	7	6	6	5												
7	78	50	17	7	5	5	5	5												
7.5	70	41	14	7	4	3	5													
8	63	37	11	5	4	4	4													
8.5	60	37	10	5	4	4	4													
9	56	32	9	5	4	3	4													
9.5	53	28	9	4	4	4														
10	52	27	7	5	4	3														
10.5	45	23	7	4	3	3														
11	43	22	7	3	3	3														
11.5	53	22	7	3	3															
12	42	20	7	4	3															

The calculation of luminance is more complex as it is necessary to calculate the angles between the observer's eye and the lantern at the point of interest on the road (β), the angle between the normal to the road surface at the point of interest and the lantern (i) and use them to look up the appropriate value in the r-table of the road surface. The angles β and i are shown in Figure 14.6.

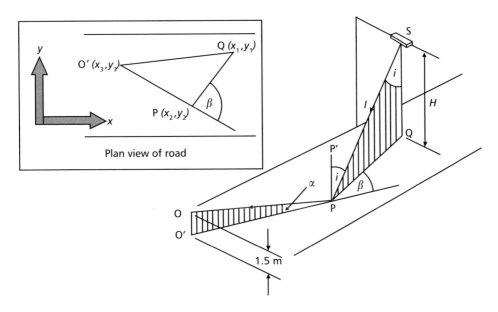

Figure 14.6 Angles β and i

Angle β may be calculated using the cosine rule by considering the lengths O′P, the distance from the road surface below the observer location to the point of interest, PQ, the distance from the point of interest to the point on the road directly beneath the luminaire and O′Q, the distance from the road surface below the observer location to the point on the road directly beneath the luminaire. β may be calculated using equation 14.17.

$$\beta = 180° - \arccos\frac{O'P^2 + PQ^2 - O'Q^2}{2 \times O'P \times PQ} \tag{14.17}$$

It is then necessary to use interpolation in the r-table to find an appropriate value of r for values of $\tan i$ and β lying between those given in the r-table. It is necessary to use quadratic interpolation. This requires three values in the r-table for each interpolated value. The process is similar to that used for quadratic interpolation in section 14.1.2.

To reduce interpolation inaccuracies as far as possible, the following rule shall be followed in selecting the values for insertion in the interpolation equations.

> *The two tabular values adjacent to the value for interpolation shall be selected. The third tabular value shall be the next greatest. Linear interpolation shall be used at the boundaries of the table.*

Once the r value has been determined, the luminance may be calculated with equation 14.18

$$L = \frac{I(C,\gamma) \times r \times 10^{-4}}{H^2} \tag{14.18}$$

Note that the 10^{-4} is in the calculation as, by convention, r-tables are listed multiplied by 10^4 to make them easier to read. If relative photometry of the luminaire, based on a nominal lamp flux of 1000 lumens, is used, it will be necessary to multiply by the total lamp flux of the luminaire in kilolumens. If the maintained road surface luminance is needed then it will be necessary to multiply the result by the maintenance factor.

14.4 Calculation of glare

The main metric for the control of glare on main roads is Threshold Increment ($TI\%$). $TI\%$ is a measure of disability glare and it is based on a measure of light scatter in the eye called veiling luminance (L_V). The actual veiling luminance experienced by any given observer is a function of a number of aspects of their eye and thus is very variable between subjects. In general, the older the subject is the worse problems they are likely to have with disability glare. In general, equation 14.19 is used for the calculation of veiling luminance and it gives the amount of veiling luminance typically found with a subject with good eyesight in the age range 20 to 30.

$$L_V = 10 \sum_{i-1}^{n} \frac{E_{eye_i}}{\theta_i^2} \tag{14.19}$$

Where there are n light sources, the illuminance at the observer's eye due to the ith source is $E_{eye,i}$ and the angle between the ith source and the direction of view is θ_i.

$TI\%$ is calculated using equation 14.20:

$$TI\% = 65 \frac{L_V}{L_B^{0.8}} \tag{14.20}$$

where L_B is the background luminance. This is usually taken as the average luminance of the road. Note that equation 14.20 is only valid for values of L_B up to 5 cd·m^{-2}. This is generally not a problem for road lighting as it is unusual for road luminance to exceed 2 cd·m^{-2}. Where L_B does exceed 5 cd·m^{-2}, then equation 14.21 should be used.

$$TI\% = 95 \frac{L_V}{L_B^{1.05}} \tag{14.21}$$

14.5 Calculations in other outdoor areas

The same basic calculation principle is used in all outdoor areas and so it is possible to calculate illuminance in outdoor workplaces and sports facilities. There are European standards BS EN 12193: 2007: *Light and lighting. Sports lighting* and BS EN 12464-2:2007: *Lighting of work places. Outdoor work places* (BSI 2007b, 2007c) that set out the number and locations of points where illuminance needs to be calculated. The main difference in these areas is that it is common to control glare with a metric called Glare Rating (GR). The system of glare rating is defined by CIE (1994). The degree of glare in any area may vary with observer position and viewing direction. The calculation is based upon the veiling luminance calculation and the basic equation for GR is given in equation 14.22:

$$GR = 27 + 24 \log_{10} \left(\frac{L_{VL}}{L_{VE}^{0.9}} \right) \tag{14.22}$$

where L_{VL} is the veiling luminance due to the luminaires and L_{VE} is the veiling luminance due to the rest of the environment.

The calculation of the veiling luminance due to the rest of the environment is a complex process involving breaking down the luminous field into a series of small elements, calculating the illuminance that they cause at the eye of the observer and dividing by the angle between them and the direction of view squared.

To overcome this complexity, the CIE (1994) also gave a simplified method that gives an approximate value of the veiling luminance due to the environment. The method works well when the background has a fairly uniform luminance. Equation 14.23 may be used to estimate the veiling luminance due to the environment:

$$L_{VE} = 0.035 \, L_{av} \tag{14.23}$$

where L_{av} is the average luminance of the horizontal area being viewed by the observer. The average luminance of a surface may be calculated by equation 14.24 provided the surface has diffuse reflection properties.

$$L_{av} = \frac{E_{hor\,av} \, \rho}{\pi} \tag{14.24}$$

where $E_{hor\,av}$ is the average horizontal illuminance and ρ is the reflectance of the area.

In the calculation of GR, the selection of observer locations and viewing directions is critical to calculating a GR value appropriate to the application; in most applications, it is normal to assume that the observer is looking 2 degrees below the horizontal.

Chapter 15: Measurement of lighting installations and interpreting the results

As with any engineering project, the results of any lighting design are prone to a level of uncertainty and it is often necessary to measure the lighting performance of a design to ensure that it has been correctly executed. However, the process of measuring light gives rise to a different set of uncertainties and so care is needed in both the process of measurement and the interpretation of the results. This chapter looks at the properties of the instruments used for measurement and the process of performing and interpreting the results of measurement.

15.1   Light measuring equipment

There are two main types of instrument used to measure light, illuminance and luminance meters. The performance of both types of instrument is covered by British Standards (BSI, 2005b,c) and they are discussed in the following sections.

15.1.1   Illuminance meters
The performance requirements for illuminance meters are set out in BS 667: 2005. The standard defines two types of meter, Type L – which are of high accuracy and generally used in a laboratory, and type F – in which a certain amount of accuracy has been sacrificed in order to make the instrument portable. The standard considers a number of potential sources of error and puts limits on them. Table 15.1 details errors and their limits.

Table 15.1 Tolerances for illuminance meters

Source of error	Maximum error over effective range	
	Type L	Type F
Calibration uncertainty[1], %		
10–10 000 lux	1.0	2.5
10 000–100 000 lux	1.0	3.0
Non-linearity, %		
10–10 000 lux	0.2	1.0
10 000–100 000 lux	0.2	2.0
Spectral correction factor,%	1.5	3.5
Infra-red response, %	0.2	0.2
Ultraviolet response	0.2	0.5
Cosine correction (unless marked as uncorrected), %	1.5	2.5
Fatigue, %	0.1	0.4
Temperature change, % per K	0.2	0.25
Range change	0.5	1.0

[1]The standard used and errors involved should be quoted.

Note 1:   For digital displays, there is a permitted tolerance of ±1 on the least significant digit.

Note 2:   A meter, which just meets the requirement of this standard, would have a best measurement capability of ±4% (type L) or ±6% (type F) when used on any of its calibrated ranges.

Calibration

Calibration should be done with a stable tungsten light source with a colour temperature of 2856 ± 20 K. The luminance meter is calibrated using a standard intensity source, or against a standard meter. Usually, the procedure is carried out on an optical bench where the illuminance at the photometer head can be varied by changing the distance between the light source and the photocell. Standard lamps are available from the National Physical Laboratory and other national standards bodies.

Linearity error

To assess the linearity error, expose a meter to an illuminance close to the illuminance used to calibrate the meter, then expose the meter to a series of illuminances covering the range. The linearity error may then be calculated using equation 15.1:

$$N = \left(1 - \left|\frac{AB}{CD}\right|\right) \tag{15.1}$$

where: A meter reading at the test point
  C meter reading close to the calibration point
  B/D ratio of illuminance that caused A and C

Spectral correction factor

As a tungsten lamp with a colour temperature of 2856 K ± 20 K has been adopted as the standard by which a meter is calibrated, it is necessary to check the error when the meter is used with other light sources. The error when using a particular light source is given by equation 15.2.

$$f_1' = \frac{\sum\limits_{380}^{780} \left| s^\star(\lambda)_{rel} - V(\lambda) \right|}{\sum\limits_{380}^{780} V(\lambda)} \times 100 \tag{15.2}$$

where $s^\star(\lambda)_{rel}$ is the normalised relative spectral responsivity as given by the following equation:

$$s^\star(\lambda)_{rel} = \frac{\sum\limits_{380}^{780} S(\lambda)_A V(\lambda)}{\sum\limits_{380}^{780} S(\lambda)_A s(\lambda)_{rel}} \; s(\lambda)_{rel}$$

  $S(\lambda)_A$ is the spectral distribution of the illuminant used in the calibration (standard illuminant A in accordance with CIE 15 (CIE, 2004b));
  $s(\lambda)_{rel}$ is the relative spectral responsivity normalised at an arbitrary wavelength;
  $V(\lambda)$ is the spectral luminous efficiency of the human eye for photopic vision.

Cosine error

It is important that light coming at high angles away from the normal is given the correct weighting according to the cosine formula used in calculating plane illuminance. Using a small source to illuminate the photometer head and then rotating the head, checking the angle of rotation and meter reading and dividing it by the reading at normal to give $f(\theta)$, the error may be assessed using equation 15.3:

$$\rho = \sum_{-85^0 - step\,5^0}^{+85^0} \left| F(\theta) - \cos(\theta) \right|$$ (15.3)

The total error may then be expressed as a percentage using equation 15.4:

$$T = \frac{\rho}{22.9 \times 2} \times 100$$ (15.4)

15.1.2 Luminance meters

Luminance meters have a similar set of errors to illuminance meters as given in Table 15.2.

Table 15.2 Tolerances for luminance meters

Source of error	Maximum error over effective range	
	Type L	Type F
Calibration uncertainty[1], %		
0.1–1000 cd m^{-2}	1.5	3.0
1000–10 000 cd m^{-2}	1.5	3.5
Non-linearity, %		
0.1–1000 cd m^{-2}	0.2	0.5
1000–10 000 cd m^{-2}	0.2	1.0
Spectral correction factor,%	2.0	4.0
Infra-red response, %	0.2	0.2
Ultraviolet response	0.2	0.2
Fatigue, %	0.1	0.4
Temperature change, % per K	0.2	0.2
Directional response	2.0	4.0
Effect of surrounding field	1.0	1.0
Errors of focus	0.4	0.6
Range change	0.1	0.6

[1]*The standard used and errors involved should be quoted.*

Note 1:   For digital displays, there is a permitted tolerance of ±1 on the least significant digit.

Note 2:   A meter, which just meets the requirement of this standard, would have a best measurement capability of ±5% (type L) or ±7% (type F) when used on any of its calibrated ranges.

Whilst there is no cosine error to consider, the following spatially related errors must be assessed.

Directional response

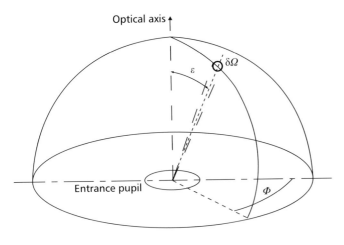

Figure 15.1 Directional response of luminance meters

Luminance meters should have a uniform response across their designed field of view. The response of the meter is characterised by the function f_2 (ε, ϕ) shown as equation 15.3:

$$f_2(\varepsilon, \phi) = \frac{Y(\varepsilon, \phi)}{Y(\varepsilon = 0)} \times 100\% \qquad (15.3)$$

where $Y (\varepsilon, \phi)$ reading at angle of incidence ε, ϕ
      $Y (\varepsilon = 0)$ reading on axis of photometer

The uniformity inside the cone of acceptance (E) can be characterised by equation 15.4:

$$E = (1 - \frac{Y_{min}}{Y_{max}}) \times 100 \qquad (15.4)$$

Effect of surrounding field
A luminance meter should not respond to light outside its field of measurement. This can be tested using a gloss trap, which is slightly larger than the acceptance area of the meter.

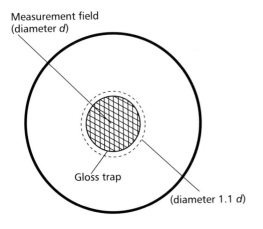

Measurement field
(diameter d)

Gloss trap

(diameter 1.1 d)

Figure 15.2 Testing effect of surrounding field

The meter is exposed to a large uniform luminance source, and the error characterised using equation 15.5:

$$f_2(U) = \frac{Y(surround)}{Y(total) - Y(surround)}$$

(15.5)

where: Y(surround) Output for the measurement with the gloss trap (black field)
Y(total) Output for the measurement without gloss trap (measurement field and surrounding field).

Focus error
This may be due to changes in the light transmitting properties of the optical system of the meter as focus is changed. It is characterised by equation 15.6:

$$f_{12} = \left(\frac{Y_1}{Y_2}\right) \times 100\%$$

(15.6)

where: Y_1 Output signal when focused at the shortest distance
Y_2 Output signal when focused at the longest distance.

15.2 Field measurements

15.2.1 Operating conditions
Whenever a performance measurement is carried out, there are a number of factors that may impact upon the performance of the lighting installation being tested. A number of issues are associated with the lighting equipment and there are others that relate to the environment.

It is important to ensure that the output of the lamps is stable. In general, this requires that they have been run for at least 100 h and they have had time to run up to and reach thermal stability, which may take half an hour or so. Also if the lamps have been in use for a long time, then their output may be lower than the nominal output. Similarly, if the luminaires have been running for a long time, the build-up of dirt on the optical surfaces may be reducing the light output. For these reasons, it is good practice to record the condition of the lighting equipment when taking measurements.

Many lamp types are sensitive to changes in temperature so it is important to record the temperature when taking readings. For example, when taking readings inside a building that has not been fully commissioned, the temperature might be quite cold. This may cause a reduction in light output from fluorescent lamps, which are designed to operate at about 25°C.

The supply voltage can significantly change the output of some lamps so it is a good idea to measure the supply voltage. As there can be a voltage drop in the supply cabling, it is best to measure the voltage in the supply network as close to the luminaires as possible.

If the measurement is being carried out to demonstrate that a lighting installation performs the way it was predicted to, then it is necessary to check that the lighting installed is the same as that which the design calculation assumed. It is also necessary to check that the room geometry and surface finishes are the same as those used in the design calculations.

15.2.2 Grids and illuminance measurement
When measuring illuminance, it is most common to evaluate the average illuminance and uniformity over some specified area. To do this, it is normal to measure the illuminance on a

regular grid of points with the maximum distance between the points (p) being given by equation 15.7

$$p = 0.2 \times 5^{\log d} \qquad (15.7)$$

where d is the length of the longer dimension of the area being measured.

For example, if it is required to calculate the illuminance on a 6 m by 4 m area (see Figure 15.3), then the following equation may be applied:

$$p = 0.2 \times 5^{\log 6} = 0.6997$$

However, if you divide the length of 6 m by 0.6997, you get 8.575, so rounding up to the nearest whole number, nine measurement points are needed. Dividing 6 by 9 gives a true spacing of 0.666 m. Then it is necessary to find the number of points in the width of the area that gives nearly the same spacing; in the case of the 6 m by 4 m area, this is easy as a spacing of 0.666 m means that six points are needed across the width. Once the number of points and the spacing have been calculated, it is simple to arrange the points, with the first point starting a half spacing from the edge (see Figure 15.3).

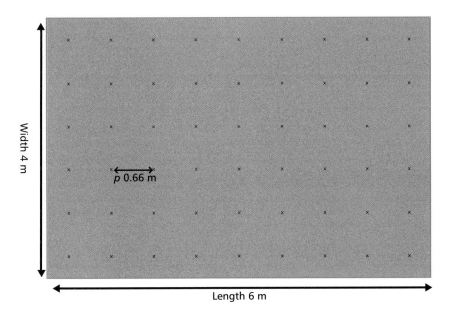

Figure 15.3 Typical measurement grid

This type of grid is suitable for most applications, however, for road lighting and some sports lighting applications, particular grids are defined in the relevant standards (BSI, 2003c, 2007b). When the area of a grid is a room or part of a room, then a band of 0.5 m from the walls is excluded from the calculation area except when a task area is in or extends into this border area.

Once the grid has been defined, it is relatively easy to take illuminance measurements at each point. This is achieved by placing a light meter at each point at the correct height and taking a measurement. The orientation of the light meter is a function of the type of measurement being taken. For horizontal illuminance, it is necessary to ensure that the illuminance meter is flat. For vertical illuminance, the meter must be kept both vertical and parallel to the correct reference plane. In some sports lighting installations, it is necessary to measure the illuminance towards a camera location; this involves orientating the illuminance meter so that it faces the camera location at each point of measurement.

15.2.3   Averages and uniformities

Once measurements have been made at each grid point as set out in section 15.2.2, it is relatively simple to calculate the average and the uniformity. The average illuminance is calculated by adding up all of the illuminance values and dividing by the number of measurements. The most common measure of uniformity is the ratio of minimum to average which is calculated by dividing the minimum value by the average. However, the minimum to maximum value is sometimes used.

15.2.4   Measurement of road luminance

To measure the luminance of a road surface, it is necessary to have access to the road in safe conditions; this usually requires the road to be closed. If it is necessary to work on a road that is only partly closed then it is sensible to do a risk assessment of the measurement procedure and take steps to reduce any possible risks to the people conducting the measurement.

Details of the measurement procedure and the luminance meter to be used are given in the European Standard on methods of measuring lighting performance for road lighting (BSI, 2003d); the key points to consider are:

● Ensure that the luminance meter is located at the correct observer position.

● Measure and record the geometry of the installation.

● Mark out the grid points to be measured with moveable markers so that the marker may be removed once the meter has been aimed at the spot but before the measurement is taken.

● Check the supply voltage to the lighting and note the age and condition of the lamps.

● Note the condition and state of the road surface, particularly if it is wet or dry.

15.2.5   Other measures of spatial illuminance

Whilst there are meters specially designed to measure hemispherical, cylindrical and semi-cylindrical illuminance, they can be expensive and the number of times each instrument is used may be limited. It is usual to measure the illuminance on six faces of a cube and derive the required illuminance value using the methods set out in Chapter 13.

Chapter 16: Colour

16.1 Introduction

Colour is a very important property of both light sources and illuminated surfaces. This section of the *Code* provides detailed information on the various ways to characterise colours, together with their strengths and limitations. In all situations, the actual impact in any use of coloured surfaces with particular light sources will be subjective. Whilst the methods detailed in this chapter may aid the prediction of the impact of a particular colour scheme, there is no real substitute for physical assessment. This chapter is divided into two sections: the first deals with the colour properties of light sources and the second is concerned with the colour properties of surfaces. There is, however, significant overlap between the two topics and some of the calculation methods are equally applicable to both the colours of surfaces and the colours of light sources.

16.2 Colour properties of light sources

All colour properties of a source are a function of its spectral power distribution. There are two main colour properties: colour appearance and colour rendering. Colour appearance is commonly expressed in terms of the position of the colour of a light source in one of several colour systems or the colour of the full radiator (sometimes known as a blackbody or a Planckian radiator) that most nearly matches it. This latter measure is known as the correlated colour temperature (CCT) of the source. It should be noted that, to truly describe the appearance of a colour, it is also necessary to consider the luminance of the source or surface, however, many of the metrics of colour appearance used in the lighting industry do not include luminance. The way that a collection of coloured surfaces appears under a given light source is more complex and whilst the two metrics described in this section give a single number metric that represents the average performance over a range of surfaces, such metrics are no guarantee of perfect colour appearance with a given light source.

16.2.1 Colour appearance in the CIE chromaticity (1931) diagram

The CIE define three colour-matching functions $\bar{x}(\lambda)$, $\bar{y}(\lambda)$, and $\bar{z}(\lambda)$, which are specified at 5 nm intervals through the visible spectrum from 380 nm to 780 nm. The X, Y, Z tristimulus values can be calculated by multiplying the power recorded at each wavelength with each of the colour-matching functions in turn and summing, thus:

$$X = \sum \varphi(\lambda)\bar{x}(\lambda)$$
$$Y = \sum \varphi(\lambda)\bar{y}(\lambda)$$
$$Z = \sum \varphi(\lambda)\bar{z}(\lambda)$$

where $\varphi(\lambda)$ is the power in each wavelength band. Tabulated values of the colour-matching functions are given in Table 16.1.

Table 16.1 Tabulated values of $\bar{x}(\lambda)$, $\bar{y}(\lambda)$, and $\bar{z}(\lambda)$

λ / nm	$\bar{x}(\lambda)$	$\bar{y}(\lambda)$	$\bar{z}(\lambda)$	λ / nm	$\bar{x}(\lambda)$	$\bar{y}(\lambda)$	$\bar{z}(\lambda)$
380	0.0014	0.0000	0.0065	585	0.9786	0.8163	0.0014
385	0.0022	0.0001	0.0105	590	1.0263	0.7570	0.0011
390	0.0042	0.0001	0.0201	595	1.0567	0.6949	0.0010
395	0.0076	0.0002	0.0362	600	1.0662	0.6310	0.0008
400	0.0143	0.0004	0.0679	605	1.0456	0.5668	0.0006
405	0.0232	0.0006	0.1102	610	1.0026	0.5030	0.0003
410	0.0435	0.0012	0.2074	615	0.9384	0.4412	0.0002
415	0.0776	0.0022	0.3713	620	0.8544	0.3810	0.0002
420	0.1344	0.0040	0.6456	625	0.7514	0.3210	0.0001
425	0.2148	0.0073	1.0391	630	0.6424	0.2650	0.0000
430	0.2839	0.0116	1.3856	635	0.5419	0.2170	0.0000
435	0.3285	0.0168	1.6230	640	0.4479	0.1750	0.0000
440	0.3483	0.0230	1.7471	645	0.3608	0.1382	0.0000
445	0.3481	0.0298	1.7826	650	0.2835	0.1070	0.0000
450	0.3362	0.0380	1.7721	655	0.2187	0.0816	0.0000
455	0.3187	0.0480	1.7441	660	0.1649	0.0610	0.0000
460	0.2908	0.0600	1.6692	665	0.1212	0.0446	0.0000
465	0.2511	0.0739	1.5281	670	0.0874	0.0320	0.0000
470	0.1954	0.0910	1.2876	675	0.0636	0.0232	0.0000
475	0.1421	0.1126	1.0419	680	0.0468	0.0170	0.0000
480	0.0956	0.1390	0.8130	685	0.0329	0.0119	0.0000
485	0.0580	0.1693	0.6162	690	0.0227	0.0082	0.0000
490	0.0320	0.2080	0.4652	695	0.0158	0.0057	0.0000
495	0.0147	0.2586	0.3533	700	0.0114	0.0041	0.0000
500	0.0049	0.3230	0.2720	705	0.0081	0.0029	0.0000
505	0.0024	0.4073	0.2123	710	0.0058	0.0021	0.0000
510	0.0093	0.5030	0.1582	715	0.0041	0.0015	0.0000
515	0.0291	0.6082	0.1117	720	0.0029	0.0010	0.0000
520	0.0633	0.7100	0.0782	725	0.0020	0.0007	0.0000
525	0.1096	0.7932	0.0573	730	0.0014	0.0005	0.0000
530	0.1655	0.8620	0.0422	735	0.0010	0.0004	0.0000
535	0.2257	0.9149	0.0298	740	0.0007	0.0002	0.0000
540	0.2904	0.9540	0.0203	745	0.0005	0.0002	0.0000
545	0.3597	0.9803	0.0134	750	0.0003	0.0001	0.0000
550	0.4334	0.9950	0.0087	755	0.0002	0.0001	0.0000
555	0.5121	1.0000	0.0057	760	0.0002	0.0001	0.0000
560	0.5945	0.9950	0.0039	765	0.0001	0.0000	0.0000
565	0.6784	0.9786	0.0027	770	0.0001	0.0000	0.0000
570	0.7621	0.9520	0.0021	775	0.0001	0.0000	0.0000
575	0.8425	0.9154	0.0018	780	0.0000	0.0000	0.0000
580	0.9163	0.8700	0.0017				

Note that these values are used where the colour source being viewed individually subtends an angle between 1° and 4° at the eye. For a larger field of view, the CIE has developed a further set of functions $\bar{x}_{10}(\lambda)$, $\bar{y}_{10}(\lambda)$, and $\bar{z}_{10}(\lambda)$ which are known as the CIE 1964 observer (details of the CIE 1964 colour matching functions are available at http://www.cie.co.at/main/freepubs.html).

The tristimulus values X, Y and Z are then normalised to create chromaticity coordinates x, y and z using the following formulae

$$x = \frac{X}{X+Y+Z} \qquad y = \frac{Y}{X+Y+Z} \qquad z = \frac{Z}{X+Y+Z}$$

Note that the sum of the chromaticity coordinates x, y and z is always equal to unity. The relative colour appearance of the light may then be plotted on the CIE chromaticity diagram: it is only necessary to plot x and y as z may be inferred. This chart is sometimes called the CIE chromaticity (1931) diagram after the year in which it was introduced. Figure 16.1 shows an example of the diagram.

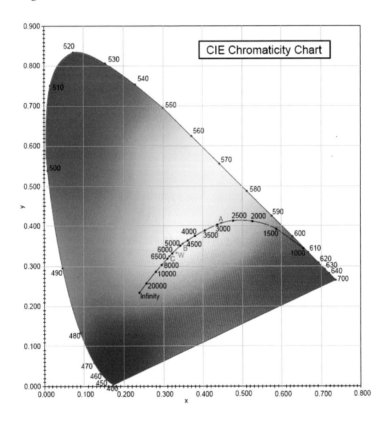

Figure 16.1 The CIE chromaticity (1931) diagram

The outer curved boundary of the CIE chromaticity diagram is called the spectrum locus. All colours that are made up of only a single wavelength plot on this line. The straight line joining the two ends of the spectrum locus is known as the purple boundary. The line running near the middle of the diagram is the full radiator (Planckian or blackbody) locus and it passes through chromaticity coordinates of objects that produce radiation by thermal emission or incandescence. See section 16.2.3 for more information on calculations with full radiators. Any point on the chromaticity diagram shows the unique colour of a given light source and so the diagram is a useful method for specifying such things as the colours of signal lights.

16.2.2 CIE UCS (1976) diagram

The CIE chromaticity (1931) diagram is very useful for providing a simple plot of all available source colours, however, it suffers from the weakness that it is not perceptually uniform. Green colours cover a large area while blue and red colours are compressed in the bottom left and right corners, respectively. The ability of a given person to discriminate colour differences depends on many personal factors including their genetics, age and tiredness; as well as a range of environmental conditions including the luminance and spatial separation of the colours being compared. In describing the comparison of colours, it is often important to establish the minimum difference between two colours necessary for the colours to be seen as different. Research into the perception of colour difference was carried out during the 1940s by MacAdam and whilst the methodology was questionable and the number of subjects used very small, the results have been used to derive a series of ellipses on the CIE chromaticity (1931) diagram. A MacAdam ellipse joins points that are just perceptually different from the centre of the ellipse and the ellipses in Figure 16.2 are drawn 10 times correct size.

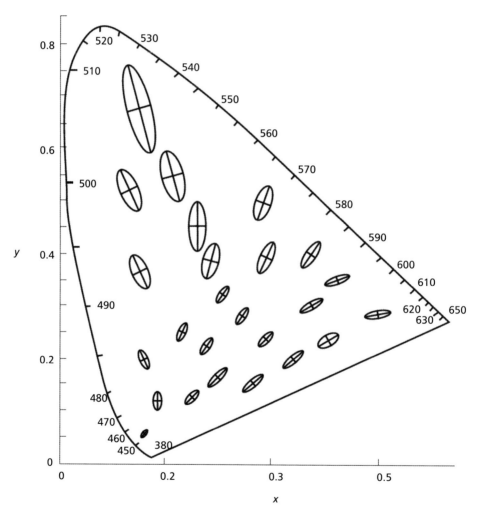

Figure 16.2 The CIE chromaticity (1931) diagram with the MacAdam ellipses displayed, multiplied ten times (after MacAdam (1942) from the IESNA Lighting Handbook)

A MacAdam ellipse about any point on the diagram can be described by the equation of an ellipse:

$$g_{11}(\Delta x)^2 + g_{12}(\Delta x)(\Delta y) + g_{22}(\Delta y)^2 = 1$$

where Δx and Δy are the differences in the x and y co-ordinates of the colours from the centre of the ellipse and g_{11}, g_{12} and g_{22} are coefficients that may be obtained from Figure 16.3 below by multiplying the value interpolated for the x and y co-ordinates for the centre of the ellipse by 10 000.

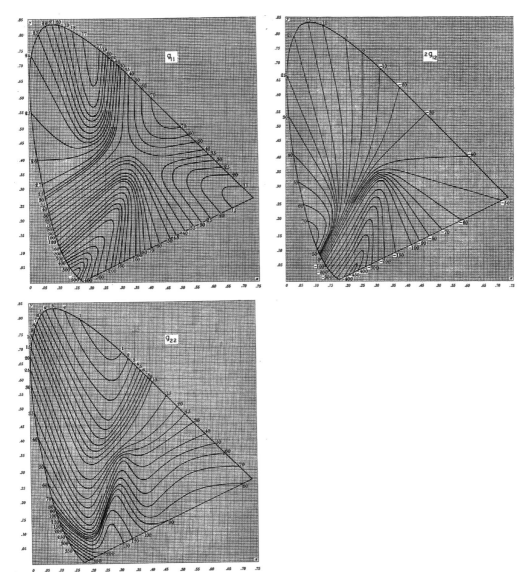

Figure 16.3 Charts used to obtain values of g_{11}, g_{12} and g_{22}

The same coefficients (g_{11}, g_{12} and g_{22}) may be used to define a measure of chromaticity difference ΔC:

$$\Delta C = \sqrt{g_{11}(\Delta x)^2 + g_{12}(\Delta x)(\Delta y) + g_{22}(\Delta y)^2}$$

To get round the problem of the non-uniformity of the CIE chromaticity (1931) diagram, in 1960, the CIE recommended the u, v chromaticity diagram for applications where a more uniform colour space was needed. For this diagram, u and v are calculated as follows:

$$u = \frac{4X}{(X+15Y+3Z)} \quad \text{or} \quad u = \frac{4x}{(-2x+12y+3)}$$

$$v = \frac{6Y}{(X+15Y+3Z)} \quad \text{or} \quad v = \frac{6y}{(-2x+12y+3)}$$

To convert these coordinate systems into a three-dimensional colour space for studying differences in surface colours which may vary in luminance as well as colour, the $U\star\ V\star\ W\star$ system was developed and recommended by the CIE in 1963. In this system, the values of $U\star$, $V\star$ and $W\star$ are calculated with the following formulae:

$$W^* = 25\ Y^{\frac{1}{3}} - 17 \quad \text{where} \quad (1 \leq Y \leq 100)$$
$$U^* = 13\ W^* (u - u_0)$$
$$V^* = 13\ W^* (v - v_0)$$

The chromaticity coordinates (u_0, v_0) refer to a nominal achromatic colour, usually that of the light source.

In 1976, the CIE recommended a revised chromaticity diagram space based on u' and v'. The u' and v' coordinates are calculated using the following formulae:

$$u' = \frac{4X}{(X+15Y+3Z)} \quad \text{or} \quad u' = \frac{4x}{(-2x+12y+3)}$$

$$v' = \frac{9Y}{(X+15Y+3Z)} \quad \text{or} \quad v' = \frac{9y}{(-2x+12y+3)}$$

When MacAdam ellipses are plotted on the 1976 colour diagram, they have more similar areas and are more nearly circular (Figure 16.4).

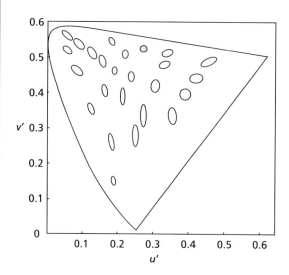

Figure 16.4 MacAdam ellipses plotted on the CIE 1976 UCS diagram

There are also other colour spaces which are sufficiently perceptually uniform to permit the calculation of colour difference. One of these is the CIE L⋆a⋆b⋆ space: see section 16.3.6 for more details.

16.2.3 Colour temperature

When an object is heated to a high temperature, the atoms within the material become excited by the many interactions between them and energy is radiated in a continuous spectrum. The exact nature of the radiation produced by an idealised radiator, known as a full radiator, was studied by Max Planck at the end of the 19th century and he developed the following formula that predicts the radiation produced:

$$M_{e\lambda}^{th} = \frac{c_1}{\lambda^5 \left[\exp(c_2 / \lambda T) - 1\right]}$$

where $M_{e\lambda}^{th}$ is the spectral radiant exitance, c_1 and c_2 are constants, with values of 3.7814×10^{-16} W m^{-2} and 1.4388×10^{-2} mK, respectively. λ is the wavelength and T the temperature in kelvin. The values of the spectral radiant exitance are plotted for different values of T in Figure 16.5.

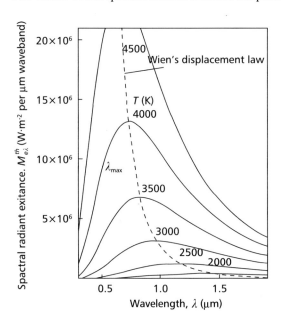

Figure 16.5 Spectral power distribution of radiation according to Planck's law

The wavelength for maximum power (λ_{max}) is inversely proportional to the temperature (T). The following formula was developed by Planck's co-worker at the University of Berlin and is known as Wien's displacement law.

$$\lambda_{max} = \frac{c_3}{T}$$

where c_3 has a value of 2.90×10^{-3} mK.

For light sources that are not full radiators, it may still be useful to categorise their colour in terms of their correlated colour temperature. The correlated colour temperature for a light source is defined as the temperature corresponding to the point on the full radiator locus which is nearest to the point representing the chromaticity of the light source in a plot of $2v'/_3$ against u'. To perform this mathematically, it is necessary to minimise the function below by changing the value of the temperature of the full radiator.

$$\left\{ \left[\frac{2\left(v'_s - v'_{fr} \right)}{3} \right]^2 + \left[u'_s - u'_{fr} \right]^2 \right\}$$

where v'_s is the v' value for the source, v'_{fr} is the v' value for the full radiator, u'_s is the u' value for the source, and u'_{fr} is the u' value for the full radiator. It is thus possible to plot a series of lines on the CIE chromaticity diagram that correspond to a particular value of correlated colour temperature, see Figure 16.6.

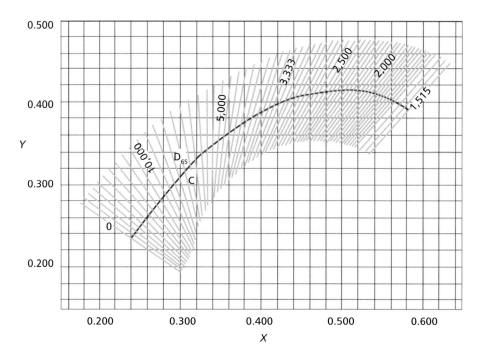

Figure 16.6 The full radiator (Planckian) locus and lines of constant correlated colour temperature plotted on the CIE 1931 (x, y) chromaticity diagram. Also shown are the chromaticity coordinates of CIE Standard Illuminants C and D65 (from the *IESNA Lighting Handbook*; IESNA, 2000)

16.2.4 Colour rendering

The ability of a light source to render colours faithfully is assessed by the calculation of the extent to which colours illuminated by the source have their colour appearance changed compared to those under a reference source. Two metrics for the colour rendering of light sources are discussed in this section: the first is the CIE Colour Rendering Index (CRI) which was developed in the 1960s and was largely designed to deal with the then new improvements in fluorescent lamps. However, the recent emergence of LEDs into mainstream lighting applications has shown up some of the weaknesses in the CRI system and an alternative system known as the Colour Quality Scale (CQS) has been developed. Whilst at present, CQS does not have the same worldwide acceptance as CRI, it does provide a useful metric that works well with most light sources. However, it should be noted that there is currently a lot of interest in new metrics for colour and no conclusion has been reached as to which method to adopt.

16.2.5 Colour Rendering Index (CRI)

The CIE (1995b) published the method to calculate colour rendering index. They put forward two metrics, Ra_8 which is based on the assessment of light source performance on eight colour samples and Ra_{14} which is based on 14 colour samples. The most commonly used version is Ra_8. Figure 16.7 gives an approximate rendering of the colours used. The top row colours are those used for the calculation of Ra_8, the bottom row gives the additional colours that are used in the calculation of Ra_{14}.

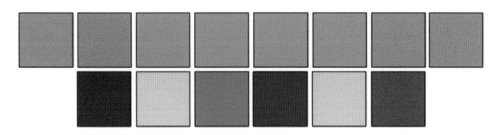

Figure 16.7 Colours used for the calculation of CRI

The CIE colour rendering system compares the colour of test samples under the test light source with the samples under a reference source. If the correlated colour temperature of the test source is less than 5000 K, then the full radiator of the same colour temperature is used for the reference, if the colour temperature is higher, then a simulated daylight spectrum is used.

To calculate the daylight spectrum, the CIE has defined a reference form of daylight that is defined by the following formula:

$$S_\lambda = S_0(\lambda) + M_1 S_1(\lambda) + M_2 S_2(\lambda)$$

where $S_0(\lambda)$, $S_1(\lambda)$ and $S_2(\lambda)$ are functions of the wavelength λ and are listed in Table 16.2. M_1 and M_2 are functions of the colour co-ordinates of the daylight x_D and y_D and may be calculated using the following formulae:

$$M_1 = \frac{-1.3515 - 1.7703\,x_D + 5.9114\,y_D}{0.0241 + 0.2562\,x_D - 0.7341\,y_D}$$

$$M_2 = \frac{-0.03 - 31.4424\,x_D + 30.0717\,y_D}{0.0241 + 0.2562\,x_D - 0.7341\,y_D}$$

For correlated colour temperatures in the range 4000 to 7000 K, the x coordinate of daylight in terms of the correlated colour temperature T_C is given by the following formula:

$$x_D = -\frac{4.607 \times 10^9}{T_C^3} + \frac{2.9678 \times 10^6}{T_C^2} + \frac{99.11}{T_C} + 0.244063$$

For colour temperatures in the range 7000 to 25 000 K, the following formula may be used:

$$x_D = -\frac{2.0064 \times 10^9}{T_C^3} + \frac{1.9018 \times 10^6}{T_C^2} + \frac{247.48}{T_C} + 0.233704$$

The y chromaticity coordinate (y_D) may then be calculated:

$$y_D = -3\,x_D^2 + 2.87\,x_D - 0.275$$

Table 16.2 Tabulated values of $S_0(\lambda)$, $S_1(\lambda)$ and $S_2(\lambda)$

λ	$S_0(\lambda)$	$S_1(\lambda)$	$S_2(\lambda)$	λ	$S_0(\lambda)$	$S_1(\lambda)$	$S_2(\lambda)$
380	63.40	38.50	3.00	585	92.10	−3.50	1.30
385	64.60	36.75	2.10	590	89.10	−3.50	2.10
390	65.80	35.00	1.20	595	89.80	−4.65	2.65
395	80.30	39.20	0.05	600	90.50	−5.80	3.20
400	94.80	43.40	−1.10	605	90.40	−6.50	3.65
405	99.80	44.85	−0.80	610	90.30	−7.20	4.10
410	104.80	46.30	−0.50	615	89.35	−7.90	4.40
415	105.35	45.10	−0.60	620	88.40	−8.60	4.70
420	105.90	43.90	−0.70	625	86.20	−9.05	4.90
425	101.25	40.50	−0.95	630	84.00	−9.50	5.10
430	96.80	37.10	−1.20	635	84.55	−10.20	5.90
435	105.35	36.90	−1.90	640	85.10	−10.90	6.70
440	113.90	36.70	−2.60	645	83.50	−10.80	7.00
445	119.75	36.30	−2.75	650	81.90	−10.70	7.30
450	125.60	35.90	−2.90	655	82.50	−11.35	7.95
455	125.55	34.25	−2.85	660	82.60	−12.00	8.60
460	125.50	32.60	−2.80	665	83.75	−13.00	9.20
465	123.40	30.25	−2.70	670	84.90	−14.00	9.80
470	121.30	27.90	−2.60	675	83.10	−13.80	10.00
475	121.30	26.10	−2.60	680	81.30	−13.60	10.20
480	121.30	24.30	−2.60	685	76.60	−12.80	9.25
485	117.40	22.20	−2.20	690	71.90	−12.00	8.30
490	113.50	20.10	−1.80	695	73.10	−12.65	8.95
495	113.30	18.15	−1.65	700	74.60	−13.30	9.60
500	113.10	16.20	−1.50	705	75.35	−13.10	9.05
505	111.95	14.70	−1.40	710	76.40	−12.90	8.50
510	110.80	13.20	−1.30	715	69.85	−11.75	7.75
515	108.65	10.90	−1.25	720	63.30	−10.60	7.00
520	106.50	8.60	−1.20	725	67.50	−11.10	7.30
525	107.65	7.35	−1.10	730	71.70	−11.60	7.60
530	108.80	6.10	−1.00	735	74.35	−11.90	7.80
535	107.05	5.15	−0.75	740	77.00	−12.20	8.00
540	105.30	4.20	−0.50	745	71.70	−11.20	7.35
545	104.85	3.05	−0.40	750	65.20	−10.20	6.70
550	104.40	1.90	−0.30	755	56.45	−9.00	5.95
555	102.20	0.95	−0.15	760	47.70	−7.80	5.20
560	100.00	0.00	0.00	765	58.15	−9.50	6.30
565	98.00	−0.80	0.10	770	68.60	−11.20	7.40
570	96.00	−1.60	0.20	775	66.80	−10.80	7.10
575	95.55	−2.55	0.35	780	65.00	−10.40	6.80
580	95.10	−3.50	0.50				

Once the spectral power distributions of the reference and test sources have been calculated, and then the tristimulus values calculated, it is necessary to normalise the tristimulus values for the sources so that they both have a Y value of 100. The mathematical process of calculating the colour rendering index is as follows:

1. The tristimulus values for the colour sample must be calculated for both the test and reference source. Tabulated data for the reflectance of the samples used is given in Table 16.3.

2. Colorimetric data must be transformed from CIE 1931 values (X, Y, Z, x, y) to the (u, v) coordinates of the 1960 diagram by the following:

$$u = 4X/(X + 15Y + 3Z) \tag{16.1a}$$

$$v = 6Y/(X + 15Y + 3Z) \tag{16.1b}$$

3. To account for the adaptive colour shift due to the different state of chromatic adaptation under the lamp to be tested, k, and under the reference illuminant, r, use the following formula:

$$u'_{k,i} = \frac{10.872 + 0.404 \frac{c_r}{c_k} c_{k,i} - 4 \frac{d_r}{d_k} d_{k,i}}{16.518 + 1.481 \frac{c_r}{c_k} c_{k,i} - \frac{d_r}{d_k} d_{k,i}} \tag{16.2}$$

$$v'_{k,i} = \frac{5.520}{16.518 + 1.481 \frac{c_r}{c_k} c_{k,i} - \frac{d_r}{d_k} d_{k,i}}$$

The values $u'_{k,i}$ and $v'_{k,i}$ are the chromaticity coordinates of a test colour sample, i, after the adaptive colour shift, obtained by moving the light source to be tested to the reference illuminant, i.e. $u'k = ur$, $v'k = vr$ and should not be confused with CIE 1976 u', v' coordinates.

4. The functions c and d for use in equation 16.2 are calculated for the light source to be tested uk, vk, and the test colour samples i under the light source to be tested uk,i, vk,i according to the following:

$$c = (4 - u - 10v)/v$$
$$d = (1.708v + 0.404 - 1.481u)/v \tag{16.3}$$

5. Colorimetric data must now be transformed into the 1964 Uniform Space by the following:

$$W^*_{r,i} = 25 (Y_{r,i})^{\frac{1}{3}} - 17; \quad W^*_{k,i} = 25 (Y_{k,i})^{\frac{1}{3}} - 17$$

$$U^*_{r,i} = 13 W^*_{r,i} (u_{r,i} - u_r); \quad U^*_{k,i} = 13 W^*_{k,i} (u'_{k,i} - u'_k) \tag{16.4}$$

$$V^*_{r,i} = 13 W^*_{r,i} (v_{r,i} - v_r); \quad V^*_{k,i} = 13 W^*_{k,i} (v'_{k,i} - v'_k)$$

The values $u'_k = u_r$, $v'_k = v_r$ are the chromaticity coordinates of the light source to be tested after consideration of the adaptive colour shift. The values $Y_{r,i}$ and $Y_{k,i}$ must be normalised so that $Y_r = Y_k = 100$.

Table 16.3 Spectral reflectance values of the colours used for the calculation of CRI

λ	TCS01	TCS02	TCS03	TCS04	TCS05	TCS06	TCS07	TCS08	TCS09	TCS10	TCS11	TCS12	TCS13	TCS14
380	0.219	0.070	0.065	0.074	0.295	0.151	0.378	0.104	0.066	0.05	0.111	0.12	0.104	0.036
385	0.239	0.079	0.068	0.083	0.306	0.203	0.459	0.129	0.062	0.054	0.121	0.103	0.127	0.036
390	0.252	0.089	0.070	0.093	0.310	0.265	0.524	0.170	0.058	0.059	0.127	0.09	0.161	0.037
395	0.256	0.101	0.072	0.105	0.312	0.339	0.546	0.240	0.055	0.063	0.129	0.082	0.211	0.038
400	0.256	0.111	0.073	0.116	0.313	0.410	0.551	0.319	0.052	0.066	0.127	0.076	0.264	0.039
405	0.254	0.116	0.073	0.121	0.315	0.464	0.555	0.416	0.052	0.067	0.121	0.068	0.313	0.039
410	0.252	0.118	0.074	0.124	0.319	0.492	0.559	0.462	0.051	0.068	0.116	0.064	0.341	0.04
415	0.248	0.120	0.074	0.126	0.322	0.508	0.560	0.482	0.05	0.069	0.112	0.065	0.352	0.041
420	0.244	0.121	0.074	0.128	0.326	0.517	0.561	0.490	0.05	0.069	0.108	0.075	0.359	0.042
425	0.240	0.122	0.073	0.131	0.330	0.524	0.558	0.488	0.049	0.07	0.105	0.093	0.361	0.042
430	0.237	0.122	0.073	0.135	0.334	0.531	0.556	0.482	0.048	0.072	0.104	0.123	0.364	0.043
435	0.232	0.122	0.073	0.139	0.339	0.538	0.551	0.473	0.047	0.073	0.104	0.16	0.365	0.044
440	0.230	0.123	0.073	0.144	0.346	0.544	0.544	0.462	0.046	0.076	0.105	0.207	0.367	0.044
445	0.226	0.124	0.073	0.151	0.352	0.551	0.535	0.450	0.044	0.078	0.106	0.256	0.369	0.045
450	0.225	0.127	0.074	0.161	0.360	0.556	0.522	0.439	0.042	0.083	0.11	0.3	0.372	0.045
455	0.222	0.128	0.075	0.172	0.369	0.556	0.506	0.426	0.041	0.088	0.115	0.331	0.374	0.046
460	0.220	0.131	0.077	0.186	0.381	0.554	0.488	0.413	0.038	0.095	0.123	0.346	0.376	0.047
465	0.218	0.134	0.080	0.205	0.394	0.549	0.469	0.397	0.035	0.103	0.134	0.347	0.379	0.048
470	0.216	0.138	0.085	0.229	0.403	0.541	0.448	0.382	0.033	0.113	0.148	0.341	0.384	0.05
475	0.214	0.143	0.094	0.254	0.410	0.531	0.429	0.366	0.031	0.125	0.167	0.328	0.389	0.052
480	0.214	0.150	0.109	0.281	0.415	0.519	0.408	0.352	0.03	0.142	0.192	0.307	0.397	0.055
485	0.214	0.159	0.126	0.308	0.418	0.504	0.385	0.337	0.029	0.162	0.219	0.282	0.405	0.057
490	0.216	0.174	0.148	0.332	0.419	0.488	0.363	0.325	0.028	0.189	0.252	0.257	0.416	0.062
495	0.218	0.190	0.172	0.352	0.417	0.469	0.341	0.310	0.028	0.219	0.291	0.23	0.429	0.067
500	0.223	0.207	0.198	0.370	0.413	0.450	0.324	0.299	0.028	0.262	0.325	0.204	0.443	0.075
505	0.225	0.225	0.221	0.383	0.409	0.431	0.311	0.289	0.029	0.305	0.347	0.178	0.454	0.083
510	0.226	0.242	0.241	0.390	0.403	0.414	0.301	0.283	0.03	0.365	0.356	0.154	0.461	0.092
515	0.226	0.253	0.260	0.394	0.396	0.395	0.291	0.276	0.03	0.416	0.353	0.129	0.466	0.1

Table 16.3 Continued

λ	TCS01	TCS02	TCS03	TCS04	TCS05	TCS06	TCS07	TCS08	TCS09	TCS10	TCS11	TCS12	TCS13	TCS14
520	0.225	0.260	0.278	0.395	0.389	0.377	0.283	0.270	0.031	0.465	0.346	0.109	0.469	0.108
525	0.225	0.264	0.302	0.392	0.381	0.358	0.273	0.262	0.031	0.509	0.333	0.09	0.471	0.121
530	0.227	0.267	0.339	0.385	0.372	0.341	0.265	0.256	0.032	0.546	0.314	0.075	0.474	0.133
535	0.230	0.269	0.370	0.377	0.363	0.325	0.260	0.251	0.032	0.581	0.294	0.062	0.476	0.142
540	0.236	0.272	0.392	0.367	0.353	0.309	0.257	0.250	0.033	0.61	0.271	0.051	0.483	0.15
545	0.245	0.276	0.399	0.354	0.342	0.293	0.257	0.251	0.034	0.634	0.248	0.041	0.49	0.154
550	0.253	0.282	0.400	0.341	0.331	0.279	0.259	0.254	0.035	0.653	0.227	0.035	0.506	0.155
555	0.262	0.289	0.393	0.327	0.320	0.265	0.260	0.258	0.037	0.666	0.206	0.029	0.526	0.152
560	0.272	0.299	0.380	0.312	0.308	0.253	0.260	0.264	0.041	0.678	0.188	0.025	0.553	0.147
565	0.283	0.309	0.365	0.296	0.296	0.241	0.258	0.269	0.044	0.687	0.17	0.022	0.582	0.14
570	0.298	0.322	0.349	0.280	0.284	0.234	0.256	0.272	0.048	0.693	0.153	0.019	0.618	0.133
575	0.318	0.329	0.332	0.263	0.271	0.227	0.254	0.274	0.052	0.698	0.138	0.017	0.651	0.125
580	0.341	0.335	0.315	0.247	0.260	0.225	0.254	0.278	0.06	0.701	0.125	0.017	0.68	0.118
585	0.367	0.339	0.299	0.229	0.247	0.222	0.259	0.284	0.076	0.704	0.114	0.017	0.701	0.112
590	0.390	0.341	0.285	0.214	0.232	0.221	0.270	0.295	0.102	0.705	0.106	0.016	0.717	0.106
595	0.409	0.341	0.272	0.198	0.220	0.220	0.284	0.316	0.136	0.705	0.1	0.016	0.729	0.101
600	0.424	0.342	0.264	0.185	0.210	0.220	0.302	0.348	0.19	0.706	0.096	0.016	0.736	0.098
605	0.435	0.342	0.257	0.175	0.200	0.220	0.324	0.384	0.256	0.707	0.092	0.016	0.742	0.095
610	0.442	0.342	0.252	0.169	0.194	0.220	0.344	0.434	0.336	0.707	0.09	0.016	0.745	0.093
615	0.448	0.341	0.247	0.164	0.189	0.220	0.362	0.482	0.418	0.707	0.087	0.016	0.747	0.09
620	0.450	0.341	0.241	0.160	0.185	0.223	0.377	0.528	0.505	0.708	0.085	0.016	0.748	0.089
625	0.451	0.339	0.235	0.156	0.183	0.227	0.389	0.568	0.581	0.708	0.082	0.016	0.748	0.087
630	0.451	0.339	0.229	0.154	0.180	0.233	0.400	0.604	0.641	0.71	0.08	0.018	0.748	0.086
635	0.451	0.338	0.224	0.152	0.177	0.239	0.410	0.629	0.682	0.711	0.079	0.018	0.748	0.085
640	0.451	0.338	0.220	0.151	0.176	0.244	0.420	0.648	0.717	0.712	0.078	0.018	0.748	0.084
645	0.451	0.337	0.217	0.149	0.175	0.251	0.429	0.663	0.74	0.714	0.078	0.018	0.748	0.084
650	0.450	0.336	0.216	0.148	0.175	0.258	0.438	0.676	0.758	0.716	0.078	0.019	0.748	0.084

Table 16.3 Continued

λ	TCS01	TCS02	TCS03	TCS04	TCS05	TCS06	TCS07	TCS08	TCS09	TCS10	TCS11	TCS12	TCS13	TCS14
655	0.450	0.335	0.216	0.148	0.175	0.263	0.445	0.685	0.77	0.718	0.078	0.02	0.748	0.084
660	0.451	0.334	0.219	0.148	0.175	0.268	0.452	0.693	0.781	0.72	0.081	0.023	0.747	0.085
665	0.451	0.332	0.224	0.149	0.177	0.273	0.457	0.700	0.79	0.722	0.083	0.024	0.747	0.087
670	0.453	0.332	0.230	0.151	0.180	0.278	0.462	0.705	0.797	0.725	0.088	0.026	0.747	0.092
675	0.454	0.331	0.238	0.154	0.183	0.281	0.466	0.709	0.803	0.729	0.093	0.03	0.747	0.096
680	0.455	0.331	0.251	0.158	0.186	0.283	0.468	0.712	0.809	0.731	0.102	0.035	0.747	0.102
685	0.457	0.330	0.269	0.162	0.189	0.286	0.470	0.715	0.814	0.735	0.112	0.043	0.747	0.11
690	0.458	0.329	0.288	0.165	0.192	0.291	0.473	0.717	0.819	0.739	0.125	0.056	0.747	0.123
695	0.460	0.328	0.312	0.168	0.195	0.296	0.477	0.719	0.824	0.742	0.141	0.074	0.746	0.137
700	0.462	0.328	0.340	0.170	0.199	0.302	0.483	0.721	0.828	0.746	0.161	0.097	0.746	0.152
705	0.463	0.327	0.366	0.171	0.200	0.313	0.489	0.720	0.83	0.748	0.182	0.128	0.746	0.169
710	0.464	0.326	0.390	0.170	0.199	0.325	0.496	0.719	0.831	0.749	0.203	0.166	0.745	0.188
715	0.465	0.325	0.412	0.168	0.198	0.338	0.503	0.722	0.833	0.751	0.223	0.21	0.744	0.207
720	0.466	0.324	0.431	0.166	0.196	0.351	0.511	0.725	0.835	0.753	0.242	0.257	0.743	0.226
725	0.466	0.324	0.447	0.164	0.195	0.364	0.518	0.727	0.836	0.754	0.257	0.305	0.744	0.243
730	0.466	0.324	0.460	0.164	0.195	0.376	0.525	0.729	0.836	0.755	0.27	0.354	0.745	0.26
735	0.466	0.323	0.472	0.165	0.196	0.389	0.532	0.730	0.837	0.755	0.282	0.401	0.748	0.277
740	0.467	0.322	0.481	0.168	0.197	0.401	0.539	0.730	0.838	0.755	0.292	0.446	0.75	0.294
745	0.467	0.321	0.488	0.172	0.200	0.413	0.546	0.730	0.839	0.755	0.302	0.485	0.75	0.31
750	0.467	0.320	0.493	0.177	0.203	0.425	0.553	0.730	0.839	0.756	0.31	0.52	0.749	0.325
755	0.467	0.318	0.497	0.181	0.205	0.436	0.559	0.730	0.839	0.757	0.314	0.551	0.748	0.339
760	0.467	0.316	0.500	0.185	0.208	0.447	0.565	0.730	0.839	0.758	0.317	0.577	0.748	0.353
765	0.467	0.315	0.502	0.189	0.212	0.458	0.570	0.730	0.839	0.759	0.323	0.599	0.747	0.366
770	0.467	0.315	0.505	0.192	0.215	0.469	0.575	0.730	0.839	0.759	0.33	0.618	0.747	0.379
775	0.467	0.314	0.510	0.194	0.217	0.477	0.578	0.730	0.839	0.759	0.334	0.633	0.747	0.39
780	0.467	0.314	0.516	0.197	0.219	0.485	0.581	0.730	0.839	0.759	0.338	0.645	0.747	0.399

6. The difference between the resultant colour shift of the test colour sample under the test lamp k and illuminated by the reference r may be calculated using the following 1964 Colour Difference Formula:

$$\Delta E_i = \sqrt{(\Delta U_i^\star)^2 + (\Delta V_i^\star)^2 + (\Delta W_i^\star)^2}$$ (16.5)

7. Calculate the Special Colour Rendering Index, R_i, for each test colour sample by the following:

$$Ra_8 = 100 - 4.6\Delta E_i$$ (16.6)

8. Calculate the general Colour Rendering Index, Ra, by the following formulae, 16.7a for Ra_8 and 16.7b for Ra_{14}:

$$Ra_8 = \frac{1}{8}\sum_{i=1}^{8} R_i$$ (16.7a)

$$Ra_{14} = \frac{1}{14}\sum_{i=1}^{14} R_i$$ (16.7b)

16.2.6 Colour Quality Scale (CQS)

The colour quality scale has been proposed by Davis and Ohno (2010) as a response to the perceived shortcomings of the CIE Colour Rendering Index system when applied to LED sources. The system is similar to the CRI in that it is based on the difference in the colour of the light reflected from a series of samples when illuminated by a test and a reference source. The reference source used is the same as the CRI system but CQS is based on the use of 15 colour samples that are more saturated in colour than the samples used in the CRI system (Figure 16.8). The colour calculations are all carried out in CIELAB colour space and the colour transform used to allow for the adaptation to a new white point uses the CMCCAT2000 transform to convert X, Y and Z to RGB, a linear shift in RGB space and then an inverse CMCCAT2000 transform back to X, Y and Z. As well as assessing colour differences between the colours of samples under a test and a reference source, an allowance is made for any increase in chroma of the colour. Thus, changes to higher chroma are scored better in the system than those that reduce chroma. The overall colour differences are summed using a root mean square method, thus making it harder for light sources that do very badly on one or two colours to obtain a good score on a simple average. The CQS score is scaled so that it always produces a value between 0 and 100 (in the CRI system, it is possible to get negative values). Finally, there is a de-rating factor applied for sources that have a colour temperature below 3500 K. A full explanation and justification of this process is given in the Davis and Ohno paper.

Figure 16.8 Colours used for the calculation of CQS

The calculation process is as follows:

1. Obtain normalised ($Y=100$) test and reference spectra in the same way as used in the CRI method.

2. Calculate the tristimulus values for the colour sample with both the test and reference sources; these need to be compared with the tristimulus values of the same sources (known as their white values); the following symbols are used to represent these values.
$X_{i,ref}\,Y_{i,ref}\,Z_{i,ref}$ – the tristimulus values of the colour sample i when illuminated with the reference source.
$X_{i,test}\,Y_{i,test}\,Z_{i,test}$ – the tristimulus values of the colour sample i when illuminated with the test source.
$X_{w,ref}\,Y_{w,ref}\,Z_{w,ref}$ – the tristimulus values of the white point of the reference source.
$X_{w,test}\,Y_{w,test}\,Z_{w,test}$ – the tristimulus values of the white point of the test source.
The values of reflectance of the 15 colour samples are given in Table 16.4.

3. Transform the tristimulus values for the white points and the colour sample illuminated by the test source using equations 16.8 to 16.10.

$$\begin{pmatrix} R_{i,test} \\ G_{i,test} \\ B_{i,test} \end{pmatrix} = M \begin{pmatrix} X_{i,test} \\ Y_{i,test} \\ Z_{i,test} \end{pmatrix} \tag{16.8}$$

$$\begin{pmatrix} R_{w,test} \\ G_{w,test} \\ B_{w,test} \end{pmatrix} = M \begin{pmatrix} X_{w,test} \\ Y_{w,test} \\ Z_{w,test} \end{pmatrix} \tag{16.9}$$

$$\begin{pmatrix} R_{w,ref} \\ G_{w,ref} \\ B_{w,ref} \end{pmatrix} = M \begin{pmatrix} X_{w,ref} \\ Y_{w,ref} \\ Z_{w,ref} \end{pmatrix} \tag{16.10}$$

where $\quad M = \begin{pmatrix} 0.7982 & 0.3389 & -0.1371 \\ -0.5918 & 1.5512 & 0.0406 \\ 0.0008 & 0.0239 & 0.9753 \end{pmatrix}$

4. The corresponding shifted RGB values that allow for white point adaptation are calculated using equations 16.11 to 16.13.

$$R_{i,test,c} = R_{i,test}\left({R_{w,ref}} \middle/ {R_{w,test}} \right) \tag{16.11}$$

$$G_{i,test,c} = G_{i,test}\left({G_{w,ref}} \middle/ {G_{w,test}} \right) \tag{16.12}$$

$$B_{i,test,c} = B_{i,test}\left({B_{w,ref}} \middle/ {B_{w,test}} \right) \tag{16.13}$$

Table 16.4 Values of reflectance of the 15 colour samples

λ (nm)	7.5P 4/10	10PB 4/10	5PB 4/2	7.5B 5/10	10BG 6/8	2.5BG 6/10	2.5G 6/12	7.5GY 7/10	2.5GY 8/10	5Y 8.5/12	10YR 7/12	5YR 7/12	10R 6/12	5R 4/14	7.5RP 4/12
380	0.1086	0.1053	0.0858	0.0790	0.1167	0.0872	0.0726	0.0652	0.0643	0.0540	0.0482	0.0691	0.0829	0.0530	0.0908
385	0.1380	0.1323	0.0990	0.0984	0.1352	0.1001	0.0760	0.0657	0.0661	0.0489	0.0456	0.0692	0.0829	0.0507	0.1021
390	0.1729	0.1662	0.1204	0.1242	0.1674	0.1159	0.0789	0.0667	0.0702	0.0548	0.0478	0.0727	0.0866	0.0505	0.1130
395	0.2167	0.2113	0.1458	0.1595	0.2024	0.1339	0.0844	0.0691	0.0672	0.0550	0.0455	0.0756	0.0888	0.0502	0.1280
400	0.2539	0.2516	0.1696	0.1937	0.2298	0.1431	0.0864	0.0694	0.0715	0.0529	0.0484	0.0770	0.0884	0.0498	0.1359
405	0.2785	0.2806	0.1922	0.2215	0.2521	0.1516	0.0848	0.0709	0.0705	0.0521	0.0494	0.0806	0.0853	0.0489	0.1378
410	0.2853	0.2971	0.2101	0.2419	0.2635	0.1570	0.0861	0.0707	0.0727	0.0541	0.0456	0.0771	0.0868	0.0503	0.1363
415	0.2883	0.3042	0.2179	0.2488	0.2702	0.1608	0.0859	0.0691	0.0731	0.0548	0.0470	0.0742	0.0859	0.0492	0.1363
420	0.2860	0.3125	0.2233	0.2603	0.2758	0.1649	0.0868	0.0717	0.0745	0.0541	0.0473	0.0766	0.0828	0.0511	0.1354
425	0.2761	0.3183	0.2371	0.2776	0.2834	0.1678	0.0869	0.0692	0.0770	0.0531	0.0486	0.0733	0.0819	0.0509	0.1322
430	0.2674	0.3196	0.2499	0.2868	0.2934	0.1785	0.0882	0.0710	0.0756	0.0599	0.0501	0.0758	0.0822	0.0496	0.1294
435	0.2565	0.3261	0.2674	0.3107	0.3042	0.1829	0.0903	0.0717	0.0773	0.0569	0.0480	0.0768	0.0818	0.0494	0.1241
440	0.2422	0.3253	0.2949	0.3309	0.3201	0.1896	0.0924	0.0722	0.0786	0.0603	0.0490	0.0775	0.0822	0.0480	0.1209
445	0.2281	0.3193	0.3232	0.3515	0.3329	0.2032	0.0951	0.0737	0.0818	0.0643	0.0468	0.0754	0.0819	0.0487	0.1137
450	0.2140	0.3071	0.3435	0.3676	0.3511	0.2120	0.0969	0.0731	0.0861	0.0702	0.0471	0.0763	0.0807	0.0468	0.1117
455	0.2004	0.2961	0.3538	0.3819	0.3724	0.2294	0.1003	0.0777	0.0907	0.0715	0.0486	0.0763	0.0787	0.0443	0.1045
460	0.1854	0.2873	0.3602	0.4026	0.4027	0.2539	0.1083	0.0823	0.0981	0.0798	0.0517	0.0752	0.0832	0.0440	0.1006
465	0.1733	0.2729	0.3571	0.4189	0.4367	0.2869	0.1203	0.0917	0.1067	0.0860	0.0519	0.0782	0.0828	0.0427	0.0970
470	0.1602	0.2595	0.3511	0.4317	0.4625	0.3170	0.1383	0.1062	0.1152	0.0959	0.0479	0.0808	0.0810	0.0421	0.0908
475	0.1499	0.2395	0.3365	0.4363	0.4890	0.3570	0.1634	0.1285	0.1294	0.1088	0.0494	0.0778	0.0819	0.0414	0.0858
480	0.1414	0.2194	0.3176	0.4356	0.5085	0.3994	0.1988	0.1598	0.1410	0.1218	0.0524	0.0788	0.0836	0.0408	0.0807
485	0.1288	0.1949	0.2956	0.4297	0.5181	0.4346	0.2376	0.1993	0.1531	0.1398	0.0527	0.0805	0.0802	0.0400	0.0752
490	0.1204	0.1732	0.2747	0.4199	0.5243	0.4615	0.2795	0.2445	0.1694	0.1626	0.0537	0.0809	0.0809	0.0392	0.0716
495	0.1104	0.1560	0.2506	0.4058	0.5179	0.4747	0.3275	0.2974	0.1919	0.1878	0.0577	0.0838	0.0838	0.0406	0.0688
500	0.1061	0.1436	0.2279	0.3882	0.5084	0.4754	0.3671	0.3462	0.2178	0.2302	0.0647	0.0922	0.0842	0.0388	0.0678
505	0.1018	0.1305	0.2055	0.3660	0.4904	0.4691	0.4030	0.3894	0.2560	0.2829	0.0737	0.1051	0.0865	0.0396	0.0639
510	0.0968	0.1174	0.1847	0.3433	0.4717	0.4556	0.4201	0.4180	0.3110	0.3455	0.0983	0.1230	0.0910	0.0397	0.0615
515	0.0941	0.1075	0.1592	0.3148	0.4467	0.4371	0.4257	0.4433	0.3789	0.4171	0.1396	0.1521	0.0920	0.0391	0.0586

Table 16.4 Continued

λ (nm)	7.5P 4/10	10PB 4/10	5PB 4/2	7.5B 5/10	10BG 6/8	2.5BG 6/10	2.5G 6/12	7.5GY 7/10	2.5GY 8/10	5Y 8.5/12	10YR 7/12	5YR 7/12	10R 6/12	5R 4/14	7.5RP 4/12
520	0.0881	0.0991	0.1438	0.2890	0.4207	0.4154	0.4218	0.4548	0.4515	0.4871	0.1809	0.1728	0.0917	0.0405	0.0571
525	0.0842	0.0925	0.1244	0.2583	0.3931	0.3937	0.4090	0.4605	0.5285	0.5529	0.2280	0.1842	0.0917	0.0394	0.0527
530	0.0808	0.0916	0.1105	0.2340	0.3653	0.3737	0.3977	0.4647	0.5845	0.5955	0.2645	0.1897	0.0952	0.0401	0.0513
535	0.0779	0.0896	0.0959	0.2076	0.3363	0.3459	0.3769	0.4626	0.6261	0.6299	0.2963	0.1946	0.0983	0.0396	0.0537
540	0.0782	0.0897	0.0871	0.1839	0.3083	0.3203	0.3559	0.4604	0.6458	0.6552	0.3202	0.2037	0.1036	0.0396	0.0512
545	0.0773	0.0893	0.0790	0.1613	0.2808	0.2941	0.3312	0.4522	0.6547	0.6661	0.3545	0.2248	0.1150	0.0395	0.0530
550	0.0793	0.0891	0.0703	0.1434	0.2538	0.2715	0.3072	0.4444	0.6545	0.6752	0.3950	0.2675	0.1331	0.0399	0.0517
555	0.0790	0.0868	0.0652	0.1243	0.2260	0.2442	0.2803	0.4321	0.6473	0.6832	0.4353	0.3286	0.1646	0.0420	0.0511
560	0.0793	0.0820	0.0555	0.1044	0.2024	0.2205	0.2532	0.4149	0.6351	0.6851	0.4577	0.3895	0.2070	0.0410	0.0507
565	0.0806	0.0829	0.0579	0.0978	0.1865	0.1979	0.2313	0.4039	0.6252	0.6964	0.4904	0.4654	0.2754	0.0464	0.0549
570	0.0805	0.0854	0.0562	0.0910	0.1697	0.1800	0.2109	0.3879	0.6064	0.6966	0.5075	0.5188	0.3279	0.0500	0.0559
575	0.0793	0.0871	0.0548	0.0832	0.1592	0.1610	0.1897	0.3694	0.5924	0.7063	0.5193	0.5592	0.3819	0.0545	0.0627
580	0.0803	0.0922	0.0517	0.0771	0.1482	0.1453	0.1723	0.3526	0.5756	0.7104	0.5273	0.5909	0.4250	0.0620	0.0678
585	0.0815	0.0978	0.0544	0.0747	0.1393	0.1234	0.1528	0.3288	0.5549	0.7115	0.5359	0.6189	0.4690	0.0742	0.0810
590	0.0842	0.1037	0.0519	0.0726	0.1316	0.1172	0.1355	0.3080	0.5303	0.7145	0.5431	0.6343	0.5067	0.0937	0.1004
595	0.0912	0.1079	0.0520	0.0682	0.1217	0.1045	0.1196	0.2829	0.5002	0.7195	0.5449	0.6485	0.5443	0.1279	0.1268
600	0.1035	0.1092	0.0541	0.0671	0.1182	0.0954	0.1050	0.2591	0.4793	0.7183	0.5493	0.6607	0.5721	0.1762	0.1595
605	0.1212	0.1088	0.0537	0.0660	0.1112	0.0913	0.0949	0.2388	0.4517	0.7208	0.5526	0.6648	0.5871	0.2449	0.2012
610	0.1455	0.1078	0.0545	0.0661	0.1071	0.0873	0.0868	0.2228	0.4340	0.7228	0.5561	0.6654	0.6073	0.3211	0.2452
615	0.1785	0.1026	0.0560	0.0660	0.1059	0.0846	0.0797	0.2109	0.4169	0.7274	0.5552	0.6721	0.6141	0.4050	0.2953
620	0.2107	0.0991	0.0560	0.0653	0.1044	0.0829	0.0783	0.2033	0.4060	0.7251	0.5573	0.6744	0.6170	0.4745	0.3439
625	0.2460	0.0995	0.0561	0.0644	0.1021	0.0814	0.0732	0.1963	0.3989	0.7274	0.5620	0.6723	0.6216	0.5335	0.3928
630	0.2791	0.1043	0.0578	0.0653	0.0991	0.0835	0.0737	0.1936	0.3945	0.7341	0.5607	0.6811	0.6272	0.5776	0.4336
635	0.3074	0.1101	0.0586	0.0669	0.1000	0.0833	0.0709	0.1887	0.3887	0.7358	0.5599	0.6792	0.6287	0.6094	0.4723
640	0.3330	0.1187	0.0573	0.0660	0.0980	0.0801	0.0703	0.1847	0.3805	0.7362	0.5632	0.6774	0.6276	0.6320	0.4996
645	0.3542	0.1311	0.0602	0.0677	0.0963	0.0776	0.0696	0.1804	0.3741	0.7354	0.5644	0.6796	0.6351	0.6495	0.5279

Table 16.4 Continued

λ (nm)	7.5P 4/10	10PB 4/10	5PB 4/2	7.5B 5/10	10BG 6/8	2.5BG 6/10	2.5G 6/12	7.5GY 7/10	2.5GY 8/10	5Y 8.5/12	10YR 7/12	5YR 7/12	10R 6/12	5R 4/14	7.5RP 4/12
650	0.3745	0.1430	0.0604	0.0668	0.0997	0.0797	0.0673	0.1766	0.3700	0.7442	0.5680	0.6856	0.6362	0.6620	0.5428
655	0.3920	0.1583	0.0606	0.0693	0.0994	0.0801	0.0677	0.1734	0.3630	0.7438	0.5660	0.6853	0.6348	0.6743	0.5601
660	0.4052	0.1704	0.0606	0.0689	0.1022	0.0810	0.0682	0.1721	0.3640	0.7440	0.5709	0.6864	0.6418	0.6833	0.5736
665	0.4186	0.1846	0.0595	0.0676	0.1005	0.0819	0.0665	0.1720	0.3590	0.7436	0.5692	0.6879	0.6438	0.6895	0.5837
670	0.4281	0.1906	0.0609	0.0694	0.1044	0.0856	0.0691	0.1724	0.3648	0.7442	0.5657	0.6874	0.6378	0.6924	0.5890
675	0.4395	0.1983	0.0605	0.0687	0.1073	0.0913	0.0695	0.1757	0.3696	0.7489	0.5716	0.6871	0.6410	0.7030	0.5959
680	0.4440	0.1981	0.0602	0.0698	0.1069	0.0930	0.0723	0.1781	0.3734	0.7435	0.5729	0.6863	0.6460	0.7075	0.5983
685	0.4497	0.1963	0.0580	0.0679	0.1103	0.0958	0.0727	0.1829	0.3818	0.7460	0.5739	0.6890	0.6451	0.7112	0.6015
690	0.4555	0.2003	0.0587	0.0694	0.1104	0.1016	0.0757	0.1897	0.3884	0.7518	0.5714	0.6863	0.6432	0.7187	0.6054
695	0.4612	0.2034	0.0573	0.0675	0.1084	0.1044	0.0767	0.1949	0.3947	0.7550	0.5741	0.6893	0.6509	0.7214	0.6135
700	0.4663	0.2061	0.0606	0.0676	0.1092	0.1047	0.0810	0.2018	0.4011	0.7496	0.5774	0.6950	0.6517	0.7284	0.6200
705	0.4707	0.2120	0.0613	0.0662	0.1074	0.1062	0.0818	0.2051	0.4040	0.7548	0.5791	0.6941	0.6514	0.7327	0.6287
710	0.4783	0.2207	0.0618	0.0681	0.1059	0.1052	0.0837	0.2071	0.4072	0.7609	0.5801	0.6958	0.6567	0.7351	0.6405
715	0.4778	0.2257	0.0652	0.0706	0.1082	0.1029	0.0822	0.2066	0.4065	0.7580	0.5804	0.6950	0.6597	0.7374	0.6443
720	0.4844	0.2335	0.0647	0.0728	0.1106	0.1025	0.0838	0.2032	0.4006	0.7574	0.5840	0.7008	0.6576	0.7410	0.6489
725	0.4877	0.2441	0.0684	0.0766	0.1129	0.1008	0.0847	0.1998	0.3983	0.7632	0.5814	0.7020	0.6576	0.7417	0.6621
730	0.4928	0.2550	0.0718	0.0814	0.1186	0.1036	0.0837	0.2024	0.3981	0.7701	0.5874	0.7059	0.6656	0.7491	0.6662
735	0.4960	0.2684	0.0731	0.0901	0.1243	0.1059	0.0864	0.2032	0.3990	0.7667	0.5885	0.7085	0.6641	0.7516	0.6726
740	0.4976	0.2862	0.0791	0.1042	0.1359	0.1123	0.0882	0.2074	0.4096	0.7735	0.5911	0.7047	0.6667	0.7532	0.6774
745	0.4993	0.3086	0.0828	0.1228	0.1466	0.1175	0.0923	0.2160	0.4187	0.7720	0.5878	0.7021	0.6688	0.7567	0.6834
750	0.5015	0.3262	0.0896	0.1482	0.1617	0.1217	0.0967	0.2194	0.4264	0.7739	0.5896	0.7071	0.6713	0.7600	0.6808
755	0.5044	0.3483	0.0980	0.1793	0.1739	0.1304	0.0996	0.2293	0.4370	0.7740	0.5947	0.7088	0.6657	0.7592	0.6838
760	0.5042	0.3665	0.1063	0.2129	0.1814	0.1330	0.1027	0.2378	0.4424	0.7699	0.5945	0.7055	0.6712	0.7605	0.6874
765	0.5073	0.3814	0.1137	0.2445	0.1907	0.1373	0.1080	0.2448	0.4512	0.7788	0.5935	0.7073	0.6745	0.7629	0.6955
770	0.5112	0.3974	0.1238	0.2674	0.1976	0.1376	0.1115	0.2489	0.4579	0.7801	0.5979	0.7114	0.6780	0.7646	0.7012
775	0.5147	0.4091	0.1381	0.2838	0.1958	0.1384	0.1118	0.2558	0.4596	0.7728	0.5941	0.7028	0.6744	0.7622	0.6996
780	0.5128	0.4206	0.1505	0.2979	0.1972	0.1390	0.1152	0.2635	0.4756	0.7793	0.5962	0.7105	0.6786	0.7680	0.7023

5. Then calculate the tristimulus values for the test source reflected on the colour allowing for white point adaptation using equation 16.14.

$$
\begin{pmatrix} X_{i,test,c} \\ Y_{i,test,c} \\ Z_{i,test,c} \end{pmatrix} = M^{-1} \begin{pmatrix} R_{i,test,c} \\ G_{i,test,c} \\ B_{i,test,c} \end{pmatrix}
$$
(16.14)

where $M^{-1} = \begin{pmatrix} 1.076450 & -0.237662 & 0.161212 \\ 0.410964 & 0.554342 & 0.034694 \\ -0.010954 & -0.013389 & 1.024343 \end{pmatrix}$

6. Calculate the position of the colour of the sample under the test and reference sources in CIE L*a*b* colour space, using equations 16.15 to 16.17 for the test source and 16.18 to 16.20 for the reference source.

$$
L^{*}_{i,test} = 116 \left(\frac{Y_{i,test,c}}{Y_{w,test}} \right)^{\frac{1}{3}} - 16
$$
(16.15)

$$
a^{*}_{i,test} = 500 \left[\left(\frac{X_{i,test,c}}{X_{w,test}} \right)^{\frac{1}{3}} - \left(\frac{Y_{i,test,c}}{Y_{w,test}} \right)^{\frac{1}{3}} \right]
$$
(16.16)

$$
b^{*}_{i,test} = 500 \left[\left(\frac{Y_{i,test,c}}{Y_{w,test}} \right)^{\frac{1}{3}} - \left(\frac{Z_{i,test,c}}{Z_{w,test}} \right)^{\frac{1}{3}} \right]
$$
(16.17)

$$
L^{*}_{i,ref} = 116 \left(\frac{Y_{i,ref}}{Y_{w,ref}} \right)^{\frac{1}{3}} - 16
$$
(16.18)

$$
a^{*}_{i,ref} = 500 \left[\left(\frac{X_{i,ref}}{X_{w,ref}} \right)^{\frac{1}{3}} - \left(\frac{Y_{i,ref}}{Y_{w,ref}} \right)^{\frac{1}{3}} \right]
$$
(16.19)

$$
b^{*}_{i,ref} = 500 \left[\left(\frac{Y_{i,ref}}{Y_{w,ref}} \right)^{\frac{1}{3}} - \left(\frac{Z_{i,ref}}{Z_{w,ref}} \right)^{\frac{1}{3}} \right]
$$
(16.20)

7. Calculate the chroma value for each sample under both the test ($C^{*}_{i,test}$) and the reference ($C^{*}_{i,ref}$) sources using equations 16.21 and 16.22:

$$
C^{*}_{i,test} = \sqrt{(a^{*}_{i,test})^2 + (b^{*}_{i,test})^2}
$$
(16.21)

$$
C^{*}_{i,ref} = \sqrt{(a^{*}_{i,ref})^2 + (b^{*}_{i,ref})^2}
$$
(16.22)

8. Calculate the colour differences between the adjusted appearance under the test source and the reference source using equations 16.23 to 16.25:

$$\Delta L_i^* = L_{i,\,test}^* - L_{i,\,ref}^* \tag{16.23}$$

$$\Delta a_i^* = a_{i,\,test}^* - a_{i,\,ref}^* \tag{16.24}$$

$$\Delta b_i^* = b_{i,\,test}^* - b_{i,\,ref}^* \tag{16.25}$$

9. Calculate the chroma difference in a similar manner using equation 16.26:

$$\Delta C_i^* = C_{i,\,test}^* - C_{i,\,ref}^* \tag{16.26}$$

10. The colour difference of the sample under the reference and test illuminants ΔE_i^* is: calculated using equation 16.27:

$$\Delta E_i^* = \sqrt{(\Delta L_i^*)^2 + (\Delta a_i^*)^2 + (\Delta b_i^*)^2} \tag{16.27}$$

11. Rather than simply calculating the colour difference of each reflective sample as above, a saturation factor is introduced into the calculations of the CQS. The saturation factor serves to negate any contribution to the colour difference that arises from an increase in object chroma from test source illumination (relative to the reference illuminant). Thus, equations 16.28a and 16.28b are used to calculate a colour difference that allows for a change in colour saturation $\Delta E_{i,sat}^*$

$$\Delta E_{i,\,sat}^* = \Delta E_i^* \quad if \quad \Delta C_i^* \le 0 \tag{16.28a}$$

$$\Delta E_{i,\,sat}^* = \sqrt{(\Delta E_i^*)^2 - (\Delta C_i^*)^2} \quad if \quad \Delta C_i^* > 0 \tag{16.28b}$$

12. The next step is to combine the individual colour differences for each of the 15 colour samples using a RMS method to ensure that it is not possible to get a good overall rating if one sample produces a large colour difference. The RMS average colour difference ΔE_{RMS}^* may be calculated using equation 16.29

$$\Delta E_{RMS} = \sqrt{\frac{1}{15} \sum_{i=1}^{15} (\Delta E_{i,\,sat}^*)^2} \tag{16.29}$$

13. This RMS error is then converted into a quality score $Q_{a,RMS}$ using equation 16.30

$$Q_{a,\,RMS} = 100 - (3.1 \times \Delta E_{RMS}) \tag{16.30}$$

14. To ensure that the final CQS score lies in the range 0 to 100, then the value of $Q_{a,RMS}$ is scaled using equation 16.31 to give $Q_{a,0-100}$

$$Q_{a,\,0-100} = 10 \ln\left[\exp\left(\frac{Q_{a,\,RMS}}{10}\right) + 1\right] \tag{16.31}$$

The action of this formula is illustrated in Figure 16.9

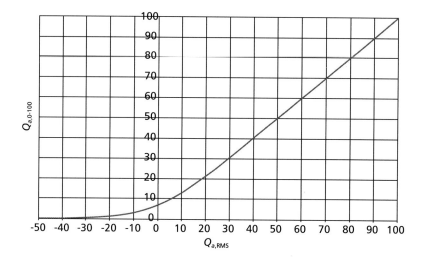

Figure 16.9 Plot of $Q_{a,0\text{-}100}$ for different values of $Q_{a,\text{RMS}}$

15. If the test light source has a low colour temperature (below 3500 K), then the CQS value is reduced. This is done by multiplying the $Q_{a,0\text{-}100}$ value by CCT factor M_{CCT}. When the colour temperature is above 3500 K, M_{CCT} is 1; below 3500 K, its value is given by equation 16.32:

$$M_{\text{CCT}} = T^3(9.2672 \times 10^{-11}) - T^2(8.3959 \times 10^{-7})$$
$$+ T(0.00255) - 1.612 \quad \text{for} \quad T \le 3500 \tag{16.32}$$

A plot of M_{CCT} against CCT is given in Figure 16.10.

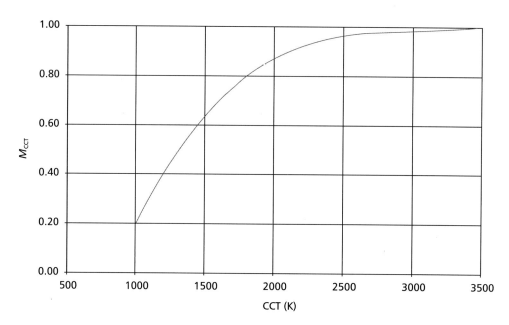

Figure 16.10 Plot of M_{CTT} against CCT

The CQS system may be compared to the CRI system. Table 16.5 gives the Ra_8 and Ra_{14} values for a series of common light sources together with the CQS values.

Table 16.5 Ra_8, Ra_{14} and CQS values for common light sources

Lamp type	CCT	Ra_8	Ra_{14}	CQS
Tungsten Halogen	3134	99.89	99.86	98.06
Ceramic Metal Halide	4230	88.97	84.06	85.21
Compact Fluorescent	2848	79.81	67.90	71.56
White LED	4301	76.56	69.31	69.98
Multi Phosphor Fluorescent	6621	97.97	97.08	96.79

16.3 Colour properties of surfaces

There are a number of systems for categorising the colour of surfaces and they are reviewed in LG11: *Surface Reflectance and Colour* (SLL, 2001). The systems are reviewed in this section and Table 16.6 gives the values for a number of colours in each system. There are also colour management systems that allow for the conversion of the colours from one environment to another. These systems are used to ensure that images appear the same in different conditions such as being viewed on a display and on a printed document. This section reviews these systems.

16.3.1 Munsell system

The Munsell system was originally devised in 1905 by A.H. Munsell, however, the system has been refined over the years. The system gives three different dimensions to colour space: Munsell Hue, Munsell Value and Munsell Chroma.

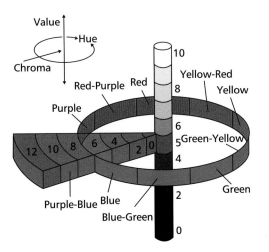

Figure 16.11 A schematic representation of the Munsell colour system

Hue describes the apparently dominant part of the spectrum occupied by the colour, the main hues are red (R), yellow (Y), green (G), blue (B) and purple (P). There are also intermediate hues of yellow red (YR), green yellow (GY), blue green (BG), purple blue (PB) and red purple(RP). Figure 16.11 shows the Munsell hues and Figure 16.12 illustrates the Munsell hues for chroma 6 and value 6.

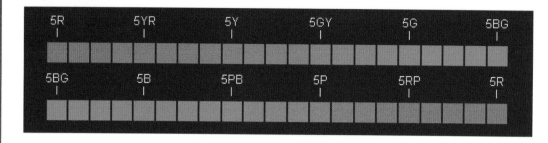

Figure 16.12 The Munsell hues for chroma 6 and value 6

Chroma is the strength of a colour that increases from neutral grey to fully saturated colour: neutral grey has a chroma of 0, however there is no intrinsic upper limit to chroma. Different areas of the colour space have different maximal values of chroma.

Value varies between perfect black (0) and perfect white (10). There is an approximate relationship between value (*V*) and the reflectance of the colour (*R*) given in equation 16.33

$$R \approx V\,(V-1)$$

(16.33)

Figure 16.13 shows three views of the Munsell colour space.

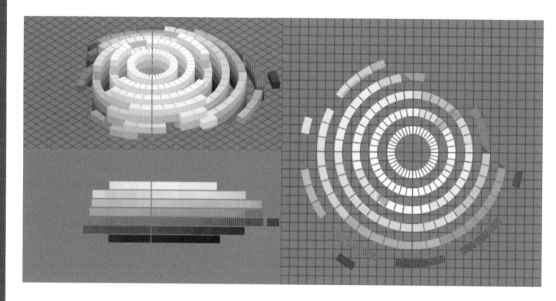

Figure 16.13 The Munsell colour solid

16.3.2 Natural Colour System (NCS)

The Swedish NCS colour system is based on the colour opponency description of colour vision, first proposed by Ewald Hering. The system is proprietary and owned by the Scandinavian Colour Institute (see http://www.ncscolour.com/ or http://www.ncscolour.co.uk). The system uses six elementary colours, white, black, red, yellow, green and blue. The system can be thought of as a colour space as illustrated in Figure 16.14.

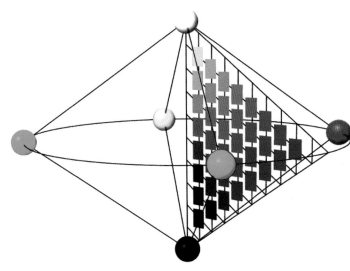

Figure 16.14 The NCS colour space (courtesy NCS Colour AB)

If you take a horizontal slice through the space, then the resulting section is a circle containing a series of hues as illustrated in Figure 16.15.

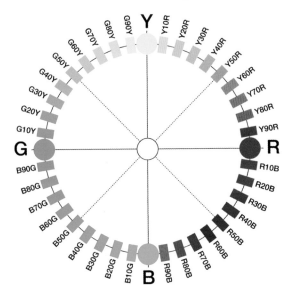

Figure 16.15 A hue circle in NCS (courtesy NCS Colour AB)

All of the colours in the NCS have a percentage of Whiteness or Blackness, and this is best illustrated using the NCS Colour Triangle (Figure 16.16). The NCS Colour Triangle is a vertical slice through the NCS Colour Solid. C stands for maximum colour intensity or Chromaticness, W stands for White and S for Black. The numbers in the system refer to the percentage of the distance across the colour space.

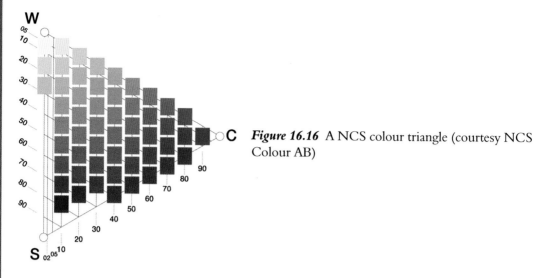

Figure 16.16 A NCS colour triangle (courtesy NCS Colour AB)

16.3.3 DIN system

The DIN (Deutsches Institut für Normung; http://www.din.de) colour system uses three variables to define the colour space. They are hue (T), saturation (S) and darkness (D). Figure 16.17 shows the DIN colour space.

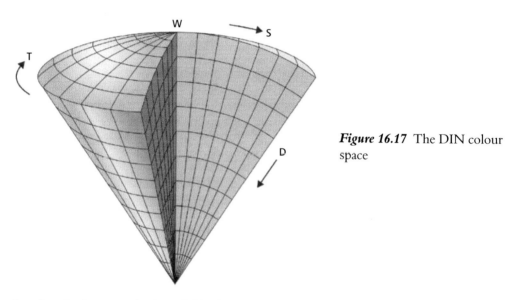

Figure 16.17 The DIN colour space

An atlas of colour samples is available showing the individual colours defined by the system, see Figure 16.18.

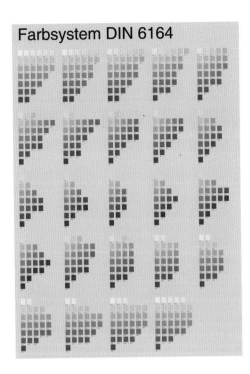

Farbsystem DIN 6164

Figure 16.18 DIN colour atlas

16.3.4 BS 5252

BS 5252: 1976: *Framework for colour co-ordination for building purposes* (BSI, 1976), provides a list of 237 surface colours. The system uses three dimensions of colour: they are hue, designated by a number (00 to 24), greyness designated by a letter (A to E), and weight given by an additional number. Figure 16.19 shows the colours in the system.

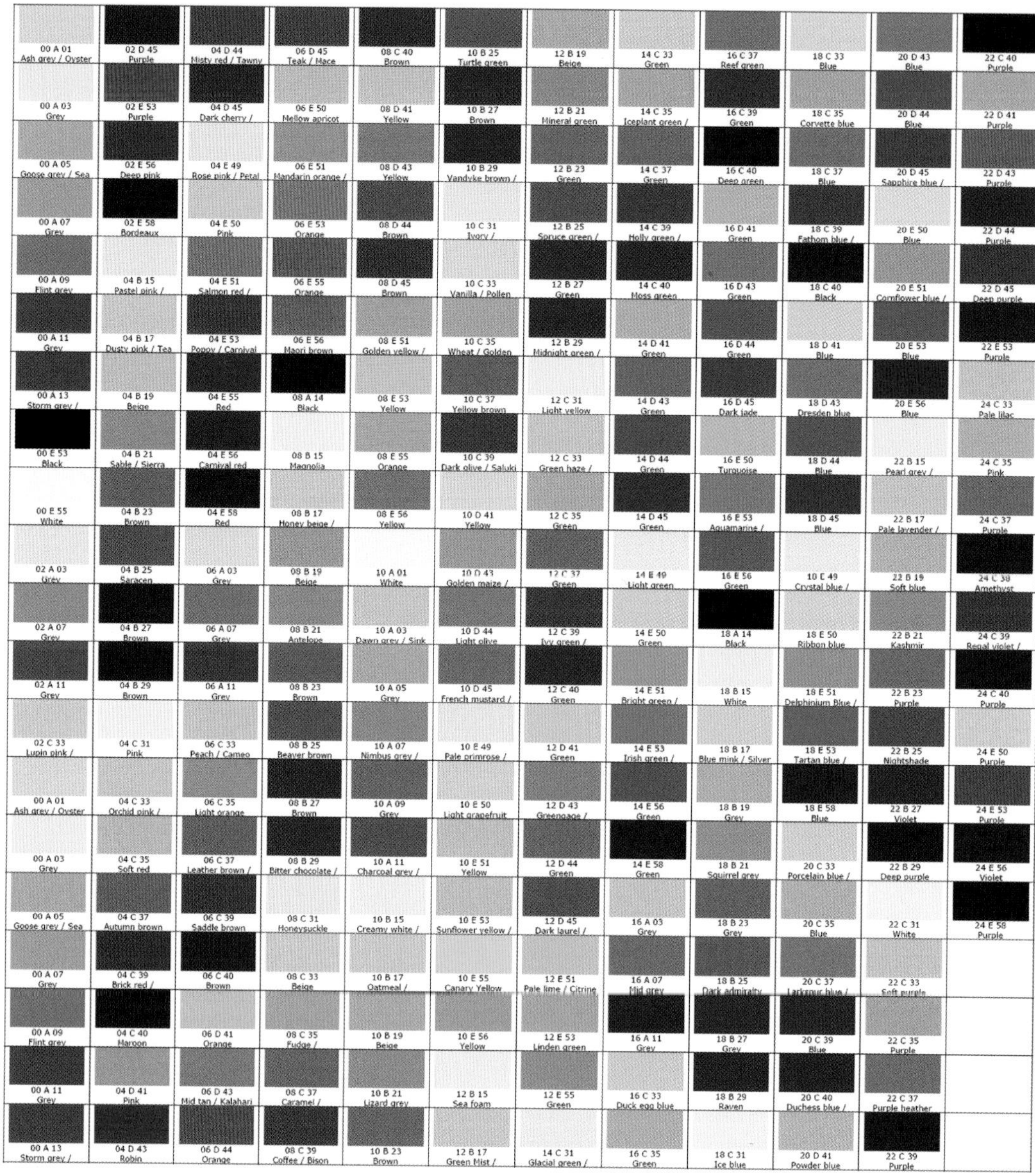

Figure 16.19 Colours defined by BS 5252 (note colours are approximate)

The colours in the standard are referenced in a number of other British Standards:

BS 381C: 1996 Specification for colours for identification, coding and special purposes
BS 4800: 1981 Specification for paint colours for building purposes
BS 4900: 1976 Specification for vitreous enamel colours for building purposes
BS 4901: 1976 Specification for plastics colours for building purposes
BS 4902: 1976 Specification for sheet and tile flooring colours for building purposes
BS 4903: 1979 Specification for external colours for farm buildings
BS 4904: 1978 Specification for external cladding colours for building purposes
BS 6770: 1988 Guide for exterior colours for park homes (mobile homes), holiday caravans and transportable accommodation units

In BS 4800, there is a table of colours that lists BS colour designations together with approximate values of the colours in the NCS and Munsell systems. This is shown here in Table 16.6.

Table 16.6 Approximate NCS and Munsell references for colours in BS 4800

Greyness group	BS colour designation	Hue	Approximate NCS reference	Approximate Munsell reference
A	00 A 01	Neutral	1501-Y03R	N 8.5
	00 A 05	Neutral	3101-Y26R	N 7
	00 A 09	Neutral	5301-R46B	N 5
	00 A 13	Neutral	7501-R97B	N 3
	10 A 03	Yellow	2002-Y03R	5Y 8/0.5
	10 A 07	Yellow	4302-Y09R	5Y 6/0.5
	10 A 11	Yellow	6702-G98Y	5Y 4/0.5
B	04 B 15	Red	0906-Y78R	10R 9/1
	04 B 17	Red	1409-Y83R	10R 8/2
	04 B 21	Red	3810-Y76R	10R 6/2
	08 B 15	Yellow-Red	0606-Y41R	10YR9.25/1
	08 B 17	Yellow-Red	1607-Y41R	8.75YR 8/2
	08 B 21	Yellow-Red	4107-Y41R	8.75YR 6/2
	08 B 25	Yellow-Red	6308-Y40R	8.75YR 4/2
	08 B 29	Yellow-Red	8105-Y53R	8.75YR 2/2
	10 B 15	Yellow	0504-Y21R	5Y 9.25/1
	10 B 17	Yellow	1811-Y01R	5Y 8/2
	10 B 21	Yellow	4011-G99Y	5Y 6/2
	10 B 25	Yellow	6211-G90Y	5Y 4/2
	10 B 29	Yellow	8305-G89Y	5Y 2/2
	12 B 15	Green-Yellow	0807-G73Y	5GY 9/1
	12 B 17	Green-Yellow	1812-G75Y	2.5GY 8/2
	12 B 21	Green-Yellow	3915-G65Y	2.5GY 6/2
	12 B 25	Green-Yellow	6313-G57Y	2.5GY 4/2
	12 B 29	Green-Yellow	8207-G53Y	2.5GY 2/2
	18 B 17	Blue	1704 B78G	5B 8/1
	18 B 21	Blue	4004-B57G	5B 6/1
	18 B 25	Blue	6405-B14G	5B 4/1
	18 B 29	Blue	8205-B06G	7.5B 2/1
	22 B 15	Violet	1000-N	10PB 9/1
	22 B 17	Violet	1804-R58B	10PB 8/2
C	02 C 33	Red-Purple	1118-R07B	7.5 RP 8/4
	02 C 37	Red-Purple	3531-R17B	7.5 RP 5/6
	02 C 39	Red-Purple	5331-R21B	7.5 RP 3/6
	02 C 40	Red-Purple	7315-R24B	7.5 RP 2/4
	04 C 33	Red	1019-Y86R	7.5 R 8/4
	04 C 37	Red	3632-Y85R	7.5 R 5/6
	04 C 39	Red	5136-Y87R	7.5 R 3/6
	06 C 33	Yellow-Red	1517-Y35R	7.5 YR 8/4
	06 C 37	Yellow-Red	4034-Y45R	5 YR 5/6
	06 C 39	Yellow-Red	6525-Y40R	7.5 YR 3/6

Table 16.6 Continued

Greyness group	BS colour designation	Hue	Approximate NCS reference	Approximate Munsell reference
C	08 C 31	Yellow-Red	0809-Y32R	10 YR 9/2
	08 C 33	Yellow-Red	2430-Y24R	10 YR 7/6
	08 C 37	Yellow-Red	4340-Y18R	10 YR 5/6
	08 C 39	Yellow-Red	6724-Y22R	10 YR 3/6
	10 C 31	Yellow	0811-Y16R	5Y 9/2
	10 C 33	Yellow	1122-Y03R	5Y 8.5/4
	10 C 37	Yellow	2536-G99Y	5Y 7/6
	10 C 39	Yellow	6921-G95Y	5Y 3/4
	12 C 33	Green-Yellow	1623-G72Y	2.5GY 8/4
	12 C 29	Green-Yellow	6626-G49Y	2.5GY 3/4
	14 C 31	Green	0609-G12Y	5G 9/1
	14 C 35	Green	2601-G06Y	5G 7/2
	14 C 39	Green	6520-G	5G 3/4
	14 C 40	Green	8007-G05Y	5G 2/2
	16 C 33	Blue-Green	1613-B68G	7.5BG 8/2
	16 C 37	Blue-Green	4326-B57G	7.5BG 5/4
	18 C 31	Blue	0704-B97G	5B 9.25/1
	18 C 35	Blue	2156-B05G	7.5B 7/3
	18 C 39	Blue	6126-B08G	7.5B 3/4
	20 C 33	Blue-Purple	1117-R83B	5PB 8/4
	20 C 37	Blue-Purple	3827-R87B	5PB 5/6
	20 C 40	Blue-Purple	7415-R82B	5PB 2/4
	22 C 37	Violet	3928-R60B	10PB 5/6
	24 C 33	Purple	1514-R35B	7.5P 4/10
	24 C 39	Purple	5431-R49B	7.5P 3/10
D	04 D 44	Red	2858-Y88R	7.5R 4/10
	04 D 45	Red	3657-Y93R	7.5R 3/10
	06 D 43	Yellow-Red	2560-Y27R	7.5YR 6/10
	06 D 45	Yellow-Red	4644-Y47R	5YR 5/8
	10 D 43	Yellow	2163-G97Y	5Y 7/10
	10 D 45	Yellow	3952-G98Y	5Y 5/8
	12 D 43	Green- Yellow	2954-G64Y	2.5GY 6/8
	12 D 45	Green- Yellow	5043-G54Y	2.5GY 4/6
	16 D 45	Blue-Green	5536-B51G	7.5BG 3/6
	18 D 45	Blue	3536-B09G	7.5B 5/6
	20 D 45	Purple-Blue	4938-R88B	5PB 3/8
	22 D 45	Violet	4542-R63B	10PB 3/8

Table 16.6 Continued

Greyness group	BS colour designation	Hue	Approximate NCS reference	Approximate Munsell reference
E	04 E 49	Red	0314-Y91B	7.5R 9/3
	04 E 51	Red	0963-Y81R	7.5R 6/12
	04 E 53	Red	1777-Y81R	7.5R 4.5/16
	06 E 50	Yellow-Red	0742-Y32R	7.5YR 8/8
	06 E 51	Yellow-Red	0860-Y50R	2.5YR 7/11
	06 E 56	Yellow-Red	2960-Y43R	5YR 5/12
	08 E 51	Yellow-Red	1178-Y16R	10YR 7.5/12
	10 E 49	Yellow	0823-G87Y	10Y 9/4
	10 E 50	Yellow	0848-Y	5Y 8.5/8
	10 E 53	Yellow	0875-G97Y	6.25Y 8.5/13
	12 E 51	Green- Yellow	0963-G66Y	2.5GY 8/10
	12 E 53	Green- Yellow	1266-G45Y	5GY 7/11
	14 E 51	Green	1854-G09Y	2.5G 6.5/8
	14 E 53	Green	2854-G5G	5G 5/10
	16 E 53	Blue- Green	3049-B50G	7.5BG 5/8
	18 E 49	Blue	0710-B64G	5B 9/2
	18 E50	Blue	0822-B11G	7.5B 8/4
	18 E51	Blue	1847-B06G	7.5B 6/8
	18 E53	Blue	2959-B	10B 4/10
	20 E 51	Purple-Blue	1548-R89B	5 PB 6/10
	00 E 53	Black	9500-N	N 1.5
	00 E 55	White	0000-N	N 9.5

16.3.5 RAL design system

The RAL system is based on an atlas of 1688 colours, the most used of which are shown in Figure 16.20. The system is similar to the Munsell system in that it uses three parameters to define each colour: hue, lightness and chroma.

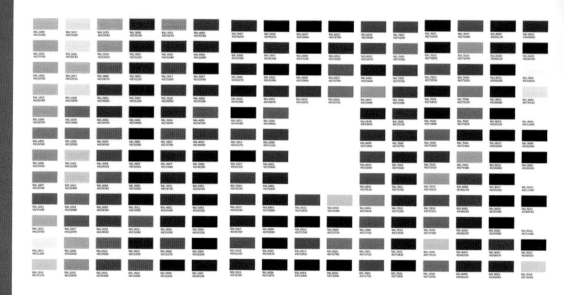

Figure 16.20 The RAL colour chart (note: colours are approximate)

16.3.6 CIE L*a*b*

The CIE L★a★b★ space is a method of describing colours that has been derived as a uniform colour space, thus not only is it possible to use the space to characterise a colour but it is also possible to assess differences between colours with some confidence.

The system is based on the tristimulus values of the light reflected from a coloured surface compared to the light reflected from a perfect white diffuser when illuminated by the same source. In the following formulae, X, Y and Z are values for the light reflected from a surface and X_n, Y_n and Z_n for the light reflected from a white surface.

$$L^\star = 116\left(\frac{Y}{Y_n}\right)^{\frac{1}{3}} - 16$$

$$a^\star = 500\left[\left(\frac{X}{X_n}\right)^{\frac{1}{3}} - \left(\frac{Y}{Y_n}\right)^{\frac{1}{3}}\right]$$

$$b^\star = 500\left[\left(\frac{Y}{Y_n}\right)^{\frac{1}{3}} - \left(\frac{Z}{Z_n}\right)^{\frac{1}{3}}\right]$$

Note these formulae are not valid for very dark surfaces where $\frac{Y}{Y_n} \leq 0.008856$.

In the CIELAB system, $L^\star$ is the lightness of the sample, $a^\star$ and $b^\star$ relate to colour. Other terms are also defined:

a, b chroma

$$C_{ab}^\star = \left(a^{\star 2} + b^{\star 2}\right)^{\frac{1}{2}}$$

a, b hue-angle

$$h_{ab} = \tan^{-1}\left(\frac{b^\star}{a^\star}\right)$$

The L★a★b★ colour space is difficult to visualise but Figure 16.21 gives sections through the diagram at $L^\star$ values of 25, 50 and 75.

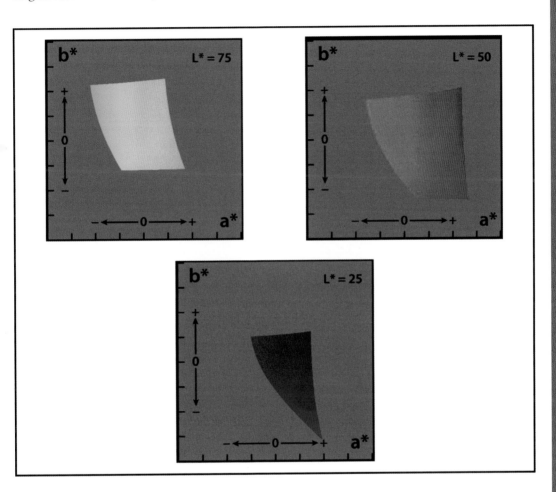

Figure 16.21 Sections through the CIE L★a★b★ colour space

Chapter 17: Daylight calculations

In the UK, the two most common cases of daylight considered are the overcast sky and clear sky. The performance of buildings under an overcast sky is often characterised by daylight factor. Calculating daylight factor at a given point in a room is often quite complex although there have been a series of tabular and graphic methods used to do this. These methods have generally been superseded by software calculation tools that generally give better results in less time.

This chapter focuses on two main calculations: that of average daylight factor and the formula for calculation of the sun position. There is a lot of further information on daylight given in SLL Lighting Guide 10: *Daylighting and window design* (SLL, 1999) and the British Standard *Code of practice for daylighting* (BSI, 2008).

17.1 Average daylight factor

Daylight factor is the ratio, expressed as a percentage, of illuminance at a point on a given plane due to light received directly or indirectly from a sky of known or assumed luminance distribution, to illuminance on a horizontal plane due to an unobstructed hemisphere of the sky. It is most common to use this ratio only for an overcast sky; however, in principle, it may be used with a sky of any luminance distribution. There are a number of sky distributions listed in the ISO/CIE Joint Standard, *Spatial distribution of daylight – CIE standard general sky* (ISO/CIE, 2003), however, many of the distributions are a function of sun position, thus, the daylight factor may be a function of time and date. The distribution of light under an overcast sky is such that the luminance of the sky at the zenith is three times the luminance of the sky at the horizon. The distribution of luminance is given in equation 17.1:

$$L_{\gamma} = \frac{L_z\left(1+2\sin\gamma\right)}{3} \tag{17.1}$$

where L_{γ} is the luminance at angle γ above the horizon and L_z is the luminance of the zenith. Note that this formula is different to the one given in the CIE Standard, however for all practical purposes, the distribution of luminance is the same. The CIE formula provides for sky luminance distributions for 15 different types of sky. The CIE formula uses five parameters to fit each sky type, however, as the basic formula is exponential in nature, it is not quite a perfect fit to the trigonometric function used in this *Code*.

The average daylight factor ($\bar{D}$) on the working plane of a room may be calculated using equation 17.2

$$\bar{D} = \frac{TA_W\theta}{A\left(1-R^2\right)} \tag{17.2}$$

where: T is the diffuse light transmittance of the glazing, including the effects of dirt;
A_w is the net glazed area of the window in m² (the net area of glazing is the area of windows less any lost to glazing bars or window frame);
θ is the angle subtended by the visible sky (degrees). It is measured in a vertical plane normal to the glass, from the window reference point (geometric centre of the window), as illustrated in Figure 17.1;
A is the total area of the ceiling, floor and walls, including windows, in m²;
R is the area-weighted average reflectance of the interior surfaces. In initial calculations for rooms with white ceilings and mid-reflectance walls, this may be taken as 0.5.

When two or more windows in a room face different obstructions, or differ in transmittance, the average daylight factor should be found separately for each window, and the results summed.

Note: The window area below the working plane does not significantly increase the amount of light falling onto the working plane. This is because the light from the lower part of the windows has to bounce off at least two room surfaces before it reaches the working plane. A study has shown that the area of the window below the working plane is only about 15 per cent as effective at letting light onto the working plane as an equivalent area above the working plane.

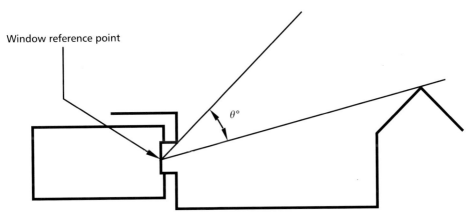

Figure 17.1 Angle of visible sky

17.2 Calculation of the sun position

The sun position is generally described by two angles, the solar altitude (a) and the solar azimuth (g); the two angles are shown in Figure 17.2.

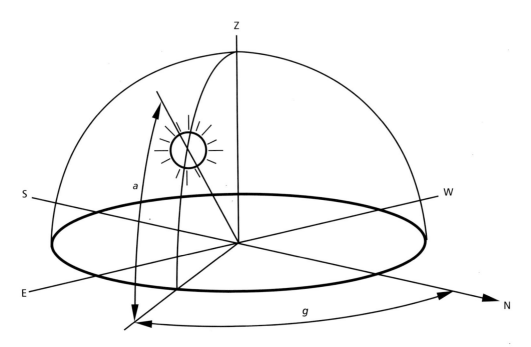

Figure 17.2 Solar altitude (a) and azimuth (g)

The calculation of solar altitude and azimuth is a complex multistep process with a number of intermediate variables being invoked. The calculations given in equations 17.3 to 17.10 may be used to calculate the sun position to ±1°.

Day angle

The day angle, τ_d, in radians, is given by equation 17.3:

$$\tau_d = \frac{2\pi(J-1)}{365} \tag{17.3}$$

where J is the day number; J is 1 for 1st January and 365 for 31st December. February is taken to have 28 days.

Solar declination

The solar declination is the angle between the sun's rays arriving at the earth and the earth's equatorial plane. The solar declination, δ_s, in radians may be calculated using equation 17.4:

$$\delta_s = 0.006918 - 0.399912\cos\tau_d + 0.070257\sin\tau_d - 0.006758\cos^2\tau_d$$
$$+ 0.000907\sin^2\tau_d - 0.002697\cos^3\tau_d + 0.00148\sin^3\tau_d \tag{17.4}$$

A plot of solar declination over the course of a year is shown in Figure 17.3.

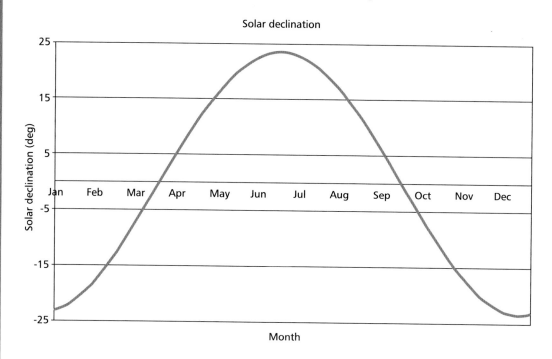

Figure 17.3 Solar declination

Equation of time

The equation of time gives the variation between solar time and clock time that is due to the eccentricity of the earth's orbit around the sun. The equation of time, ET, in hours, depends on the day number J and is given in equation 17.5

$$ET = 0.17\sin\left\{\frac{4\pi(J-80)}{373}\right\} - 0.129\sin\left\{\frac{2\pi(J-8)}{355}\right\} \tag{17.5}$$

Figure 17.4 is a plot of ET values over the course of a year.

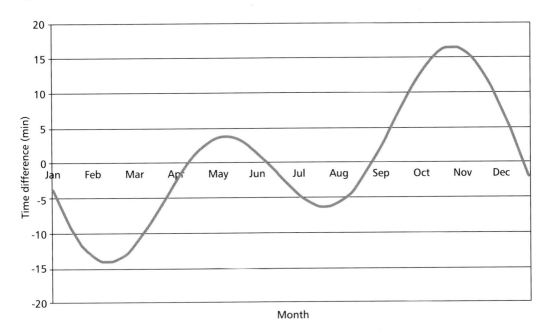

Figure 17.4 Equation of time

True solar time

The true solar time differs from clock time due to a number of factors: the difference in longitude between the site and the standard meridian for the time zone, the equation of time and the adoption of clock changes between summer and winter. True solar time, TST, in hours is given by equation 17.6:

$$TST = LT + \frac{\lambda - \lambda_s}{15} + ET + TD \tag{17.6}$$

where: LT is the local clock time in hours from midnight
        ET is the equation of time in hours (see equation 17.5)
        TD is the daylight savings time difference in hours (for example, –1 for the change from GMT to BST in the UK)
        λ is the longitude of the site in degrees (west positive, east negative)
        λ_s is the longitude of the standard meridian of the time zone in degrees

Hour angle

The hour angle, ξ, in radians is given by equation 17.7:

$$\xi = \frac{\pi \times TST}{12} \tag{17.7}$$

Solar altitude

Solar altitude, a, may be given either in radians or degrees, depending on which version of the arcsine function is used. Equation 17.8 may be used to calculate solar altitude:

$$a = \sin^{-1}\left(\sin\varphi \sin\delta_s - \cos\varphi \cos\delta_s \cos\xi\right) \tag{17.8}$$

where φ is the latitude (φ is positive in the northern hemisphere and negative in the southern) of the site. Figure 17.5 gives a plot of solar altitude at Birmingham for 3 days in the year.

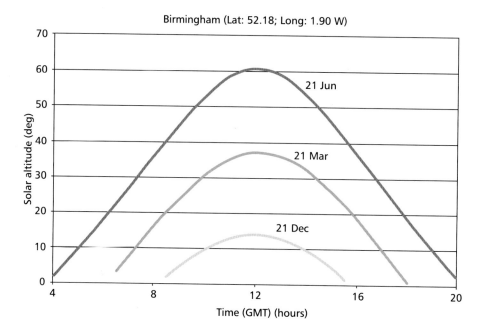

Figure 17.5 Solar altitude

Solar azimuth

Solar azimuth g, may be given either in radians or degrees from true north, depending on which version of the arccosine function is used. The azimuth angle is shown in Figure 17.2. Solar azimuth may be calculated using equation 17.9 for the northern hemisphere and equation 17.10 for the southern. Note the equations as they stand are in radians, however when solar azimuth is calculated in degrees the first 2π in some of the equations should be replaced by 360.

$$g = \cos^{-1}\left(\frac{-\sin\varphi\sin a + \sin\delta_s}{\cos\varphi\cos a}\right), \qquad 0 < \xi \leq \pi$$

$$g = 2\pi - \cos^{-1}\left(\frac{-\sin\varphi\sin a + \sin\delta_s}{\cos\varphi\cos a}\right), \qquad \pi < \xi \leq 2\pi \qquad (17.9)$$

$$g = 2\pi - \cos^{-1}\left(\frac{-\sin\varphi\sin a + \sin\delta_s}{\cos\varphi\cos\alpha}\right), \qquad 0 < \xi \leq \pi$$

$$g = \cos^{-1}\left(\frac{-\sin\varphi\sin a + \sin\delta_s}{\cos\varphi\cos a}\right), \qquad \pi < \xi \leq 2\pi \qquad (17.10)$$

where φ is the latitude of the site. Figure 17.6 gives a plot of solar azimuth at Birmingham for 3 days in the year.

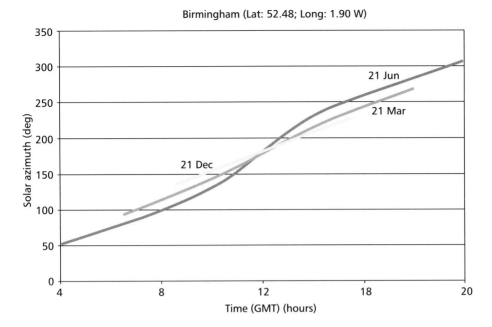

Figure 17.6 Solar azimuth

Chapter 18: Predicting maintenance factor

During the life of a lighting installation, the light available for the task progressively decreases due to accumulation of dirt on surfaces and the aging of equipment. The rate of reduction is influenced by the equipment choice and the environmental and operating conditions. In lighting scheme design, it is necessary to take account of this fall using a maintenance factor and plan suitable maintenance schedules to limit the decay. A high maintenance factor together with an effective maintenance programme promotes energy efficiency, good design of lighting schemes and limits the installed lighting power requirements.

This chapter of the *Code* describes the parameters influencing the depreciation process and gives a method for estimating the maintenance factor for electric lighting systems.

The definition of maintenance factor is 'the ratio of maintained illuminance to initial illuminance', i.e. taking account of all losses including lamp lumen maintenance.

18.1 Determination of maintenance factor

18.1.1 Indoor lighting
The maintenance factor (MF) is a multiple of four factors:

$$MF = LLMF \times LSF \times LMF \times RSMF$$

where $LLMF$ is the lamp lumen maintenance factor; LSF is the lamp survival factor (used only if spot-replacement of lamps is not carried out); LMF is the luminaire maintenance factor; $RSMF$ is the room surface maintenance factor.

For more information on indoor lighting maintenance factors, see CIE 97:2005 (CIE, 2005).

18.1.2 Outdoor lighting
The maintenance factor in outdoor lighting depends on factors similar to those for indoor lighting; however, as outdoor lighting does not depend on inter-reflected light, then there is no need to consider the term for room surface maintenance factor. The $LLMF$ and LSF values for outdoor lighting are the same as for indoor lighting, however, due to the harsher outdoor environment, a separate table is used for LMF for outdoor use.

$$MF = LLMF \times LSF \times LMF$$

For more information on outdoor lighting maintenance factors, see CIE 154:2003 (CIE, 2003).

18.2 Lamp lumen maintenance factor and survival factor

Lamp lumen (luminous flux) maintenance factor (LLMF)
The lumen output from all lamp types decreases with time of operation. The rate of fall-off varies for different lamp types and it is essential to consult the manufacturer's data. From such data, it is possible to obtain the lamp lumen maintenance factor for a specific number of hours of operation. The lamp lumen maintenance factor is therefore the proportion of the initial light output that is produced after a specified time and, where the rate of fall-off is regular, may be quoted as a percentage reduction per thousand hours of operation.

Manufacturers' data will normally be based on British Standards test procedures which specify the ambient temperature in which the lamp will be tested, with a regulated voltage applied to the lamp and, if appropriate, a reference set of control gear. If any of the aspects of the proposed design are unusual, e.g. high ambient temperature, vibration, switching cycle, operating attitude, etc, the manufacturer should be made aware of the conditions and will advise if they affect the life and/or light output of the lamp.

Lamp survival factor (LSF)
As with lamp lumen maintenance factor, it is essential to consult manufacturer's data. These will give the percentage of lamp failures for a specific number of hours operation and is only applicable where group lamp replacement, without spot replacement, is to be carried out. These data will also be based on assumptions such as switching cycle, supply voltage and control gear. Manufacturers should be made aware of these aspects and should advise if these will affect the lamp life or lamp survival. Typical lumen maintenance and lamp survival data are given in Table 18.1.

18.3   Luminaire maintenance factor (LMF) – Indoor

Dirt deposited on or in the luminaire will cause a reduction in light output from the luminaire. The rate at which dirt is deposited depends on the construction of the luminaire and on the extent to which dirt is present in the atmosphere, which in turn is related to the nature of the dirt generated in the specific environment. Table 18.2 gives a list of the luminaire classes and Table 18.3 gives a list of typical locations where the various environmental conditions may be found. Note that some of the descriptions in Table 18.2 refer to ingress protection classes; these are discussed in section 18.6.

Table 18.1 LLMF and LSF values

Burning hours in thousand hours

Luminaire type[1]		Differences[2]	0.1	0.5	1	2	4	6	8	10	12	15	20	30
Incandescent	LLMF	Moderate	1.00	0.97	0.93	–	–	–	–	–	–	–	–	–
	LSF	Large	1.00	0.98	0.50	–	–	–	–	–	–	–	–	–
Halogen	LLMF	Large	1.00	0.99	0.97	0.95	–	–	–	–	–	–	–	–
	LSF	Large	1.00	1.00	0.78	0.50	–	–	–	–	–	–	–	–
Fluorescent tri-phosphor (HF gear)	LLMF	Moderate	1.00	0.99	0.98	0.97	0.93	0.92	0.90	0.90	0.90	0.90	0.90	–
	LSF	Moderate	1.00	1.00	1.00	1.00	1.00	0.99	0.98	0.98	0.97	0.94	0.50	–
Fluorescent tri-phosphor	LLMF	Moderate	1.00	0.99	0.98	0.97	0.93	0.92	0.90	0.90	0.90	0.90	–	–
	LSF	Moderate	1.00	1.00	1.00	1.00	1.00	0.99	0.98	0.98	0.92	0.50	–	–
Fluorescent halophosphate	LLMF	Moderate	1.00	0.98	0.96	0.95	0.87	0.84	0.81	0.79	0.77	0.75	–	–
	LSF	Moderate	1.00	1.00	1.00	1.00	1.00	0.99	0.98	0.98	0.92	0.50	–	–
Compact fluorescent	LLMF	Large	1.00	0.98	0.97	0.94	0.91	0.89	0.87	0.85	–	–	–	–
	LSF	Large	1.00	0.99	0.99	0.98	0.97	0.94	0.86	0.50	–	–	–	–
Mercury	LLMF	Moderate	1.00	0.99	0.97	0.93	0.85	0.82	0.80	0.79	0.78	0.77	0.76	–
	LSF	Moderate	1.00	1.00	0.99	0.98	0.97	0.94	0.90	0.86	0.79	0.69	0.50	–
Metal halide (250/400 W)[3]	LLMF	Large	1.00	0.98	0.95	0.90	0.87	0.83	0.79	0.65	0.63	0.58	0.50	–
	LSF	Large	1.00	0.99	0.99	0.98	0.97	0.92	0.86	0.80	0.73	0.66	0.50	–
Ceramic metal halide (50/150 W)	LLMF	Large	1.00	0.95	0.87	0.75	0.72	0.68	0.64	0.60	0.56	–	–	–
	LSF	Large	1.00	0.99	0.99	0.98	0.98	0.98	0.95	0.80	0.50	–	–	–
High pressure sodium (250/400 W)	LLMF	Moderate	1.00	1.00	0.98	0.98	0.98	0.97	0.97	0.97	0.97	0.96	0.94	0.90
	LSF	Moderate	1.00	1.00	1.00	1.00	0.99	0.99	0.99	0.99	0.97	0.95	0.92	0.50

[1]Data for LEDs are changing rapidly and no values can be given.
[2]Indicates differences in LLMF and LSF among lamps which belong to the same lamp type category.
[3]Differences in the group of metal halides are extremely remarkable. Very high and very low wattage lamps live significantly shorter than the values given here.

Table 18.2 Luminaire classes

Class	Description
A	Bare lamp batten
B	Open top reflector (ventilated self-cleaning)
C	Closed top housing (unventilated)
D	Enclosed (IP2X)
E	Dustproof (IP5X)
F	Indirect uplighter
G	Air handling and forced ventilated

Table 18.3 Locations of environmental conditions

Environment	Typical locations
Very clean (VC)	Clean rooms, semiconductor plants, hospital clinical areas, computer centres
Clean (C)	Offices, schools, hospital wards
Normal (N)	Shops, laboratories, restaurants, warehouses, assembly areas, workshops
Dirty (D)	Steelworks, chemical works, foundries, welding, polishing, woodwork

Table 18.4 shows typical changes in light output from a luminaire caused by dirt deposition, for a number of luminaire types and environment classes.

18.4 Luminaire maintenance factor (LMF) – Outdoor

Table 18.5 shows typical luminaire maintenance factors, which may be used in design calculations. It takes into account luminaire IP rating, pollution category and cleaning interval. See section 18.6 for information on IP classes.

Table 18.4 LMF values for indoor use

Elapsed time between cleanings / years	0.5				1.0				1.5				2.0				2.5				3.0			
										Environment														
Luminaire type	VC	C	N	D	VC	C	N	D	VC	C	N	D	VC	C	N	D	VC	C	N	D	VC	C	N	D
A	0.98	0.95	0.92	0.88	0.96	0.93	0.89	0.83	0.95	0.91	0.87	0.80	0.94	0.89	0.84	0.78	0.93	0.87	0.82	0.75	0.92	0.85	0.79	0.73
B	0.96	0.95	0.91	0.88	0.95	0.90	0.86	0.83	0.94	0.87	0.83	0.79	0.92	0.84	0.80	0.75	0.91	0.82	0.76	0.71	0.89	0.79	0.74	0.68
C	0.95	0.93	0.89	0.85	0.94	0.89	0.81	0.75	0.93	0.84	0.74	0.66	0.91	0.80	0.69	0.59	0.89	0.77	0.64	0.54	0.87	0.74	0.61	0.52
D	0.94	0.92	0.87	0.83	0.94	0.88	0.82	0.77	0.93	0.85	0.79	0.73	0.91	0.83	0.77	C.71	0.90	0.81	0.75	0.68	0.89	0.79	0.73	0.65
E	0.94	0.96	0.93	0.91	0.96	0.94	0.90	0.86	0.92	0.92	0.88	0.83	0.93	0.91	0.86	0.81	0.92	0.90	0.85	0.80	0.92	0.90	0.84	0.79
F	0.94	0.92	0.89	0.85	0.93	0.86	0.81	0.74	0.91	0.81	0.73	0.65	0.88	0.77	0.66	0.57	0.86	0.73	0.60	0.51	0.85	0.70	0.55	0.45
G	1.00	1.00	0.99	0.98	1.00	0.99	0.96	0.93	0.99	0.97	0.94	0.89	0.99	0.96	0.92	0.87	0.98	0.95	0.91	0.86	0.98	0.95	0.90	0.85

Table 18.5 LMF values for outdoor use

Cleaning interval / months	Luminaire maintenance factor								
	IP2X minimum (a)			IP5X minimum (a)			IP6X minimum (a)		
	High pollution (b)	Medium pollution (c)	Low pollution (d)	High pollution (b)	Medium pollution (c)	Low pollution (d)	High pollution (b)	Medium pollution (c)	Low pollution (d)
12	0.53	0.62	0.82	0.89	0.90	0.92	0.91	0.92	0.93
18	0.48	0.58	0.80	0.87	0.88	0.91	0.90	0.91	0.92
24	0.45	0.56	0.79	0.84	0.86	0.90	0.88	0.89	0.91
36	0.42	0.53	0.78	0.76	0.82	0.88	0.83	0.87	0.90

(a) Ingress protection code number of lamp housing; see BS EN 60529 (BSI, 1992).
(b) High pollution generally occurs in the centre of large urban areas and heavy industrial areas.
(c) Medium pollution generally occurs in semi-urban, residential and light industrial areas.
(d) Low pollution generally occurs in rural areas.

18.5   Room surface maintenance factor (RSMF)

Changes in room surface reflectance caused by dirt deposition will cause changes in the illuminance produced by the lighting installation. The magnitude of these changes is governed by the extent of dirt deposition and the importance of inter-reflection to the illuminance produced. Inter-reflection is closely related to the distribution of light from the luminaire. For luminaires that have a strongly downward distribution, i.e. direct luminaires, inter-reflection has little effect on the illuminance produced on the horizontal working plane. Conversely, indirect lighting is completely dependent on inter-reflections. Most luminaires lie somewhere between these extremes so most lighting installations are dependent to some extent on inter-reflection.

Tables 18.6 to 18.8 show the typical changes in illuminance from an installation that occur with time due to dirt deposition on the room surfaces for very clean, clean, normal and dirty conditions lit by direct, direct/indirect and indirect luminaires. From the tables, it is possible to select a room surface maintenance factor appropriate to the circumstances. The areas in which very clean, clean, normal and dirty environments are found are given in Table 18.3.

Table 18.6 Room surface maintenance factor (RSMF) for direct flux distribution

Reflectances ceiling/walls/floor	Time / years						Room surface maintenance factor						
	Environment	0.5	1.0	1.5	2.0	2.5	3.0	3.5	4.0	4.5	5.0	5.5	6.0
0.80/0.70/0.20	Very clean	0.97	0.96	0.95	0.95	0.95	0.95	0.95	0.95	0.95	0.95	0.95	0.95
	Clean	0.93	0.92	0.91	0.91	0.91	0.91	0.91	0.91	0.91	0.91	0.91	0.91
	Normal	0.88	0.86	0.86	0.85	0.85	0.85	0.85	0.85	0.85	0.85	0.85	0.85
	Dirty	0.81	0.80	0.80	0.80	0.80	0.80	0.80	0.80	0.80	0.80	0.80	0.80
0.80/0.50/0.20	Very clean	0.98	0.97	0.97	0.97	0.97	0.97	0.97	0.97	0.97	0.97	0.97	0.97
	Clean	0.95	0.94	0.94	0.94	0.94	0.94	0.94	0.94	0.94	0.94	0.94	0.94
	Normal	0.91	0.90	0.90	0.90	0.90	0.90	0.90	0.90	0.90	0.90	0.90	0.90
	Dirty	0.86	0.85	0.85	0.85	0.85	0.85	0.85	0.85	0.85	0.85	0.85	0.85
0.80/0.30/0.20	Very clean	0.99	0.98	0.98	0.98	0.98	0.98	0.98	0.98	0.98	0.98	0.98	0.98
	Clean	0.97	0.96	0.96	0.96	0.96	0.96	0.96	0.96	0.96	0.96	0.96	0.96
	Normal	0.94	0.93	0.93	0.93	0.93	0.93	0.93	0.93	0.93	0.93	0.93	0.93
	Dirty	0.91	0.90	0.90	0.90	0.90	0.90	0.90	0.90	0.90	0.90	0.90	0.90
0.70/0.70/0.20	Very clean	0.97	0.96	0.96	0.96	0.96	0.96	0.96	0.96	0.96	0.96	0.96	0.96
	Clean	0.94	0.92	0.92	0.92	0.92	0.92	0.92	0.92	0.92	0.92	0.92	0.92
	Normal	0.89	0.87	0.87	0.87	0.87	0.87	0.87	0.87	0.87	0.87	0.87	0.87
	Dirty	0.83	0.81	0.81	0.81	0.81	0.81	0.81	0.81	0.81	0.81	0.81	0.81
0.70/0.50/0.20	Very clean	0.98	0.97	0.97	0.97	0.97	0.97	0.97	0.97	0.97	0.97	0.97	0.97
	Clean	0.96	0.95	0.94	0.94	0.94	0.94	0.94	0.94	0.94	0.94	0.94	0.94
	Normal	0.92	0.91	0.90	0.90	0.90	0.90	0.90	0.90	0.90	0.90	0.90	0.90
	Dirty	0.87	0.86	0.86	0.86	0.86	0.86	0.86	0.86	0.86	0.86	0.86	0.86

Table 18.6 Continued

Reflectances ceiling/walls/floor	Time / years	0.5	1.0	1.5	2.0	2.5	3.0	3.5	4.0	4.5	5.0	5.5	6.0
	Environment	Room surface maintenance factor											
0.70/0.30/0.20	Very clean	0.99	0.98	0.98	0.98	0.98	0.98	0.98	0.98	0.98	0.98	0.98	0.98
	Clean	0.97	0.96	0.96	0.96	0.96	0.96	0.96	0.96	0.96	0.96	0.96	0.96
	Normal	0.95	0.94	0.94	0.94	0.93	0.93	0.93	0.93	0.93	0.93	0.93	0.93
	Dirty	0.92	0.91	0.91	0.91	0.91	0.91	0.91	0.91	0.91	0.91	0.91	0.91
0.50/0.70/0.20	Very clean	0.98	0.97	0.97	0.96	0.96	0.96	0.96	0.96	0.96	0.96	0.96	0.96
	Clean	0.95	0.94	0.93	0.93	0.93	0.93	0.93	0.93	0.93	0.93	0.93	0.93
	Normal	0.91	0.89	0.89	0.89	0.89	0.89	0.89	0.89	0.89	0.89	0.89	0.89
	Dirty	0.85	0.84	0.84	0.84	0.84	0.84	0.84	0.84	0.84	0.84	0.84	0.84
0.50/0.50/0.20	Very clean	0.98	0.98	0.98	0.98	0.97	0.97	0.97	0.97	0.97	0.97	0.97	0.97
	Clean	0.97	0.96	0.95	0.95	0.95	0.95	0.95	0.95	0.95	0.95	0.95	0.95
	Normal	0.94	0.92	0.92	0.92	0.92	0.92	0.92	0.92	0.92	0.92	0.92	0.92
	Dirty	0.89	0.89	0.88	0.88	0.88	0.88	0.88	0.88	0.88	0.88	0.88	0.88
0.50/0.30/0.20	Very clean	0.99	0.99	0.98	0.98	0.98	0.98	0.98	0.98	0.98	0.98	0.98	0.98
	Clean	0.98	0.97	0.97	0.97	0.97	0.97	0.97	0.97	0.97	0.97	0.97	0.97
	Normal	0.96	0.95	0.95	0.95	0.95	0.95	0.95	0.95	0.95	0.95	0.95	0.95
	Dirty	0.93	0.92	0.92	0.92	0.92	0.92	0.92	0.92	0.92	0.92	0.92	0.92

Table 18.7 Room surface maintenance factor (RSMF) for direct/indirect flux distribution

Reflectances ceiling/walls/floor	Time / years	Room surface maintenance factor											
	Environment	0.5	1.0	1.5	2.0	2.5	3.0	3.5	4.0	4.5	5.0	5.5	6.0
0.80/0.70/0.20	Very clean	0.95	0.94	0.93	0.93	0.93	0.93	0.93	0.93	0.93	0.93	0.93	0.93
	Clean	0.90	0.88	0.87	0.87	0.87	0.87	0.87	0.87	0.87	0.87	0.87	0.87
	Normal	0.81	0.78	0.77	0.77	0.77	0.77	0.77	0.77	0.77	0.77	0.77	0.77
	Dirty	0.70	0.67	0.67	0.67	0.67	0.67	0.67	0.67	0.67	0.67	0.67	0.67
0.80/0.50/0.20	Very clean	0.96	0.95	0.95	0.95	0.95	0.95	0.95	0.95	0.95	0.95	0.95	0.95
	Clean	0.93	0.91	0.90	0.90	0.90	0.90	0.90	0.90	0.90	0.90	0.90	0.90
	Normal	0.85	0.83	0.82	0.82	0.82	0.82	0.82	0.82	0.82	0.82	0.82	0.82
	Dirty	0.76	0.73	0.73	0.73	0.73	0.73	0.73	0.73	0.73	0.73	0.73	0.73
0.80/0.30/0.20	Very clean	0.97	0.96	0.96	0.96	0.96	0.96	0.96	0.96	0.96	0.96	0.96	0.96
	Clean	0.94	0.93	0.92	0.92	0.92	0.92	0.92	0.92	0.92	0.92	0.92	0.92
	Normal	0.89	0.87	0.86	0.86	0.86	0.86	0.86	0.86	0.86	0.86	0.86	0.86
	Dirty	0.81	0.79	0.78	0.78	0.78	0.78	0.78	0.78	0.78	0.78	0.78	0.78
0.70/0.70/0.20	Very clean	0.96	0.94	0.94	0.93	0.93	0.93	0.93	0.93	0.93	0.93	0.93	0.93
	Clean	0.91	0.89	0.88	0.88	0.88	0.88	0.88	0.88	0.88	0.88	0.88	0.88
	Normal	0.83	0.80	0.79	0.79	0.79	0.79	0.79	0.79	0.79	0.79	0.79	0.79
	Dirty	0.72	0.69	0.69	0.69	0.69	0.69	0.69	0.69	0.69	0.69	0.69	0.69
0.70/0.50/0.20	Very clean	0.97	0.96	0.95	0.95	0.95	0.95	0.95	0.95	0.95	0.95	0.95	0.95
	Clean	0.93	0.91	0.91	0.91	0.91	0.91	0.91	0.91	0.91	0.91	0.91	0.91
	Normal	0.87	0.84	0.84	0.83	0.83	0.83	0.83	0.83	0.83	0.83	0.83	0.83
	Dirty	0.77	0.75	0.75	0.75	0.75	0.75	0.75	0.75	0.75	0.75	0.75	0.75

Table 18.7 Continued

Reflectances ceiling/walls/floor	Time / years												
	Environment	0.5	1.0	1.5	2.0	2.5	3.0	3.5	4.0	4.5	5.0	5.5	6.0
						Room surface maintenance factor							
0.70/0.30/0.20	Very clean	0.98	0.97	0.96	0.96	0.96	0.96	0.96	0.96	0.96	0.96	0.96	0.96
	Clean	0.95	0.93	0.93	0.93	0.93	0.93	0.93	0.93	0.93	0.93	0.93	0.93
	Normal	0.90	0.88	0.87	0.87	0.87	0.87	0.87	0.87	0.87	0.87	0.87	0.87
	Dirty	0.82	0.80	0.80	0.80	0.80	0.80	0.80	0.80	0.80	0.80	0.80	0.80
0.50/0.70/0.20	Very clean	0.97	0.95	0.95	0.95	0.95	0.95	0.95	0.95	0.95	0.95	0.95	0.95
	Clean	0.93	0.91	0.90	0.90	0.90	0.90	0.90	0.90	0.90	0.90	0.90	0.90
	Normal	0.86	0.83	0.83	0.83	0.83	0.83	0.83	0.83	0.83	0.83	0.83	0.83
	Dirty	0.76	0.74	0.74	0.74	0.74	0.74	0.74	0.74	0.74	0.74	0.74	0.74
0.50/0.50/0.20	Very clean	0.97	0.96	0.96	0.96	0.96	0.96	0.96	0.96	0.96	0.96	0.96	0.96
	Clean	0.94	0.93	0.92	0.92	0.92	0.92	0.92	0.92	0.92	0.92	0.92	0.92
	Normal	0.89	0.87	0.86	0.86	0.86	0.86	0.86	0.86	0.86	0.86	0.86	0.86
	Dirty	0.81	0.79	0.79	0.79	0.79	0.79	0.79	0.79	0.79	0.79	0.79	0.79
0.50/0.30/0.20	Very clean	0.98	0.97	0.97	0.97	0.97	0.97	0.97	0.97	0.97	0.97	0.97	0.97
	Clean	0.96	0.95	0.94	0.94	0.94	0.94	0.94	0.94	0.94	0.94	0.94	0.94
	Normal	0.92	0.90	0.90	0.90	0.90	0.90	0.90	0.90	0.90	0.90	0.90	0.90
	Dirty	0.85	0.84	0.84	0.84	0.84	0.84	0.84	0.84	0.84	0.84	0.84	0.84

Table 18.8 Room surface maintenance factor (RSMF) for indirect flux distribution

Reflectances ceiling/walls/floor	Time / years Environment	Room surface maintenance factor											
		0.5	1.0	1.5	2.0	2.5	3.0	3.5	4.0	4.5	5.0	5.5	6.0
0.80/0.70/0.20	Very clean	0.93	0.91	0.90	0.90	0.90	0.90	0.89	0.89	0.89	0.89	0.89	0.89
	Clean	0.86	0.82	0.81	0.81	0.81	0.81	0.81	0.81	0.81	0.81	0.81	0.81
	Normal	0.72	0.67	0.66	0.66	0.66	0.66	0.66	0.66	0.66	0.66	0.66	0.66
	Dirty	0.54	0.50	0.49	0.49	0.49	0.49	0.49	0.49	0.49	0.49	0.49	0.49
0.80/0.50/0.20	Very clean	0.94	0.93	0.92	0.92	0.92	0.91	0.91	0.91	0.91	0.91	0.91	0.91
	Clean	0.88	0.85	0.84	0.84	0.84	0.84	0.84	0.84	0.84	0.84	0.84	0.84
	Normal	0.76	0.72	0.71	0.71	0.71	0.71	0.71	0.71	0.71	0.71	0.71	0.71
	Dirty	0.59	0.55	0.55	0.55	0.55	0.55	0.55	0.55	0.55	0.55	0.55	0.55
0.80/0.30/0.20	Very clean	0.96	0.94	0.93	0.93	0.93	0.93	0.93	0.93	0.93	0.93	0.93	0.93
	Clean	0.90	0.88	0.87	0.87	0.87	0.87	0.87	0.87	0.87	0.87	0.87	0.87
	Normal	0.80	0.76	0.75	0.75	0.75	0.75	0.75	0.75	0.75	0.75	0.75	0.75
	Dirty	0.64	0.60	0.60	0.60	0.60	0.60	0.60	0.60	0.60	0.60	0.60	0.60
0.70/0.70/0.20	Very clean	0.93	0.91	0.90	0.90	0.90	0.90	0.90	0.90	0.90	0.90	0.90	0.90
	Clean	0.86	0.83	0.82	0.81	0.81	0.81	0.81	0.81	0.81	0.81	0.81	0.81
	Normal	0.73	0.68	0.67	0.67	0.67	0.67	0.67	0.67	0.67	0.67	0.67	0.67
	Dirty	0.55	0.51	0.50	0.50	0.50	0.50	0.50	0.50	0.50	0.50	0.50	0.50
0.70/0.50/0.20	Very clean	0.95	0.93	0.92	0.92	0.92	0.92	0.92	0.92	0.92	0.92	0.92	0.92
	Clean	0.89	0.86	0.85	0.85	0.84	0.84	0.84	0.84	0.84	0.84	0.84	0.84
	Normal	0.77	0.73	0.72	0.72	0.72	0.72	0.72	0.72	0.72	0.72	0.72	0.72
	Dirty	0.60	0.56	0.55	0.55	0.55	0.55	0.55	0.55	0.55	0.55	0.55	0.55
0.70/0.30/0.20	Very clean	0.96	0.94	0.94	0.93	0.93	0.93	0.93	0.93	0.93	0.93	0.93	0.93
	Clean	0.91	0.88	0.87	0.87	0.87	0.87	0.87	0.87	0.87	0.87	0.87	0.87
	Normal	0.80	0.77	0.76	0.76	0.76	0.76	0.76	0.75	0.75	0.75	0.75	0.75
	Dirty	0.65	0.61	0.60	0.60	0.60	0.60	0.60	0.60	0.60	0.60	0.60	0.60

Table 18.8 Continued

Reflectances ceiling/walls/floor	Time / years	Room surface maintenance factor											
	Environment	0.5	1.0	1.5	2.0	2.5	3.0	3.5	4.0	4.5	5.0	5.5	6.0
0.50/0.70/0.20	Very clean	0.94	0.92	0.91	0.91	0.91	0.91	0.91	0.91	0.91	0.91	0.91	0.91
	Clean	0.87	0.84	0.83	0.83	0.83	0.83	0.83	0.83	0.83	0.83	0.83	0.83
	Normal	0.75	0.70	0.69	0.69	0.69	0.69	0.69	0.69	0.69	0.69	0.69	0.69
	Dirty	0.57	0.52	0.52	0.52	0.52	0.52	0.52	0.52	0.52	0.52	0.52	0.52
0.50/0.50/0.20	Very clean	0.95	0.93	0.93	0.93	0.92	0.92	0.92	0.92	0.92	0.92	0.92	0.92
	Clean	0.90	0.87	0.86	0.86	0.85	0.85	0.85	0.85	0.85	0.85	0.85	0.85
	Normal	0.78	0.74	0.73	0.73	0.73	0.73	0.73	0.73	0.73	0.73	0.73	0.73
	Dirty	0.61	0.57	0.57	0.57	0.57	0.57	0.57	0.57	0.57	0.57	0.57	0.57
0.50/0.30/0.20	Very clean	0.96	0.95	0.94	0.94	0.94	0.94	0.94	0.94	0.94	0.94	0.94	0.94
	Clean	0.91	0.89	0.88	0.88	0.88	0.88	0.88	0.88	0.88	0.88	0.88	0.88
	Normal	0.81	0.78	0.77	0.77	0.77	0.77	0.77	0.77	0.77	0.77	0.77	0.77
	Dirty	0.66	0.62	0.61	0.61	0.61	0.61	0.61	0.61	0.61	0.61	0.61	0.61

18.6 Ingress protection (IP) classes

The Ingress Protection (IP) system, see BS EN 60529 (BSI, 1992), classifies luminaires according to the degree of protection provided against the ingress of foreign bodies, dust and moisture. The degree of protection is indicated by the letters IP followed by two numbers. The first number indicates the degree of protection against the ingress of foreign bodies and dust. The second indicates the protection against the ingress of moisture. Table 18.9 shows the degree of protection indicated by each number. Using this table, it can be seen that a luminaire classified as IP55 is dust protected and able to withstand water jets.

Table 18.9 IP Classification of luminaires

First number	Degree of protection	Second number	Degree of protection
0	Not protected	0	Not protected
1	Protected against solid objects greater than 50 mm	1	Protected against dripping water
2	Protected against solid objects greater than 12 mm	2	Protected against dripping water when tilted up to 15 degrees
3	Protected against solid objects greater than 2.5 mm	3	Protected against spraying water
4	Protected against solid objects greater than 1.0 mm	4	Protected against splashing
5	Dust-protected	5	Protected against water jets
6	Dust-tight	6	Protected against heavy seas
		7	Protected against the effects of immersion
		8	Protected against submersion to a specified depth

Chapter 19: Glossary

The following definitions for lighting terms are taken from BS EN 12665: 2011 (BSI, 2011b). The list includes some definitions that are not used within the *SLL Code*, however, some specialised terms, mainly relating to tunnel lighting, are not given in this glossary. Note that for some terms, an additional or alternative definition is given.

Absence factor (F_A)
Factor indicating the proportion of time that a space is unoccupied

Absorptance
Ratio of the luminous flux absorbed in a body to the luminous flux incident on it

Accommodation
Adjustment of the dioptric power of the crystalline lens by which the image of an object, at a given distance, is focused on the retina

or

Adjustment of the power of the lens of the eye for focusing an image of an object on the retina

Acuity
See visual acuity

Annual operating time (t_o)
Number of hours per annum for which the lamps are operating (unit: h)

Atmospheric luminance (L_{atm})
Light veil as a result of the scatter in the atmosphere expressed as a luminance (unit: cd·m⁻²)

Average illuminance $(\bar{E})$
Illuminance averaged over the specified surface (unit: lx)

Average luminance $(\bar{L})$
Luminance averaged over the specified surface or solid angle (unit: cd·m⁻²)

Background area
Area in the workplace adjacent to the immediate surrounding area

Ballast
Device connected between the supply and one or more discharge lamps which serves mainly to limit the current of the lamp(s) to the required value
Note: A ballast may also include means for transforming the supply voltage, correcting the power factor and, either alone or in combination with a starting device, provide the necessary conditions for starting the lamp(s)

Ballast lumen factor $(F_{Ballast})$
Ratio of the luminous flux emitted by a reference lamp when operated with a particular production ballast to the luminous flux emitted by the same lamp when operated with its reference ballast
Note: Ballast lumen factor is sometimes signified by the abbreviation BLF

Brightness
Attribute of a visual sensation according to which an area appears to emit more or less light
Note: obsolete term – luminosity

Brightness contrast
Subjective assessment of the difference in brightness between two or more surfaces seen simultaneously or successively

Built-in luminaire
Fixed luminaire installed into structure or equipment to provide illumination

Carriageway
Part of the road normally used by vehicular traffic

Chromaticity
Property of a colour stimulus defined by its chromaticity coordinates, or by its dominant or complementary wavelength and purity taken together
See also CIE 15: 2004 (CIE, 2004b)

Chromaticity coordinates
Ratio of each of a set of three tristimulus values to their sum
Note 1: As the sum of the three chromaticity coordinates equals 1, two of them are sufficient to define a chromaticity
Note 2: In the CIE standard colorimetric systems, the chromaticity coordinates are presented by the symbols x, y, z and x_{10}, y_{10}, z_{10}

CIE 1974 general colour rendering index (Ra)
Mean of the CIE 1974 special colour rendering indices for a specified set of eight test colour samples

or

Value intended to specify the degree to which objects illuminated by a light source have an expected colour relative to their colour under a reference light source

Note: Ra is derived from the colour rendering indices for a specified set of 8 test colour samples. Ra has a maximum of 100, which generally occurs when the spectral distributions of the light source and the reference light source are substantially identical

Circuit luminous efficacy of a source (c)
Quotient of the luminous flux emitted by the power absorbed by the source and associated circuits (unit: $lm \cdot W^{-1}$)

Cold spot
Coldest point on lamp surface

Colorimeter
Instrument for measuring colorimetric quantities, such as the tristimulus values of a colour stimulus

Colour contrast
Subjective assessment of the difference in colour between two or more surfaces seen simultaneously or successively

Colour rendering
Effect of an illuminant on the colour appearance of objects by conscious or subconscious comparison with their colour appearance under a reference illuminant

For design purposes, colour rendering requirements shall be specified using the general colour rendering index and shall take one of the following values of Ra: 20; 40; 60; 80; 90

Colour rendering index
See CIE 1974 general colour rendering index

Colour stimulus
Visible radiation entering the eye and producing a sensation of colour, either chromatic or achromatic

Colour temperature (T_c)
Temperature of a Planckian radiator whose radiation has the same chromaticity as that of a given stimulus (unit: K)
Note: The reciprocal colour temperature is also used, unit: K^{-1}

Constant illuminance factor (F_C)
Ratio of the average input power over a given time to the initial installed power to the luminaire

Contrast
1. In the perceptual sense: assessment of the difference in appearance of two or more parts of a field seen simultaneously or successively (hence: brightness contrast, lightness contrast, colour contrast, simultaneous contrast, successive contrast, etc)

2. In the physical sense: quantity intended to correlate with the perceived brightness contrast, usually defined by one of a number of formulae which involve the luminances of the stimuli considered, for example: $\Delta L/L$ near the luminance threshold, or L_1/L_2 for much higher luminances

Contrast revealing coefficient (q_c)
Quotient between the luminance (L) of the road surface, and the vertical illuminance (E_v) at that point (unit: $cd \cdot m^{-2} \cdot lx^{-1}$)

$$q_c = \frac{L}{E_v}$$

where:

q_c	is the contrast revealing coefficient;
L	is the luminance of the road surface at the point;
E_v	is the vertical illuminance at the point

Control gear
Components required to control the electrical operation of the lamp(s)

Note: Control gear may also include means for transforming the supply voltage, correcting the power factor and, either alone or in combination with a starting device, provide the necessary conditions for starting the lamp(s)

Correction factor
Factor to modify the luminaire data as presented on a particular photometric data sheet to those of similar luminaires

Note: Examples are ballast lumen factor, length, lumen corrections

Correlated colour temperature (T_{cp})
Temperature of the Planckian radiator whose perceived colour most closely resembles that of a given stimulus at the same brightness and under specified viewing conditions (unit: K)

Note 1: The recommended method of calculating the correlated colour temperature of a stimulus is to determine on a chromaticity diagram the temperature corresponding to the point on the Planckian locus that is intersected by the agreed isotemperature line containing the point representing the stimulus (see CIE Publication No 15; CIE, 2004b)

Note 2: Reciprocal correlated colour temperature is used rather than reciprocal colour temperature whenever correlated colour temperature is appropriate

Cosine correction
Correction of a detector for the influence of the incident direction of the light

Note: For the ideal detector, the measured illuminance is proportional to the cosine of the angle of incidence of the light. The angle of incidence is the angle between the direction of the light and the normal to the surface of the detector

Critical flicker frequency
See fusion frequency

Curfew
Time period during which stricter requirements (for the control of obtrusive light) will apply

Note: It is often a condition of use of lighting applied by a government controlling authority, usually the local government

Cut-off
Technique used for concealing lamps and surfaces of high luminance from direct view in order to reduce glare

Note: In public lighting, distinction is made between full-cut-off luminaires, semi-cut-off luminaires and non-cut-off luminaires

Cut-off angle (of a luminaire)
Angle, measured up from nadir, between the vertical axis and the first line of sight at which the lamps and the surfaces of high luminance are not visible (unit: degree)

Cylindrical illuminance (at a point, for a direction) (E_z)
Total luminous flux falling on the curved surface of a very small cylinder located at the specified point divided by the curved surface area of the cylinder (unit: lx)

Daylight
Visible part of global solar radiation

Daylight dependency factor (F_D)
Level of efficiency that a control system or control strategy achieves in exploiting the saving potential of daylight in a space

Daylight factor (D or DF)
Ratio of the illuminance at a point on a given plane due to the light received directly or indirectly from a sky of assumed or known luminance distribution, to the illuminance on a horizontal plane due to an unobstructed hemisphere of this sky, excluding the contribution of direct sunlight to both illuminances

Note 1:   Glazing, dirt effects, etc are included

Note 2:   When calculating the lighting of interiors, the contribution of direct sunlight needs to be considered separately

Daylight time usage (t_D)
Annual operating hours during the daylight time, measured in hours (unit: h)

Daylight screens/daylight louvres
Devices that transmit (part of) the ambient daylight

Design speed
Speed adopted for a particular stated purpose in designing a road (unit: $km·h^{-1}$)

Diffuse sky radiation
That part of solar radiation which reaches the earth as a result of being scattered by the air molecules, aerosol particles, cloud particles or other particles

Diffused lighting
Lighting in which the light on the working plane or on an object is not incident predominantly from a particular direction

Direct lighting
Lighting by means of luminaires having a distribution of luminous intensity such that the fraction of the emitted luminous flux directly reaching the working plane, assumed to be unbounded, is 90 to 100 per cent

Direct solar radiation
That part of the extraterrestrial solar radiation which, as a collimated beam, reaches the earth's surface after selective attenuation by the atmosphere

Directional lighting
Lighting in which the light on the working plane or on an object is incident predominantly from a particular direction

Disability glare
Glare that impairs the vision of objects without necessarily causing discomfort. Disability glare can be produced directly or by reflection

Discomfort glare

Glare that causes discomfort without necessarily impairing the vision of objects. Discomfort glare can be produced directly or by reflection

Display screen equipment

Alphanumeric or graphic display screen, regardless of the display process employed

Note: Display screen equipment is sometimes signified by the abbreviation DSE

Diversity (luminance, illuminance) (U_d) (Extreme uniformity)

Ratio of minimum illuminance (luminance) to maximum illuminance (luminance) on (of) a surface
See also uniformity

Downward light output ratio (of a luminaire) (R_{DLO})

Ratio of the downward flux of the luminaire, measured under specified practical conditions with its own lamps and equipment, to the sum of the individual luminous fluxes of the same lamps when operated outside the luminaire with the same equipment, under specified conditions

Note 1:   The luminaire attitude should be declared so that appropriate corrections to the DLOR can be made if, in application, the installed attitude is different

Note 2:   Downward light output ratio is sometimes signified by the abbreviation DLOR

Efficacy

See luminous efficacy of a source

Emergency ballast lumen factor ($F_{EBallast}$)

Ratio of the luminous flux of the lamp, operated with ballast under test, at the lowest voltage which can occur during emergency mode, after failure of the normal supply (for the appropriate start time for the application requirement) and continuously to the end of rated duration of operation, to the luminous flux of the same lamp operated with the appropriate reference ballast supplied at its rated voltage and frequency

$$F_{EBallast} = F_{Ballast} \times F_{min}$$

where:

$F_{EBallast}$ is the emergency ballast lumen factor;
$F_{Ballast}$ is the ballast lumen factor;
F_{min} is the worst case of the emergency time-dependent factors

Emergency escape lighting

Part of emergency lighting that provides illumination for visibility for people leaving a location or attempting to terminate a potentially dangerous process before doing so

Emergency exit

Way out that is intended to be used during an emergency

Emergency lamp flux

See practical emergency lamp flux

Emergency lane (hard shoulder)
Lane parallel to the traffic lane(s) provided for emergency and/or broken-down vehicles only

Emergency lighting
Lighting provided for use when the supply to the normal lighting fails

Emergency lighting charge time (t_{em})
Operating hours during which the emergency lighting batteries are being charged (unit: h)

Emergency lighting charging power (P_{ei})
Input power to the charging circuit of emergency luminaires when the lamps are not operating (unit: W)

Emergency lighting, total installed charging power
See total installed charging power of the emergency lighting luminaires in the room or zone

Energy consumption used for illumination ($W_{L,t}$)
Energy consumed in period t, by the luminaires when the lamps are operating, to fulfil the illumination function and purpose in the building (unit: kW·h)

Equivalent veiling luminance (for disability glare or veiling reflections) (L_{ve})
Luminance that, when added by superposition to the luminance of both the adapting background and the object, makes the luminance threshold or the luminance difference threshold the same under the two following conditions: (1) glare present, but no additional luminance; (2) additional luminance present, but no glare (unit: cd·m^{-2})

Escape route
Route designated for escape in the event of an emergency

Escape route lighting
Part of emergency escape lighting provided to ensure that the means of escape can be effectively identified and safely used when the location is occupied

Essential data
Lamp and luminaire data required for the verification of conformity to requirements

Externally illuminated safety sign
Safety sign that is illuminated, when it is required, by an external source

Extreme uniformity
See diversity

Flicker
Impression of unsteadiness of visual sensation induced by a light stimulus whose luminance or spectral distribution fluctuates with time

Flicker frequency
See fusion frequency

Floodlighting
Lighting of a scene or object, usually by projectors, in order to increase considerably its illuminance relative to its surroundings

Flux
See luminous flux, rated lamp luminous flux

F_{min}
See minimum value emergency factor

Fusion frequency
Critical flicker frequency (for a given set of conditions)
Frequency of alternation of stimuli above which flicker is not perceptible (unit: Hz)

General colour rendering index
See CIE 1974 general colour rendering index

General lighting
Substantially uniform lighting of an area without provision for special local requirements

Glare
Condition of vision in which there is discomfort or a reduction in the ability to see details or objects, caused by an unsuitable distribution or range of luminance, or extreme contrasts

See also disability glare and discomfort glare

Glare rating limit (R_{GL})
Upper limit of glare by the CIE Glare Rating system

Global solar radiation
Combined direct solar radiation and diffuse sky radiation

Grid points for measurement and calculation
Arrangement of calculation and measurement points and their number in each dimension of the reference surface or plane

Hemispherical illuminance (at a point) (E_{hs})
Total luminous flux falling on the curved surface of a very small hemisphere located at the specified point divided by the curved surface area of the hemisphere (unit: lx)

High risk task area lighting
Part of emergency escape lighting that provides illumination for visibility for people involved in a potentially dangerous process or situation and facilitates safe termination of activities

Note: In sports lighting, it is referred to as 'Safety lighting for participants'

Illuminance (at a point of a surface) (*E*)
Quotient of the luminous flux dΦ incident on an element of the surface containing the point, by the area d*A* of that element (unit: lx = lm·m^{-2})

Note 1: Equivalent definition: Integral, taken over the hemisphere visible from the given point, of the expression $L \cdot \cos\theta \cdot d\Omega$, where L is the luminance at the given point in the various directions of the incident elementary beams of solid angle $d\Omega$, and θ is the angle between any of these beams and the normal to the surface at the given point

$$E = \frac{d\Omega}{dA} = \int_{2\pi sr} L \cos\theta d\Omega$$

where

E	is the illuminance at a point on a surface;
L	is the luminance at the given point in the various directions of the incident elementary beams of solid angle $d\Omega$;
θ	is the angle between an incident beam and the normal to the surface at the given point;
$d\Omega$	is the solid angle

Note 2: The orientation of the surface may be defined, e.g. horizontal, vertical, hence horizontal illuminance, vertical illuminance

See also average illuminance, cylindrical illuminance, hemispherical illuminance, initial illuminance, maintained illuminance, maximum illuminance, minimum illuminance, semi-cylindrical illuminance and spherical illuminance

Illuminance meter
Instrument for measuring illuminance

Immediate surrounding area
See surrounding area

Indirect lighting
Lighting by means of luminaires having a distribution of luminous intensity such that the fraction of the emitted luminous flux directly reaching the working plane, assumed to be unbounded, is 0 to 10 per cent

Initial average luminance ($\bar{L}_i$)
Average luminance when the installation is new (unit: cd·m^{-2})

Initial illuminance ($\bar{E}_i$)
Average illuminance on the specified surface when the installation is new (unit: lx)

Initial luminous flux
See rated luminous flux

Installed loading
Installed power of the lighting installation per unit area (for interior and exterior areas) or per unit length (for road lighting) (unit: W·m^{-2} (for areas) or kW·km^{-1} (for road lighting))

Integral lighting system (of a machine)
Lighting system consisting of lamp(s), luminaire(s) and associated mechanical and electrical control devices which forms a permanent part of the machine, designed to provide illumination in and/or at the machine

Intensity
See luminous intensity

Intensity distribution
See luminous intensity distribution

Internally illuminated safety sign
Safety sign that is illuminated, when it is required, by an internal source

Lamp
Source made in order to produce an optical radiation, usually visible

Note: This term is also sometimes used for certain types of luminaires

Lamp code
Any combination of letters and numbers by which the lamp type is identified

Lamp dimensions
All dimensions of the lamp that are relevant for the luminaire

Lamp lumen maintenance factor (F_{LLM})
Ratio of the luminous flux of a lamp at a given time in its life to the initial luminous flux

Note: Lamp lumen maintenance factor is sometimes signified by the abbreviation LLMF

Lamp luminous flux
See rated luminous flux

Lamp survival factor (F_{LS})
Fraction of the total number of lamps which continue to operate at a given time under defined conditions and switching frequency

Note: Lamp survival factor is sometimes signified by the abbreviation LSF

Lamp wattage
See nominal lamp wattage

LENI
See Lighting Energy Numeric Indicator

Life of lighting installation
Period after which the installation cannot be restored to satisfy the required performance because of non-recoverable deteriorations

Light centre
Point used as origin for photometric measurements and calculations

Light loss factor
See maintenance factor

Light output ratio (of a luminaire) (R_{LO})

Ratio of the total flux of the luminaire, measured under specified practical conditions with its own lamps and equipment, to the sum of the individual luminous fluxes of the same lamps when operated outside the luminaire with the same equipment, under specified conditions

Note 1:   For luminaires using incandescent lamps only, the optical light output ratio and the light output ratio are the same in practice

Note 2:   Light output ratio is sometimes signified by the abbreviation LOR

See also downward light output ratio and upward light output ratio

Light output ratio working (of a luminaire) (R_{LOW})

Ratio of the total flux of the luminaire, measured under specified practical conditions with its own lamps and equipment, to the sum of the individual luminous fluxes of the same lamps when operating outside the luminaire with a reference ballast, under reference conditions

Light source

See source

Light source colour

The colour of a light source can be expressed by its correlated colour temperature

Loading

See installed loading

Lighting Energy Numeric Indicator (LENI)

A numerical indicator that expresses the total amount of energy used by a lighting system per square metre per year

Local lighting

Lighting for a specific visual task, additional to and controlled separately from the general lighting

Localised lighting

Lighting designed to illuminate an area with a higher illuminance at certain specified positions, for instance, those at which work is carried out

Longitudinal uniformity (of road surface luminance of a carriageway) (U_l)

Ratio of the minimum to the maximum road surface luminance found in a line in the centre along a driving lane

Note: The longitudinal uniformity is considered for each driving lane

Louvres

See daylight screens

Luminaire

Apparatus which distributes, filters or transforms the light transmitted from one or more lamps and which includes, except the lamps themselves, all of the parts necessary for fixing and protecting the lamps and, where necessary, circuit auxiliaries together with the means for connecting them to the electric supply

Luminaire code
Any combination of letters and numbers by which the luminaire type is identified

Luminaire maintenance factor (F_{LM})
Ratio of the light output ratio of a luminaire at a given time to the initial light output ratio

Note: Luminaire maintenance factor is sometimes signified by the abbreviation LMF

Luminaire luminous efficacy (l)
Quotient of the luminous flux emitted by the luminaire by the power absorbed by the lamp and associated circuits of the luminaire (unit: $lm \cdot W^{-1}$)

Luminaire parasitic energy consumption ($W_{P,t}$)
Parasitic energy consumed in period t, by the luminaire emergency lighting charging circuit plus the standby control system controlling the luminaires when the lamps are not operating (unit: $kW \cdot h$)

Luminaire parasitic power (P_{pi})
Input power consumed by the charging circuit of emergency lighting luminaires and the standby power for automatic controls in the luminaire when lamps are not operating (unit: W)

$$P_{pi} = P_{ci} + P_{ei}$$

where

P_{pi} is the luminaire parasitic power consumed by the luminaire with the lamps off, expressed in watts;

P_{ci} is the parasitic power of the controls only during the time with the lamps off, expressed in watts;

P_{ei} is the emergency lighting charging power, expressed in watts

Luminaire power (P_i)
Input power consumed by the lamp(s), control gear and control circuit in or associated with the luminaire, which includes any parasitic power when the luminaire is turned on (unit: W)

Note: The rated luminaire power (P_i) for a specific luminaire may be obtained from the luminaire manufacturer

Luminance (in a given direction, at a given point of a real or imaginary surface) (L)
Quantity defined by the equation (unit: $cd \cdot m^{-2} = lm \cdot m^{-2} \cdot sr^{-1}$)

$$L = \frac{d\Phi}{dA \cos\theta d\Omega}$$

where

L is the luminance in a given direction or at a given point of a surface;

$d\Phi$ is the luminous flux transmitted by an elementary beam passing through the given point and propagating in the solid angle $d\Omega$ containing the given direction;

dA is the area of a section of that beam containing the given point;

$d\Omega$ is the solid angle;

θ is the angle between the normal to that section and the direction of the beam

Luminance shall be specified as maintained luminance and shall take one of the following values:
1×10^N $cd \cdot m^{-2}$; 1.5×10^N $cd \cdot m^{-2}$; 2.0×10^N $cd \cdot m^{-2}$; 3.0×10^N $cd \cdot m^{-2}$; 5.0×10^N $cd \cdot m^{-2}$; 7.5×10^N $cd \cdot m^{-2}$ (where N is an integer)

The area over which the luminance is to be calculated or measured shall be specified

See also atmospheric luminance, average luminance, equivalent luminance, initial luminance, interior luminance, maintained luminance, maximum luminance and minimum luminance

Luminance contrast
Photometric quantity intended to correlate with brightness contrast, usually defined by one of a number of equations which involve the luminances of the stimuli considered

Note: Luminance contrast can be defined as luminance ratio

$$C_1 = \frac{L_2}{L_1}$$ (usually for successive stimuli)

or by the following equation:

$$C_2 = \frac{L_2 - L_1}{L_1}$$ (usually for surfaces viewed simultaneously)

When the areas of different luminance are comparable in size and it is desired to take an average, the following equation can be used instead:

$$C_2 = \frac{L_2 - L_1}{0.5 \times (L_2 + L_1)}$$

where
L_1 is the luminance of the background, or largest part of the visual field;
L_2 is the luminance of the object

Luminance meter
Instrument for measuring luminance

Luminosity
See brightness

Luminous efficacy of a source (η)
Quotient of the luminous flux emitted by the power absorbed by the source (unit: lm·W^{-1})

Luminous environment
Lighting considered in relation to its physiological and psychological effects

Luminous flux (Φ)
Quantity derived from radiant flux Φ_e by evaluating the radiation according to its action upon the CIE standard photometric observer (unit: lm)

Note 1:   For photopic vision

$$\Phi = K_m \int_0^\infty \left(\frac{d\Phi_e(\lambda)}{d\lambda} \right) \times v(\lambda)$$

where
Φ *is the luminous flux;*
$\frac{d\Phi_e(\lambda)}{d\lambda}$ *is the spectral distribution of the radiant flux;*
$V(\lambda)$ *is the spectral luminous efficiency function*

or

Quantity derived from radiant flux (radiant power) by evaluating the radiation according to the spectral sensitivity of the human eye (as defined by the CIE standard photometric observer). It is the light power emitted by a source or received by a surface (unit: lumen, lm)

See also initial luminous flux and rated luminous flux

Luminous intensity (of a source, in a given direction) (*I*)
Quotient of the luminous flux dΦ leaving the source and propagated in the element of solid angle dΩ containing the given direction, by the element of solid angle (unit: cd = lm sr^{-1})

$$I = \frac{d\Phi}{d\Omega}$$

where

I is the luminous intensity of a source in a given direction;
dΦ is the luminous flux leaving the source;
dΩ is the solid angle

or

Luminous flux per unit solid angle in the direction in question, i.e. the luminous flux on a small surface, divided by the solid angle that the surface subtends at the source

(Spatial) **Distribution of luminous intensity** (of a source)
Display, by means of curves or tables, of the value of the luminous intensity of the source as a function of direction in space

or

Luminous intensity of a source (lamp or luminaire) as a function of direction in space

Machinery, Machine
Assembly of linked parts or components, at least one of which moves, with the appropriate machine actuators, control and power circuits, etc joined together for a specific application, in particular, for the processing, treatment, moving or packaging of a material

Note: The term 'machinery' also covers an assembly of machines which, in order to achieve the same end, are arranged and controlled so that they function as an integral whole

Maintained illuminance ($\bar{E}_m$)
Minimum average illuminance (unit: lx)

Note 1:   Value below which the average illuminance on the specified area should not fall

Note 2:   It is the average illuminance at the time maintenance should be carried out

Maintained luminance ($\bar{L}_m$)
Minimum average luminance (unit: $cd \cdot m^{-2}$)

Note 1: Value below which the average luminance on the specified area should not fall

Note 2: It is the average luminance at the time maintenance should be carried out

Maintenance cycle
Repetition of lamp replacement, lamp/luminaire cleaning and room surface cleaning intervals

Maintenance factor
(Light loss factor) (obsolete)
Ratio of the average illuminance on the working plane after a certain period of use of a lighting installation to the initial average illuminance obtained under the same conditions for the installation

Note 1: The term depreciation factor has been formerly used to designate the reciprocal of the above ratio
Note 2: The light losses take into account dirt accumulation on luminaire and room surfaces and lamp depreciation

or

Ratio of maintained illuminance to initial illuminance

Note: Maintenance factor of an installation depends on lamp lumen maintenance factor, lamp survival factor, luminaire maintenance factor and (for an interior lighting installation) room surface maintenance factor

See also lamp lumen maintenance factor, luminaire maintenance factor and room surface maintenance factor

Maintenance schedule
Set of instructions specifying maintenance cycle and servicing procedures

Maximum illuminance (E_{max})
Highest illuminance at any relevant point on the specified surface (unit: lx)

Maximum luminance (L_{max})
Highest luminance of any relevant point on the specified surface (unit: $cd \cdot m^{-2}$)

Measurement field (of a photometer)
Area including all points in object space, radiating towards the acceptance area of the detector

Minimum illuminance (E_{min})
Lowest illuminance at any relevant point on the specified surface (unit: lx)

Minimum luminance (L_{min})
Lowest luminance at any relevant point on the specified surface (unit: $cd \cdot m^{-2}$)

Minimum value emergency factor (F_{min})
Worst case of the emergency time-dependent factors

Mixed traffic
Traffic that consists of motor vehicles, cyclists, pedestrians, etc

Motor traffic (motorised traffic)
Traffic that consists of motorised vehicles only

Nominal lamp wattage (W_{lamp})
Approximate wattage used to designate or identify the lamp (unit: W)

Non-daylight time usage (t_N)
Annual operating hours during the non-daylight time (unit: h)

Obtrusive light
Spill light which because of quantitative, directional or spectral attributes in a given context gives rise to annoyance, discomfort, distraction or reduction in the ability to see essential information

Note 1:   In the case of outdoor sports lighting installations, obtrusive light is considered around the installation and not for spectators, referees or players within the sports area

Note 2:   In the case of large tertiary buildings with predominantly glazed facades, interior lighting may be considered as obtrusive light if it gives rise to annoyance, discomfort, distraction or a reduction in the ability to see essential information due to light spilling outside of the building structure

Occupancy dependency factor (F_o)
Factor indicating the proportion of time that a space is occupied and lighting is required

Open area lighting (anti-panic lighting)
Part of emergency escape lighting provided to avoid panic and provide illumination allowing people to see their way to an escape route

Operating time (t)
Time period for the energy consumption (unit: h)

See also annual operating time

Parasitic energy consumption
See luminaire parasitic energy consumption

Parasitic power
See luminaire parasitic power

Parasitic power of the controls (with the lamps off) (P_{ci})
Parasitic input power to the control system in the luminaires during the period with the lamps not operating (unit: W)

Principal area ($A_{Principal}$)
Actual playing area needed for the performance of a certain sport

Note: Usually this means the actual marked out 'field' area for that sport (for instance football), but in some cases, this area comprises an extra playing area around the marked area (e.g. tennis, volleyball, table tennis). The dimensions of the particular area should be checked at the time when a lighting installation is being installed

Performance
See visual performance

Photometer
Instrument for measuring photometric quantities

Photometric observer
See luminous flux

Photometry
Measurement of quantities referring to radiation as evaluated according to a given spectral luminous efficiency function, e.g. $V(\lambda)$ or $V'(\lambda)$

Photopic vision
See luminous flux

Practical emergency lamp flux (Φ_{PEL})
Lowest luminous flux of the lamp observed during the rated duration of the emergency mode (unit: lm)

$$\Phi_{PEL} = \Phi_{LD} \times F_{EBallast}$$

where

Φ_{PEL} is the practical emergency lamp flux, expressed in lumens;
Φ_{LD} is the initial lighting design lumens at 100 h;
$F_{EBallast}$ is the emergency ballast lumen factor

Radiant flux
See luminous flux

Rated luminous flux (of a type of lamp)
Value of the initial luminous flux of a given type of lamp declared by the manufacturer or the responsible vendor, the lamp being operated under specified conditions (unit: lm)

Note 1: The initial luminous flux is the luminous flux of a lamp after a short ageing period as specified in the relevant lamp standard

Note 2: The rated luminous flux is sometimes marked on the lamp

Reference ballast
Special type ballast designed for providing comparison standards for use in testing ballasts, for the selection of reference lamps and for testing regular production lamps under standardised conditions

Reference surface
Surface on which illuminance is measured or specified

Reflectance (for incident radiation of given spectral composition, polarisation and geometrical distribution) (ρ)
Ratio of the reflected radiant or luminous flux to the incident flux in the given conditions

Reflections
See veiling reflections

Reflectometer
Instrument for measuring quantities pertaining to reflection

Rooflight
Daylight opening on the roof or on a horizontal surface of a building

Room surface maintenance factor (F_{RSM})
Ratio of room surface reflectance at a given time to the initial reflectance value

Note: Room surface maintenance factor is sometimes signified by the abbreviation RSMF

Safety sign
Sign which gives a general safety message, obtained by a combination of colour and geometric shape and which, by the addition of a graphic symbol or text, gives a particular safety message

Scene setting operation time (t_s)
Operating hours of the scene setting controls (unit: h)

Scotopic observer
See luminous flux

Screens
See daylight screens

Semi-cylindrical illuminance (at a point) (E_{sz})
Total luminous flux falling on the curved surface of a very small semi-cylinder located at the specified point, divided by the curved surface area of the semi-cylinder (unit: lx)

Note: The axis of the semi-cylinder is taken to be vertical unless stated otherwise. The direction of the curved surface should be specified

Semi-direct lighting
Lighting by means of luminaires having a distribution of luminous intensity such that the fraction of the emitted luminous flux directly reaching the working plane, assumed to be unbounded, is 60 to 90 per cent

Semi-indirect lighting
Lighting by means of luminaires having a distribution of luminous intensity such that the fraction of the emitted luminous flux directly reaching the working plane, assumed to be unbounded, is 10 to 40 per cent

Shielding angle
The angle between the horizontal plane and the first line of sight at which the luminous parts of the lamps in the luminaire are directly visible (unit: degrees)

Note: The complementary angle to the shielding angle is named cut-off angle

Skylight
Visible part of diffuse sky radiation

Source (light source)
Object that produces light or other radiant flux

Note: The term light source indicates the source is essentially intended for illuminating and signalling purposes

Solar radiation
Electromagnetic radiation from the sun

See also direct solar radiation and global solar radiation

Spacing (in an installation)
Distance between the light centres of adjacent luminaires of the installation

Spacing to height ratio
Ratio of spacing to the height of the geometric centres of the luminaires above the reference plane

Note: For indoor lighting, the reference plane is usually the horizontal working plane; for exterior lighting, the reference plane is usually the ground

Spectral luminous efficiency
See luminous flux

Spherical illuminance (at a point) (E_o)
Total luminous flux falling on the whole surface of a very small sphere located at the specified point divided by the surface area of the sphere (unit: lx)

Spill light (stray light)
Light emitted by a lighting installation which falls outside the boundaries of the property for which the lighting installation is designed

Spotlighting
Lighting designed to increase considerably the illuminance of a limited area or of an object relative to the surroundings, with minimum diffused lighting

Stroboscopic effect
Apparent change of motion and/or appearance of a moving object when the object is illuminated by a light of varying intensity

Note: To obtain apparent immobilisation or constant change of movement, it is necessary that both the object movement and the light intensity variation are periodic, and some specific relation between the object movement and light variation frequencies exists. The effect is only observable if the amplitude of the light variation is above certain limits. The motion of the object can be rotational or translational

Standard photometric observer
See luminous flux

Standard year time (t_y)
Time taken for one standard year to pass, taken as 8760 h

Standby lighting
That part of emergency lighting provided to enable normal activities to continue substantially unchanged

Stray light
See spill light

Sunlight
Visible part of direct solar radiation

Surrounding area (immediate surrounding area)
Band surrounding the task area within the field of vision

Survival factor
See lamp survival factor

Task area
Area within which the visual task is carried out

Total energy used for lighting (W_t)
Energy consumed in period t, by the luminaires, when the lamps are operating plus the parasitic loads when the lamps are not operating, in a room or zone (unit: kW·h)

Total installed charging power of the emergency lighting luminaires in the room or zone (P_{em})
Input charging power of all emergency lighting luminaires (unit: W)

$$P_{em} = \sum_i P_{e,i}$$

where
P_{em} is the total installed charging power of the emergency lighting luminaires in the room or zone, expressed in watts;
$P_{e,i}$ is the emergency lighting charging power of the individual luminaires, expressed in watts

Total installed lighting power in the room or zone (P_n)
Power of all luminaires (unit: W)

$$P_n = \sum_i P_i$$

where
P_n is the total installed lighting power in the room or zone, expressed in watts;
P_i is the luminaire power expressed in watts

Total installed parasitic power of the controls in the room or zone (P_{pc})
Input power of all control systems in luminaires when the lamps are not operating (unit: W)

$$P_{pc} = \sum_i P_{c,i}$$

where

P_{pc} is the total installed parasitic power of the controls in the room or zone, expressed in watts;

$P_{c,i}$ is the parasitic power of the controls only during the time with the lamps off, expressed in watts

Traffic lane
Strip of carriageway intended to accommodate a single line of moving vehicles

Transmittance (for incident radiation of given spectral composition, polarisation and geometrical distribution) (τ)
Ratio of the transmitted radiant or luminous flux to the incident flux in the given conditions

Tristimulus values (of a colour stimulus)

Amounts of the three reference colour stimuli, in a given trichromatic system, required to match the colour of the stimulus considered

Note 1:   In the CIE standard colorimetric systems, the tristimulus values are represented by the symbols X, Y, Z and X10, Y10, Z10
Note 2:   See also CIE 15 (CIE, 2004b)

Unified glare rating limit (R_{UGL})
Upper limit of glare by the CIE Unified Glare Rating system

Uniformity (luminance, illuminance) (U_o)
Ratio of minimum illuminance (luminance) to average illuminance (luminance) on (of) a surface

Upward flux maximum
Maximum possible value of flux in an installation that is potentially emitted above the horizontal both directly from the luminaire(s) mounted in their installed attitude, and indirectly due to reflection from lit surfaces within the space (unit: lm)

Upward flux minimum
Minimum possible value of flux in an installation that is emitted above the horizontal (unit: lm)

Upward flux ratio
Ratio between the flux from all considered luminaires above the horizontal plane passing through the luminaires in their installed position on site plus their flux reflected by the ground and the minimal irreducible flux reflected towards the sky by the sole reference surface

Note: Upward flux ratio is sometimes signified by the abbreviation UFR

Upward light output ratio (of a luminaire) (R_{ULO})
Ratio of the upward flux of the luminaire, measured under specified practical conditions with its own lamps and equipment, to the sum of the individual luminous fluxes of the same lamps when operated outside the luminaire with the same equipment, under specified conditions

Note 1: *Upward light output ratio is sometimes signified by the abbreviation ULOR*
Note 2: *The luminaire attitude should be declared so that appropriate corrections to the ULOR can be made if, in application, the installed attitude is different*

Upward light ratio (R_{UL})
Proportion of the total luminaire flux that is emitted above the horizontal by all luminaires to the total luminaire flux from all luminaires in an installation, when the luminaires are mounted in their installed attitudes

Useful area (A)
Floor area inside the outer walls excluding non-habitable cellars and un-illuminated spaces (unit: m²)

Useful data
Lamp and luminaire data beneficial to the designers and users in the planning and operation of lighting installations

Utilance (of an installation, for a reference surface) (U)
Ratio of the luminous flux received by the reference surface to the sum of the individual total fluxes of the luminaires of the installation

Utilisation factor (of an installation, for a reference surface) (F_U)
Ratio of the luminous flux received by the reference surface to the sum of the individual fluxes of the lamps of the installation

$V(\lambda)$ correction
Correction of the spectral responsivity of a detector to match the photopic spectral sensitivity of the human eye

Veiling luminance
See equivalent veiling luminance

Veiling reflections
Specular reflections that appear on the object viewed and that partially or wholly obscure the details by reducing contrast

Visual acuity
1. Qualitatively: capacity for seeing distinctly fine details that have very small angular separation

2. Quantitatively: any of a number of measures of spatial discrimination such as the reciprocal of the value of the angular separation in minutes of arc of two neighbouring objects (points or lines or other specified stimuli) which the observer can just perceive to be separate

Visual comfort
Subjective condition of visual well-being induced by the visual environment

Visual field
Area or extent of physical space visible to an eye at a given position and direction of view

Note: *It should be stated whether the visual field is monocular or binocular*

Visual performance

Performance of the visual system as measured for instance by the speed and accuracy with which a visual task is performed

Visual task

Visual elements of the activity being undertaken

Note: The main visual elements are the size of the structure, its luminance, its contrast against the background and its duration

Window

Daylight opening on a vertical or nearly vertical area of a room envelope

Work place

Place intended to house work stations on the premises of the undertaking and/or establishment and any other place within the area of undertaking and/or establishment to which the worker has access in the course of his employment

Work plane (working plane)

Reference surface defined as the plane at which work is normally done

Work station

Combination and spatial arrangement of work equipment, surrounded by the work environment under the conditions imposed by the work tasks

Chapter 20: Bibliography

20.1 Standards

British Standards Institution (1976) BS 5252: 1976: *Framework for colour co-ordination for building purposes*, London: BSI.

British Standards Institution (1992) BS EN 60529: 1992: *Specification for degrees of protection provided by enclosures (IP code)*, London: BSI.

British Standards Institution (1998) BS EN 60598-2-5: 1998: *Luminaires. Particular requirements. Floodlights*, London: BSI.

British Standards Institution (2003a) BS EN 13201-2: 2003: *Road lighting. Performance requirements*, London: BSI.

British Standards Institution (2003b) BS EN 60598-2-3: 2003: *Luminaires. Particular requirements. Luminaires for road and street lighting*, London: BSI.

British Standards Institution (2003c) BS EN 13201-3: 2003: *Road lighting. Calculation of performance*, London: BSI.

British Standards Institution (2003d) BS EN 13201-4: 2003: *Road lighting. Methods of measuring lighting performance*, London: BSI.

British Standards Institution (2003e) BS 5489-1:2003+A2: 2008: *Code of practice for the design of road lighting. Lighting of roads and public amenity areas*, London: BSI.

British Standards Institution (2004a) BS EN 13032-1: 2004: *Light and lighting. Measurement and presentation of photometric data of lamps and luminaires. Measurement and file format*, London: BSI.

British Standards Institution (2004b) BS EN 13032-2: 2004: *Light and lighting. Measurement and presentation of photometric data of lamps and luminaires. Presentation of data for indoor and outdoor work places*, London: BSI.

British Standards Institution (2005a) BS EN 14225: *Measurement and assessment of personal exposure to incoherent optical radiation, Parts 1 to 4*, London: BSI.

British Standards Institution (2005b) BS 667: *Illuminance meters. Requirements and test methods*, London: BSI.

British Standards Institution (2005c) BS 7920: *Luminance meters. Requirements and test methods*, London: BSI.

British Standards Institution (2007a) BS EN 15193: 2007: *Energy performance of buildings. Energy requirements for lighting*, London: BSI.

British Standards Institution (2007b) BS EN 12193: 2007: *Light and lighting. Sports lighting*, London: BSI.

British Standards Institution (2007c) BS EN 12464-2: 2007: *Light and lighting – Lighting of work places – Part 2: Outdoor work places*, London: BSI.

British Standards Institution (2007d) BS EN 13032-3: 2007: *Measurement and presentation of photometric data of lamps and luminaires. Presentation of data for emergency lighting of work places*, London: BSI.

British Standards Institution (2008) BS 8206-2: 2008: *Lighting for buildings – Part 2: Code of practice for daylighting*, London: BSI.

British Standards Institution (2011a) BS EN 12464-1: 2011: *Light and lighting – Lighting of work places – Part 1: Indoor work places*, London: BSI.

British Standards Institution (2011b) BS EN 12665: 2011: *Light and lighting. Basic terms and criteria for specifying lighting requirements*, London: BSI.

International Standards Organisation / Commission Internationale de l'Eclairage (2003) ISO 15469: 2004 (E) / CIE S 011/E: 2003, *Spatial distribution of daylight – CIE standard general sky*, Geneva: ISO, Vienna: CIE.

International Standards Organisation (2005) BS EN ISO 23539: 2005: *Photometry – The CIE system of physical photometry*, Geneva: ISO.

International Standards Organisation (2007) BS EN ISO 9680: 2007: *Dentistry. Operating lights*, Geneva: ISO.

International Standards Organisation (2008a) BS EN ISO 9241-307: 2008: *Ergonomics of human–system interaction. Analysis and compliance test methods for electronic visual displays*, Geneva: ISO.

International Standards Organisation (2008b) BS EN ISO 9241-302: 2008: *Ergonomics of human–system interaction. Terminology for electronic visual displays*, Geneva: ISO.

International Standards Organisation (2009) BS ISO 3864-1: 2009: *Graphical symbols. Safety colours and safety signs. Part 1. Design principles for safety signs and safety markings*, Geneva: ISO.

20.2 Guidance

Commission Internationale de l'Eclairage CIE (1978) CIE Publication 40: 1978: *Calculations for interior lighting: Basic method*, Vienna: CIE.

Commission Internationale de l'Eclairage CIE (1994) CIE Publication 112: 1994: *Glare evaluation system for use within outdoor sports and area lighting*, Vienna: CIE.

Commission Internationale de l'Eclairage CIE (1995a) CIE Publication 117: 1995: *Discomfort glare in interior lighting*, Vienna: CIE.

Commission Internationale de l'Eclairage CIE (1995b) CIE Publication 13-3: 1995: *Method of measuring and specifying colour rendering properties of light sources*, Vienna: CIE.

Commission Internationale de l'Eclairage (CIE) (1997) CIE Publication 126: 1997: *Guidelines for minimizing sky glow*, Vienna: CIE.

Commission Internationale de l'Eclairage (CIE) (2000) CIE Publication 140: 2000: *Road lighting calculations*, Vienna: CIE.

Commission Internationale de l'Eclairage CIE (2003) CIE Publication 154: 2003: *The maintenance of outdoor lighting systems*, Vienna: CIE.

Commission Internationale de l'Eclairage (CIE) (2004a) CIE Publication 158: 2004: *Ocular lighting effects on human physiology and behaviour*, Vienna: CIE.

Commission Internationale de l'Eclairage (CIE) (2004b) CIE Publication 15: 2004: *Colorimetry, 3rd edition*, Vienna: CIE.

Commission Internationale de l'Eclairage CIE (2005) CIE Publication 97: 2005: *Guide on the maintenance of indoor electric lighting systems*, Vienna: CIE.

Commission Internationale de l'Eclairage (CIE) (2006) CIE Publication S009: 2006: *Photobiologic safety of lamps and lamp systems*, Vienna: CIE.

Commission Internationale de l'Eclairage (CIE) (2010) CIE Publication 190: 2010: *Calculation and presentation of unified glare rating tables for indoor lighting luminaires*, Vienna: CIE.

The Society of Light and Lighting (SLL) (1999) SLL Lighting Guide 10: *Daylighting and window design*. ISBN 0 900953 98 5, London: CIBSE.

The Society of Light and Lighting (SLL) (2001) SLL Lighting Guide 11: *Surface reflectance and colour*. ISBN 1 903287 14 6, London: CIBSE.

The Society of Light and Lighting (SLL) (2004) SLL Lighting Guide 12: *Emergency lighting design guide*. ISBN 1 903287 51 0, London: CIBSE.

The Society of Light and Lighting (SLL) (2006) SLL Lighting Guide 4: *Sports*. ISBN 1 903287 78 2, London: CIBSE.

The Society of Light and Lighting (SLL) (2008) SLL Lighting Guide 2: *Hospital and health care buildings*. ISBN 978 1 903287 99 6, London: CIBSE.

The Society of Light and Lighting (SLL) (2009) *The SLL Lighting Handbook*. ISBN: 9781906846022, London: The Society of Light and Lighting.

The Society of Light and Lighting (SLL) (2011) SLL Lighting Guide 5: *Lighting for education*. ISBN 978 1 906846 17 6, London: CIBSE.

20.3 References

Akashi, Y., Myer, M. and Boyce, P.R. (2006) Identifying sparkle. *Lighting Research and Technology*, 38, 325–340.

Akashi, Y., Rea, M.S. and Bullough, J.D. (2007) Driver decision making in response to peripheral moving targets under mesopic lighting levels. *Lighting Research and Technology*, 39, 53–67.

American Conference of Governmental Industrial Hygienists (ACGIH) (2010) *TLVs and BEIs threshold limit values for chemical substances and physical agents, biological exposure indices*, Cincinnati, OH: ACGIH.

Badia, P., Myers, B., Boecker, M. and Culpeper, J. (1991) Bright light effects on body temperature, alertness, EEG and behavior. *Physiology and Behavior*, 50, 583–588.

Baron, R. A. (1990) Environmentally induced positive affect: Its impact on self-efficacy, task performance, negotiation, and conflict. *Journal of Applied Social Psychology*, 20(5), 368–384.

Baron, R.A. and Thomley, J. (1994) A whiff of reality: Positive affect as a potential mediator of the effects of pleasant fragrances on task performance and helping. *Environment and Behavior*, 26(6), 766–784.

Baron, R.A., Rea, M.S. and Daniels, S.G. (1992) Effects of indoor lighting (illuminance and spectral distribution) on the performance of cognitive tasks and interpersonal behaviors: The potential mediating role of positive affect. *Motivation and Emotion*, 16, 1–33.

Begley, K. and Linderson, T. (1991) Management of mercury in lighting products, *Proceedings of the 1st European Conference on Energy-Efficient Lighting*. Stockholm, Sweden: Swedish National Board for Industrial and Technical Development.

Berman, S.M. (1992) Energy efficiency consequences of scotopic sensitivity. *Journal of the Illuminating Engineering Society*, 21, 3–14.

Berman, S.M., Navvab, M., Martin, M.J., Sheedy, J. and Tithof, W. (2006) A comparison of traditional and high colour temperature lighting on the near acuity of elementary school children. *Lighting Research and Technology*, 38, 41–52.

Berson, D.M., Dunn, F.A. and Takao, M. (2002) Phototransduction by retinal ganglion cells that set the circadian clock. *Science*, 295, (5557), 1070–1073.

Boyce, P.R. (1996) Illuminance selection based on visual performance – and other fairy stories. *Journal of the Illuminating Engineering Society*, 25, 41–49.

Boyce, P.R. (2003) *Human factors in lighting*. London: Taylor and Francis.

Boyce, P.R. (2006) Lemmings, light and health. *Leukos*, 2(3), 175–184.

Boyce, P.R. and Rea, M.S. (1987) Plateau and escarpment: The shape of visual performance, *Proceedings of the CIE 21st Session, Venice*. Vienna: CIE.

Boyce, P.R., Beckstead, J.W., Eklund, N.H., Strobel, R.W. and Rea, M.S. (1997) Lighting the graveyard shift: the influence of a daylight-simulating skylight on the task performance and mood of night-shift workers. *Lighting Research and Technology*, 29, 105–142.

Boyce, P.R., Eklund, N.H., Hamilton, B.J. and Bruno, L.D. (2000) Perceptions of safety at night in different lighting conditions. *Lighting Research and Technology*, 32, 79–91.

Boyce, P.R., Akashi, Y., Hunter, C.M. and Bullough, J.D. (2003) The impact of spectral power distribution on visual performance. *Lighting Research and Technology*, 35, 141–161.

Boyce, P.R., Veitch, J.A., Newsham, G.R., Jones, C.C., Heerwagen, J., Myer, M. and Hunter, C.M. (2006a) Lighting quality and office work: Two field simulation experiments. *Lighting Research and Technology*, 38(3), 191–223.

Boyce, P. R., Veitch, J.A., Newsham, G.R., Jones, C.C., Heerwagen, J., Myer, M. and Hunter, C.M. (2006b) Occupants use of switching and dimming in offices. *Lighting Research and Technology*, 38, 358–378.

Campbell, S.S., Dawson, D. and Anderson, M.W. (1993) Alleviation of sleep maintenance insomnia with timed exposure to bright light. *Journal of the American Geriatric Society*, 41, 829–836.

Campbell, S.S., Dijk, D.J., Boulos, Z., Eastman, C.I., Lewy, A.J. and Terman, M. (1995) Light treatment for sleep disorders: Consensus report III Alerting and activating effects. *Journal of Biological Rhythms*, 10, 129–132.

Chartered Institution of Building Services Engineers (CIBSE) (1999) *Environmental factors affecting office worker performance: A review of evidence*, CIBSE Technical Memorandum TM24. London: CIBSE.

Clear, R. and Berman, S. (1994) Environmental and health aspects of lighting: Mercury. *Journal of the Illuminating Engineering Society*, 23, 138–156.

Cuttle, C. (1997) Cubic illumination. *Lighting Research and Technology*, 29, 1–14.

Cuttle, C. and Brandston, H. (1995) Evaluation of retail lighting. *Journal of the Illuminating Engineering Society*, 24(2), 33–49.

Czeisler, C.A., Rios, C.D., Sanchez, R., Brown, E.N., Richardson, G.S., Ronda, J.M. and Rogacz, S. (1988) Phase advance and reduction in amplitude of the endogenous circadian oscillator correspond with systematic changes in sleep/wake habits and daytime functioning in the elderly. *Sleep Research*, 15, 268.

Davis, W. and Ohno, Y. (2010) Color quality scale. *Optical Engineering*, 49(3), 033602.

Dijk, D-J., Boulos, Z., Eastman, C.I., Lewy, A.J., Campbell, S.S. and Terman, M. (1995) Light treatment for sleep disorders: Consensus report II Basic properties of circadian physiology and sleep regulation. *Journal of Biological Rhythms*, 10, 113–125.

Eastman, C.I., Stewart, K.T., Mahoney, M.P., Liu, L. and Fogg, L.F. (1994) Dark goggles and bright light improve circadian rhythm adaptation to night shift work. *Sleep*, 17, 535–543.

EC (2002) Directive 2002/91/EC of the European Parliament and of the Council of 16 December 2002 on the energy performance of buildings. *Official Journal of the European Communities*, 4.1.2003.

Eklund, N.H. (1999) Exit sign recognition for color normal and color deficient observers. *Journal of the Illuminating Engineering Society*, 28, 71–81.

Eklund, N.H., Boyce, P.R. and Simpson, S.N. (2001) Lighting and sustained performance: Modeling data-entry task performance. *Journal of the Illuminating Engineering Society*, 30, 126–141.

Elvik, R. (1995) Meta-analysis of evaluations of public lighting as accident countermeasure. *Transportation Research Record*, 1485, 112–123.

Environmental Protection Agency (EPA) (1997) *Mercury study report to Congress*. Washington, DC: EPA.

Farley, K.M.J. and Veitch, J.A. (2001) *A room with a view: A review of the effects of windows on work and well-being (IRC-RR-136)*. Ottawa, ON: NRC Institute for Research in Construction. Retrieved from http://irc.nrc-cnrc.gc.ca/fulltext/rr/rr136/

Figueiro, M.G., Rea, M.S. and Bullough, J.D. (2006) Does architectural lighting contribute to breast cancer? *Journal of Carcinogenesis*, 5(1), 20.

Fotios, S. and Cheal, C. (2009) Obstacle detection: A pilot study investigating the effects of lamp type, illuminance and age. *Lighting Research and Technology*, 41, 321–342.

Galasiu, A.D. and Veitch, J.A. (2006) Occupant preferences and satisfaction with the luminous environment and control systems in daylit offices: a literature review. *Energy and Buildings*, 38(7), 728–742.

Heschong Mahone Group (1999) *Skylighting and retail sales: An investigation into the relationship between daylighting and human performance*. San Francisco, CA: Pacific Gas & Electric Co. Retrieved from http://www.pge.com/mybusiness/edusafety/training/pec/daylight/daylight.shtml

HMSO (1992) The Workplace (Health, Safety and Welfare) Regulations 1992, http://www.legislation.gov.uk/uksi/1992/3004/contents/made (accessed September 2011)

HMSO (1999) The Management of Health and Safety at Work Regulations 1999, Regulation 3, http://www.legislation.gov.uk/uksi/1999/3242/regulation/3/made (accessed September 2011)

HMSO (2005) Clean Neighbourhoods and Environment Act 2005, Chapter 16, Section 102 Statutory nuisance: lighting, http://www.legislation.gov.uk/ukpga/2005/16/section/102

HMSO (2007) The Construction (Design and Management) Regulations 2007 Statutory Instruments No. 320 2007, http://www.legislation.gov.uk/uksi/2007/320/contents/made (accessed September 2011)

Hosoda, M., Stone-Romero, E.F. and Coats, G. (2003) The effects of physical attractiveness on job-related outcomes: A meta-analysis of experimental studies. *Personnel Psychology*, 56(2), 431–462.

Houser, K.W., Tiller, D.K., Bernecker, C.A. and Mistrick, R.G. (2002) The subjective response to linear fluorescent direct/indirect lighting systems. *Lighting Research and Technology*, 34(3), 243–264.

HSE (2007) Managing health and safety in construction. HSE Approved Code of Practice L144. Bootle: Health and Safety Executive, http://www.hse.gov.uk/pubns/priced/l144.pdf (accessed September 2011)

Illuminating Engineering Society of North America (IESNA) (2000) *The IESNA lighting handbook*, 9th edition. New York: IESNA.

Illuminating Engineering Society of North America (IESNA) (2005, 2007, 2009) *Photobiological safety for lamps and lamp systems – General requirements (2005), Photobiological safety for lamps and lamp systems – Risk group classification and labeling (2007) and Photobiological safety for lamps and lamp systems – Measurement techniques (2009)*. New York: IESNA.

Isen, A.M. and Baron, R.A. (1991) Positive affect as a factor in organizational behavior. In B.M. Staw and L.L. Cummings (eds), *Research in organizational behavior, 13*. Greenwich, CT: JAI Press.

Jasser, S.A., Blask, D.E. and Brainard, G.C. (2006) Light during darkness and cancer: relationships in circadian photoreception and tumor biology. *Cancer Causes and Control*, 17(4), 513–523.

Jay, P.A. (1973) The theory of practice in lighting engineering. *Light and Lighting*, 66, 303–306.

Kang, J. (2004) The effect of light on the movement of people. *Dissertation Abstracts International Section A: Humanities and Social Sciences*, 65(6-A), 2007.

Kaplan, S. and Kaplan, R. (Eds.) (1982) *Cognition and environment: Functioning in an uncertain world*. New York: Praeger.

Lack, L. and Schumacher, K. (1993) Evening light treatment of early morning insomnia. *Sleep Research*, 22, 225.

LaGiusa, F.F. and Perney, L.R. (1973) Brightness patterns influence attention spans. *Lighting Design and Application*, 3(5), 26–30.

LaGiusa, F.F. and Perney, L.R. (1974) Further studies on the effects of brightness variations on attention span in a learning environment. *Journal of the Illuminating Engineering Society*, 3, 249–252.

Lam, R.W. and Levitt, A.J. (1998) Canadian consensus guidelines for the treatment of seasonal affective disorder: A summary of the report of the Canadian consensus group on SAD. *Canadian Journal of Diagnosis*, 15(10 (October supplement)), 1–17.

Langlois, J.H., Kalakanis, L., Rubenstein, A.J., Larson, A., Hallam, M. and Smoot, M. (2000) Maxims or myths of beauty? A meta-analytic and theoretical review. *Psychological Bulletin*, 126(3), 390–423.

Liebel, B., Berman, S., Clear, R. and Lee, R. (2010) Reading performance is affected by light level and lamp spectrum, *Proceedings of the Illuminating Engineering Society of North America Annual Conference, Toronto*. New York: IESNA.

Lockley, S.W., Barger, L.K., Ayas, N.T., Rothschild, J.M., Czeisler, C.A. and Landrigan, C.P. (2007) Effects of health care provider work hours and sleep deprivation on safety and performance. *Joint Commission Journal on Quality and Patient Safety*, 33(1), 7–18.

Loe, D.L., Mansfield, K.P. and Rowlands, E. (1994) Appearance of lit environment and its relevance in lighting design: Experimental study. *Lighting Research and Technology*, 26, 119–133.

MacAdam, D.L. (1942) Visual sensitivity to color differences in daylight. *Journal of the Optical Society of America*, 32, 247–274.

Mangum, S.R. (1998) Effective constrained illumination of three-dimensional, light-sensitive objects. *Journal of the Illuminating Engineering Society*, 27, 115–131.

McCloughan, C.L.B., Aspinall, P.A. and Webb, R.S. (1999) The impact of lighting on mood. *Lighting Research and Technology*, 31, 81–88.

McNally, D. (1994) *The vanishing universe*, Cambridge, UK: Cambridge University Press.

Megaw, E.D. and Richardson, J. (1979) Eye movements and industrial inspection. *Applied Ergonomics*, 10, 145–154.

Mills, E. and Borg, N. (1999) Trends in recommended illuminance levels: An international comparison. *Journal of the Illuminating Engineering Society*, 28, 155–163.

Newsham, G.R. and Veitch, J.A. (2001) Lighting quality recommendations for VDT offices: A new method of derivation. *Lighting Research and Technology*, 33, 97–116.

Newsham, G.R., Richardson, C., Blanchet, C. and Veitch, J.A. (2005) Lighting quality research using rendered images of offices. *Lighting Research and Technology*, 37(2), 93–115.

O'Donell, B.M., Colombo, E.M. and Boyce, P.R. (2011) Colour information improves relative visual performance. *Lighting Research and Technology*, 43, 423–438.

O'Hagan, J.B., Khazova, M. and Jones, B.W. (2011) Ultra-violet emissions from HMI daylight luminaires. *Lighting Research and Technology*, 43, 249–257.

Ouellette, M.J. and Rea, M.S. (1989) Illuminance requirements for emergency lighting. *Journal of the Illuminating Engineering Society*, 18, 37–42.

Painter, K. and Farrington, D.P. (1999) Street lighting and crime: Diffusion of benefits in the Stoke-on-Trent project, in K. Painter and N. Tilley (eds), *Crime prevention studies*, Monsey, NY: Criminal Justice Press.

Painter, K.A. and Farrington, D.P. (2001) The financial benefits of improved street lighting based on crime reduction. *Lighting Research and Technology*, 33, 3–12.

Partonen, T. and Lönnqvist, J. (2000) Bright light improves vitality and alleviates distress in healthy people. *Journal of Affective Disorders*, 57(1–3), 55–61.

Ravindran, A.V., Lam, R.W., Filteau, M.J., Lespérance, F., Kennedy, S.H., Parikh, S.V. et al. (2009) Canadian Network for Mood and Anxiety Treatments (CANMAT) Clinical guidelines for the management of major depressive disorder in adults. V. Complementary and alternative medicine treatments. *Journal of Affective Disorders*, 117 (Suppl. 1).

Rea, M.S. (1986) Toward a model of visual performance: Foundations and data. *Journal of the Illuminating Engineering Society*, 15, 41–58.

Rea, M.S. and Ouellette, M.J. (1991) Relative visual performance: A basis for application. *Lighting Research and Technology*, 23, 135–144.

Rich, C. and Longcore, T. (2006) *Ecological consequences of artificial night lighting.* Washington, DC: Island Press.

Rosa, R.R. and Colligan, M.J. (1997) *Plain language on shiftwork.* Cincinnati, OH: NIIOSH.

Rosekind, M.R., Gregory K.B., Mallis, M.M., Brandt, S.L., Seal, B. and Lerner, D. (2010) The cost of poor sleep: Workplace productivity loss and associated costs. *Journal of Occupational and Environmental Medicine*, 52(1), 91–98.

Slater, A.I. and Boyce, P.R. (1990) Illuminance uniformity on desks: where is the limit? *Lighting Research and Technology*, 22, 165–174.

Sullivan, J.M. and Flannagan, M.J. (2007) Determining the potential safety benefit of improved lighting in three pedestrian crash scenarios. *Accident Analysis and Prevention*, 39, 638–647.

Summers, T.A. and Hebert, P.R. (2001) Shedding some light on store atmospherics: Influence of illumination on consumer behavior. *Journal of Business Research*, 54(2), 145–150.

Taylor, L.H. and Sucov, E.W. (1974) The movement of people towards lights. *Journal of the Illuminating Engineering Society*, 3(3), 237–241.

van Someren, E.J.W., Hagebeuk, E.E.O., Lijzenga, C., Schellens, P., de Rooij, S.E., Jonker, C., Pot, A.M., Mirmiran, M. and Swaab, D.F. (1996) Circadian rest-activity rhythm disturbances in Alzheimer's disease. *Biological Psychiatry*, 40, 259–270.

van Someren, E.J.W., Kessler, A., Mirmiran, M. and Swaab, D.F. (1997) Indirect bright light improves circadian rest-activity rhythm disturbances in demented patients. *Biological Psychiatry*, 41, 955–963.

Veitch, J.A. (2001a) Lighting quality considerations from biophysical processes. *Journal of the Illuminating Engineering Society*, 30, 3–16.

Veitch, J.A. (2001b) Psychological processes influencing lighting quality. *Journal of the Illuminating Engineering Society*, 30, 124–140.

Veitch, J.A. and Newsham, G.R. (2000) Preferred luminous conditions in open-plan offices: Research and practice recommendations. *Lighting Research and Technology*, 32, 199–212.

Veitch, J.A., Newsham, G.R., Boyce, P.R. and Jones, C.C. (2008) Lighting appraisal, well-being and performance in open-plan offices: A linked mechanisms approach. *Lighting Research and Technology*, 40, 133–151.

Weston, H.C. (1935) The relation between illumination and visual efficiency: The effect of size of work. *Industrial Health Research Board and the Medical Research Council.* London: HMSO.

Weston, H.C. (1945) The relation between illumination and visual efficiency: The effect of brightness contrast. *Industrial Health Research Board, Report No. 87.* London: HMSO.

Zhou, Y. and Boyce, P.R. (2001) Evaluation of speech intelligibility under different lighting conditions. *Journal of the Illuminating Engineering Society*, 30(1), 34–46.

Index

Note: page numbers in italics refer to figures; page numbers in bold refer to tables.